上

龙志伟　编著
Edited by Long Zhiwei

| 创意之镜圆璧合 | 革新与极致造型 |
| Perfect Concept | Innovative Shape |

广西师范大学出版社
·桂林·

图书在版编目(CIP)数据

文体建筑方案集成:汉英对照/龙志伟 编著.—桂林:广西师范大学出版社,2015.6
ISBN 978-7-5495-6534-4

Ⅰ.①文… Ⅱ.①龙… Ⅲ.①文化建筑-建筑方案-汇编 ②体育建筑-建筑方案-汇编 Ⅳ.①TU24

中国版本图书馆 CIP 数据核字(2015)第 074934 号

出 品 人:刘广汉
责任编辑:肖 莉
装帧设计:龙志杰

广西师范大学出版社出版发行

(广西桂林市中华路 22 号　邮政编码:541001)
(网址:http://www.bbtpress.com)

出版人:何林夏
全国新华书店经销
销售热线:021-31260822-882/883
上海锦良印刷厂印刷
(上海市普陀区真南路 2548 号 6 号楼　邮政编码:200331)
开本:646mm×960mm　1/8
印张:72　　　　字数:60 千字
2015 年 6 月第 1 版　2015 年 6 月第 1 次印刷
定价:598.00 元

如发现印装质量问题,影响阅读,请与印刷单位联系调换。
(电话:021-56519605)

■ 序 Preface

文化与体育,作为现代建筑造像的两副面孔,其一面记录着人文主义以来所形成的建筑艺术的不断发展,另一面则演绎着新文化基础上的对运动追求的巨大张力。

文体建筑作为文化行为、体育运动的基本物质载体,在很大程度上表现出人们对精神文化形态和形体感知运动的诉求。而渗透其中的艺术形式和社会文化心理,则共同构筑了文化属性的完整系统,反映出当代建筑多层面的品格特质。同时,文体建筑作为人们通过思想交流、形体运动进行自我完善、自我提升的物质环境,具有传播先进文化、促进科教发展、弘扬体育精神、丰富城市文化生活、完善城市环境设施等多层面的社会价值。

21世纪,一个文化思想与社会信息愈趋开放的新时代,文体建筑也日益显示出其开放性的新趋势。它不仅体现的是建筑空间形式的开放性,也是建筑内容与功能的开放性。如建筑设计通过开合屋面、自然采光等技术措施,使人工场所与自然环境有机结合;通过辅助空间的敞开设计,使城市公共空间与建筑的内部空间相互渗透,形成建筑多层面、多功能的开放性文化特征。

《文体建筑方案集成》精选了全球近70个经典的文化体育建筑案例,有学校、图书馆、博物馆、体育中心、教堂等多种建筑类型。来自阿特金斯、OMA、Perkins Eastman、UNStudio、Steven Holl Architects以及Architecture-Studio等数十家当今世界顶级设计事务所的名家巨制。本书从柔性创意、极致造型、视觉冲击、矩形立面、刚柔并济等多方面对每一个不同的案例做出细致的解读。如"辽宁大连公共图书馆"、"甘肃兰州生态馆"的拟态建筑设计风格;"白俄罗斯巴特足球场"、"山东济宁奥体中心"的圆形穹顶场馆设计;"德国史普格尔博物馆"的花园式展览布局以及"马里博尔美术馆"、"中国美术馆"的跨学科枢纽和微缩城市概念等等,让您尽赏文体建筑的典雅风格,领略文体建筑的独特魅力。

Culture and Sports, as the two faces of modern architecture statues, recorded the continuous development of the architectural art formed since the humanism on one hand, while on other hand deduced a huge tension on the movement pursuit on the basis of the new culture.

Cultural and Sports architecture, as the basic material carrier of cultural behaviors and sports, largely show people's appeal for spiritual cultural patterns and body perception movement. And the art forms and socio-cultural psychology penetrated therein work together to build a complete system of cultural property, reflecting the multifaceted character traits of the contemporary architecture. Meanwhile, cultural and sports architecture as the physical environment for people to improve and develop themselves through the exchange of ideas and physical exercise, has the multi-level social value to spread advanced culture, promote the development of science and education, carry forward the spirit of sports, enrich the cultural life of the city, improve the urban environment facilities and so on.

21st century, is the new era that the cultural ideas and social information are more open, and cultural and sports architecture is increasingly showing its new trend of openness. However, the openness is not only reflected on the architectural space form, but also reflected on the building content and functionality. For example, architectural design by opening and closing the roof as well as using natural lighting and other technical measures makes artificial spaces organically combine with the natural environment, and through the open design of the auxiliary space, makes urban public spaces and the internal spaces of architecture mutually penetrate, so as to form the multi-level and multi-functional buildings with open cultural characteristics.

Cultural and Sports Architecture Program Integration carefully selects the world's 70 classic cultural and sports building cases, including various architecture types like schools, libraries, museums, sports centers, churches, etc. In addition, the cases are all famous masterworks from dozens of the world's top design firms such as Atkins, OMA, Perkins Eastman, UNStudio, Steven Holl Architects and Architecture-Studio and so on. The book from flexible design, extreme shape, visual impact, rectangular façade, firmness and flexibility, and many aspects makes the careful interpretations for each different case. For instance, the cases of "Dalian Public Library" and "Lanzhou Bioenergy Eco-Pavilion" with the mimic architectural design style; the cases of "Belarus Bate Football Stadium" and "Ji'ning Olympic Sports Center" with the circular dome stadium design; the case of "Sprengel Museum" with garden-style exhibition layout, and the cases of "Maribor Art Gallery" and "National Art Museum of China" with the interdisciplinary hub and miniature city concept, all of these can make you feel the elegant styles of the cultural and sports architecture and appreciate the unique charm of the cultural and sports architecture.

目录
Contents

■ 创意之镜圆璧合
Perfect Concept — 8

白俄罗斯鲍里索夫足球俱乐部巴特足球场
Borisov FC Bate Borisov Football Stadium — 10

山东济宁奥体中心
Ji'ning Olympic Sports Center — 16

日本东京新国家体育场
New Tokyo National Stadium — 20

新疆喀什市图书馆
Kashi Library — 26

辽宁大连公共图书馆
Dalian Public Library — 32

土耳其伊斯坦布尔恰姆勒贾清真寺
Camlica Mosque — 40

■ 革新与极致造型
Innovative Shape — 44

阿根廷布宜诺斯艾利斯艺术博物馆
Buenos Aires Contemporary Museum of Art — 46

江苏南京汽车博物馆
Nanjing Automobile Museum — 50

阿联酋迪拜中东现代艺术博物馆 *The Museum of Middle Eastern Modern Art, Khor Dubai, UAE*	58
立陶宛维尔纽斯博物馆 *Museum in Vilnius*	64
辽宁大连城市规划博物馆 *Dalian Urban Planning Museum*	70
江苏武进第八届中国花卉博览会科技馆 & 特色馆 *Pavilion of Art & Pavilion of Science, The 8th China Flower Expo*	78
宁夏银川规划展览馆 *Yinchuan Urban Planning Exhibition Hall*	86
韩国丽水世博会展览馆 *Expo Pavillon in Yeosu, South Korea*	96
福建泉州鼎立雕刻馆 *Fujian Quanzhou Dingli Sculpture Art Museum*	102
台湾新北市立美术馆 *New Taipei City Museum*	108
台湾高雄爱的小海湾 *Love Cove*	114
荷兰斯贝克尼萨剧院 *Netherlands Theatre Spijkenisse*	120
广东广州增城大剧院 *Zengcheng Opera House*	126

台湾台北艺术表演中心 *Taipei Performing Arts Center*	**130**
山东青岛文化艺术中心 *Qingdao Culture and Art Center*	**138**
湖南长沙梅溪湖国际文化艺术中心 *Changsha Meixi Lake International Culture & Arts Center*	**142**
山东济南文化综合体 *Ji'nan Cultural Complex*	**146**
美国纽约州立大学石溪分校西门子几何物理中心 *SUNY Stony Brook Simons Center for Geometry and Physics*	**152**
美国加州洛杉矶西方学院全球事务中心 *Los Angeles Occidental College Center for Global Affairs*	**160**
伊朗扎布尔医科大学总部 *Headquarter of Zabol University of Medical Sciences*	**166**
美国纽黑文福特学院科技大厦 *The Foote School Science and Technology Building*	**170**
美国德克萨斯大学教育和商业大楼 *The University of Texas Education and Business Complex*	**174**
斯洛文尼亚卢布尔雅那国家和大学图书馆 *Ljubljana National and University Library*	**180**
荷兰阿默斯福特文化中心"Eemhuis" *Amersfoort Cultural Center "Eemhuis"*	**188**

韩国大邱 Gosan 公共图书馆（方案一） *Daegu Gosan Public Library*	**198**
韩国大邱 Gosan 公共图书馆（方案二） *Daegu Gosan Public Library*	**206**
韩国大邱 Gosan 公共图书馆（方案三） *Daegu Gosan Public Library*	**214**
内蒙古鄂尔多斯东胜基督教堂 *Dong Sheng Protestant Church*	**218**
广西南宁李宁体育园 *Lining Sports Park*	**226**
瑞典于默奥艺术馆 *Umeå Art Museum*	**236**
广东深圳当代艺术馆与城市规划展览馆 *Museum of Contemporary Art and Planning Exhibition*	**246**
西班牙毕尔巴鄂新圣马梅斯体育场 *New San Mames Stadium*	**250**
芬兰赫尔辛基古根海姆博物馆 *Helsinki Guggenheim Museum*	**262**
匈牙利布达佩斯 Pancho 体育场 *Pancho Arena*	**274**

创意之镜圆璧合

Perfect Concept

白俄罗斯鲍里索夫足球俱乐部巴特足球场
Borisov FC Bate Borisov Football Stadium

设计单位：OFIS Arhitekti	Designed by: OFIS Arhitekti
委托方：贝特足球俱乐部	Client: FC Bate Borisov
项目地址：白俄罗斯鲍里索夫	Location: Borisov, Belarus
项目面积：6 000m²	Area: 6,000m²

地区标志
圆形穹顶
静音效果

项目概况

该球场是为鲍里索夫城市在 2012 年庆祝周年纪念而设计的。整个项目占地面积达 6 000m²，其中公共项目 3 628m²，办公楼 480m²，服务区 2 000m²，体育场共设置了 13 000 个座位，可以满足广大球迷的需求。项目建成后将成为城市主导的、可识别的标志性建筑。

建筑设计

该建筑在设计时考虑到现有位置和地形的干预，所以尽可能多地保持现有的树木，维持周围环境的原状。

球场看台被设计成一个统一的圆形穹顶，肤色得圆顶给人一种脆弱的拉伸感。表层和看台之间的覆盖空间是一个与公共项目一起的前厅街道（包括商店，酒吧，服务，厕所）和上层画廊。

在内部，圆形竞技场提供了不错的静音效果，方便球员在培训时间可以集中精神。看台面的南北向，有一块面积为 85m×105m 用于表演的区域，其余区域也有足够的空间用于广告屏、摄影和摄像头的安装。全场 13 000 个座位全都围绕着运动场沿 17 行的侧边和沿 27~28 行的短边两侧进行安排。上西区画廊是留给记者的小木屋，有可以满足约 40 名记者的座位和桌子，从这里可以直接进入新闻发布厅和混合区。在东部的贵宾看台上，有一个 250 个座位的酒吧和娱乐空间。贵宾是通过景区门口的车道电梯直接进入的。在每个入口点到运动场分别设置了两个更衣室、一个混合区、一个物理治疗室和一个接受兴奋剂检查的空间。西部看台区域是围绕内部体育场舞台扩展起来的，该空间包含有游客的厕所、酒吧、急救室和停留室，这是一个"雪域高原"，自然通而不热。其他公共区域的空间则是专为各种商业活动设计的。

| 文体建筑方案集成 | CULTURAL AND SPORTS ARCHITECTURE PROGRAM INTEGRATION |

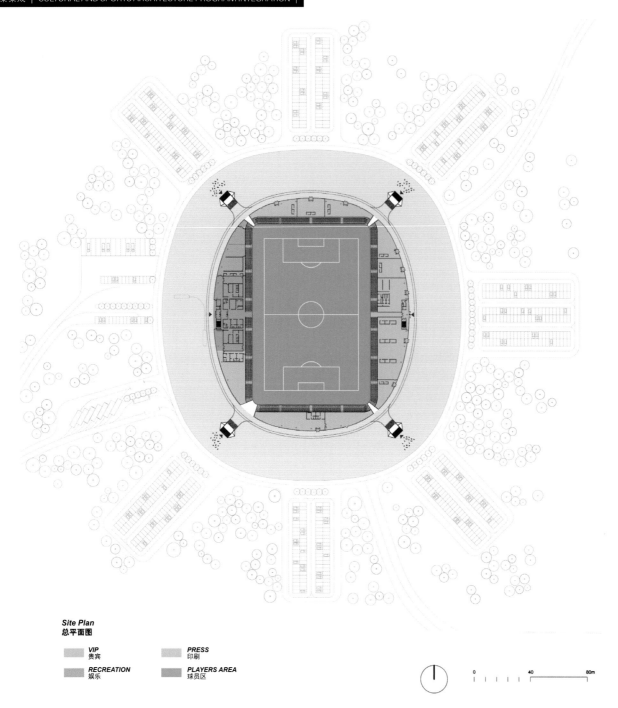

Site Plan
总平面图

- VIP / 贵宾
- RECREATION / 娱乐
- PRESS / 印刷
- PLAYERS AREA / 球员区

0　40　80m

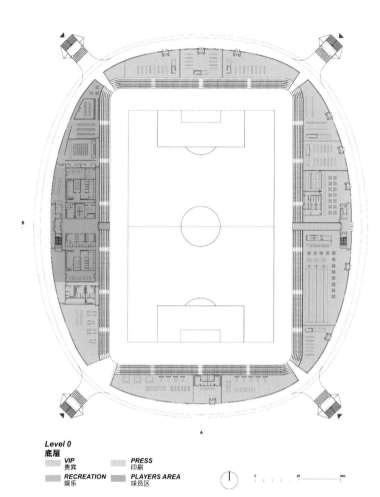

Level 0
底层

- VIP 贵宾
- RECREATION 娱乐
- PRESS 印刷
- PLAYERS AREA 球员区

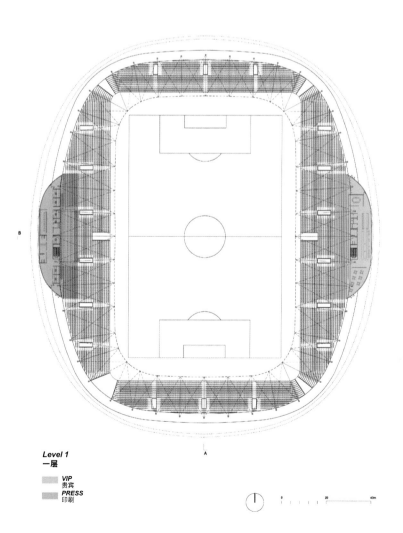

Level 1
一层

- VIP 贵宾
- PRESS 印刷

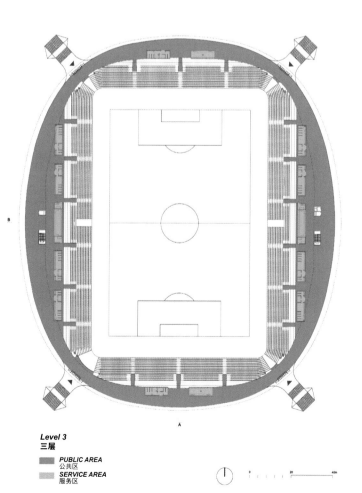

Level 3
三层

- PUBLIC AREA 公共区
- SERVICE AREA 服务区

| 12–13 |

| CULTURAL AND SPORTS ARCHITECTURE PROGRAM INTEGRATION |

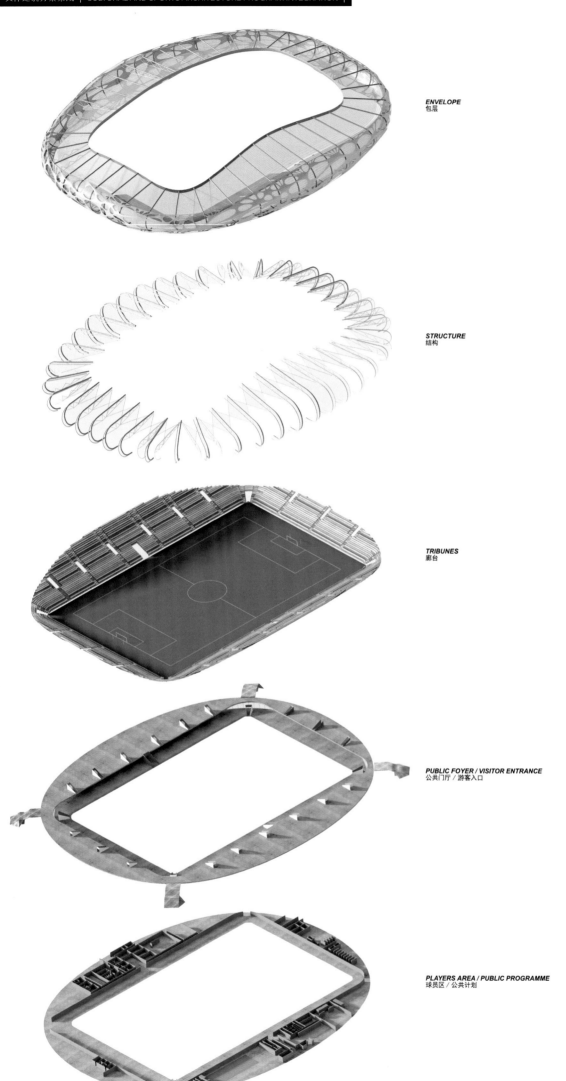

ENVELOPE 包层

STRUCTURE 结构

TRIBUNES 廊台

PUBLIC FOYER / VISITOR ENTRANCE 公共门厅 / 游客入口

PLAYERS AREA / PUBLIC PROGRAMME 球员区 / 公共计划

Regional Mark Circular Dome Mute Effect

Project Overview

The stadium is designed as the city of Borisov celebrates a significant anniversary in 2012. The site area of the whole project is 6,000m^2, including public project 3,628 m^2, office building 480m^2 and service area 2,000 m^2, and the stadium has 13,000 seats which can meet the vast soccer fans' demands. After the construction of the project, it will be the dominant, recognizable and symbol architecture for the city.

Building Design

The program takes into account the interventions of the existing location and the terrain, maintaining as many of the existing trees on site as possible and keeping the original state of the surrounding environment.

The stands of the court are designed a unified rounded dome, and the skin of the dome gives an impression of a fragile stretched perforated textile pulled over the stadium skeleton. The covered space between the skin and the tribunes is a public street—a vestibule with public program (including shops, bars, services, toilets) and galleries above.

Internally, the rounded arena provides good acoustics which can make the players focus concentration during training time. The playing surface has N-S orientation, with a total area of 85mx105m using for playing, and the remaining area allows enough space for the installation of advertising screens, photographers and cameras. The number of 13,000 seats is arranged around the playing field in rows of 17 along the sides and rows of 27 ~ 28 along the short sides. The upper west gallery is reserved for press cabins, with seats and tables for 40 journalists and direct stair access to the press room and mix zone. In the east are the VIP stands, with 250 seats and bar and entertainment spaces. The VIP is accessed directly via an elevator from the entrance area with a car driveway. At each entry point to the field are two dressing rooms, mix zone, physiotherapy and a space for doping control. The west stand area is extending all around the inner stadium arena; this space contains the visitor's toilets, bars, first-aid room and detention: it is a covered plateau, naturally ventilated and unheated, and other public area spaces are designed for various commercial activities.

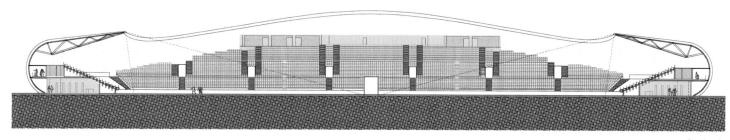

Section A
剖面 A

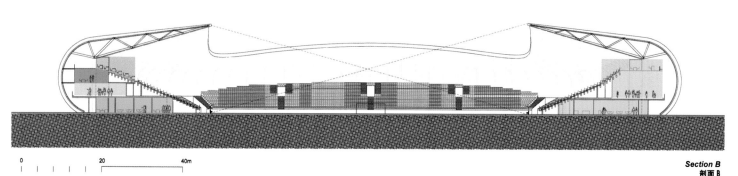

Section B
剖面 B

山东济宁奥体中心
Ji'ning Olympic Sports Center

设计单位：美国 KDG 建筑设计咨询有限公司
委托方：济宁市城乡规划局
项目地址：山东省济宁市
建筑面积：52 795m²

Designed by: Kalarch Design Group
Client: Ji'ning Urban Planning Bureau
Location: Ji'ning, Shandong
Building Area: 52,795m²

简洁 现代 实用

项目概况

济宁奥体中心是重要的城市地标，将作为山东省运动会的举办场馆。项目在现有体育场的基础上进行整体规划，并建造综合体育馆、水上运动馆、射击射箭场馆以及标志性的体育广场等设施。

建筑设计

由于该项目用地处在现有城市与自然景观带及湖面之间，周边用地包括政府行政用地、教育用地和住宅用地等。因此，项目整体设计的挑战性在于如何使奥体中心更好地融入城市，使之能够更好地为周边市民服务而不是成为市民正常生活秩序的干扰因素。

总平面以一条斜轴贯穿场地，斜轴设计有助于建立地块与周边用地、城市与自然湖面景观之间的联系。将主体建筑集中于用地一侧，另一侧布置景观用地，以期形成赛事期间的良好环境。未来可考虑作预留用地，也可用于未来城市商业开发，以便将综合功能引入该用地。

建筑的形象设计力求简洁、现代、实用，以突出济宁市的新型城市形象，并与已经建成的体育场融为一体。建筑材料平实而具有时代感，让未来的体育中心能够成为平易近人的城市公共建筑，而不是偶尔使用之后废弃的场馆。

该建筑设计重在确立了高效、省地、多功能的整体设计理念，重点在于满足大型赛事需要的前提下，能够在今后可持续性利用与维护，提高市民的参与度，使之真正服务于民。

Simplicity
Modernism
Practicality

Project Overview

Ji'ning Olympic Sports Center is an important urban landmark, the venue of Shandong's sports meeting in the future. The Olympic Sports Center is integrally planned based on the existing site including comprehensive gymnasium, water sports pavilion, shooting range and symbolic sports square.

Building Design

The site lies between the city and natural lake landscape. The surrounding areas are for administration, education and residential uses. Therefore, the challenge of the general plan design is how to better integrate the Olympic Sports Center with the city, and to provide better service for citizens but not be the interference factors for the people's normal life order.

An inclined axis penetrating the site not only facilitates to build the relationship between the site and its surrounding areas, but also build a connection between the city and natural lake landscape. The planning concentrates main buildings on one side while the other side is occupied by landscapes. The reserved area is also possible to be used as commercial land in the future.

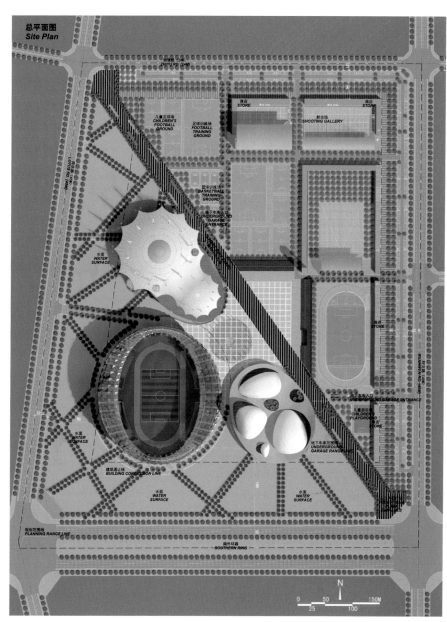

The building is expected to be concise, modern and functional, highlighting the new image of Ji'ning and integrating with the completed gymnasium. Pristine materials with sense of modern are used to create a truly hospitable urban public building but not be abandoned stadiums after the occasional use.

The architectural design focuses on establishing the general layout design with high efficiency, land-saving and multi-function. With emphasis on the premise of meeting the needs of large-scale events, it is for future sustainable use and maintenance, and it can improve the degree of participation of the public to make it truly serve the people.

游泳馆功能分析
Swimming Pool Functional Analysis

一层平面 / 1F Plan

二层平面 / 2F Plan

- 看台 / STANDS
- 大厅 / HALL
- 泳池 / SWIMMING POOL
- 中庭 / ATRIUM
- 交通 / TRAFFIC
- 公共空间 / PUBLIC SPACE
- 屋顶 / ROOF
- 更衣室/浴室 / DRESSING ROOM / BATHROOM
- 儿童乐园 / CHILDREN'S PARADISE
- 休息室 / REST ROOM
- 医疗室等辅助用房 / MEDICAL ROOM & AUXILIARY ROOMS
- 新闻广播室 / NEWS BROADCAST ROOM
- 设备间 / EQUIPMENT ROOM
- 卫生间 / TOILET

游泳馆流线分析
Swimming Pool Flow Line Analysis

一层平面 / 1F Plan

二层平面 / 2F Plan

- 观众流线 / AUDIENCE FLOW LINE
- 媒体流线 / MEDIA FLOW LINE
- 运动员、裁判员流线 / ATHLETES, REFEREES FLOW LINE
- 管理人员流线 / MANAGERS FLOW LINE
- VIP流线 / VIP FLOW LINE

篮球馆功能分析
Basketball Arena Functional Analysis

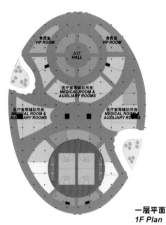
一层平面
1F Plan

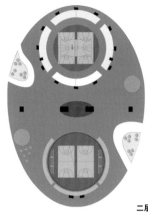
二层平面
2F Plan

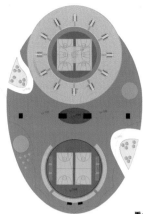
看台平面
Stands Plan

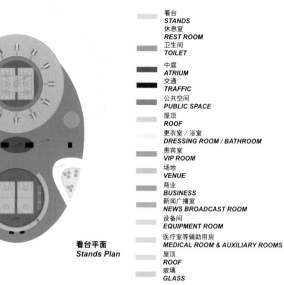

看台 STANDS
休息室 REST ROOM
卫生间 TOILET
中庭 ATRIUM
交通 TRAFFIC
公共空间 PUBLIC SPACE
屋顶 ROOF
更衣室/浴室 DRESSING ROOM / BATHROOM
贵宾室 VIP ROOM
场地 VENUE
商业 BUSINESS
新闻广播室 NEWS BROADCAST ROOM
设备间 EQUIPMENT ROOM
医疗室等辅助用房 MEDICAL ROOM & AUXILIARY ROOMS
屋顶 ROOF
玻璃 GLASS

篮球馆流线分析
Basketball Arena Flow Line Analysis

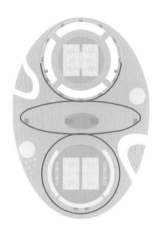

观众流线 AUDIENCE FLOW LINE
媒体流线 MEDIA FLOW LINE
运动员、裁判员流线 ATHLETES, REFEREES FLOW LINE
管理人员流线 MANAGERS FLOW LINE
VIP流线 VIP FLOW LINE

日本东京新国家体育场
New Tokyo National Stadium

设计单位: Andrea Maffei Architects
委托方: 日本体育委员会
项目地址: 日本东京
建筑面积: 301 219m²

Designed by: Andrea Maffei Architects
Client: Japan Sports Council
Location: Tokyo, Japan
Building Area: 301,219m²

矩形盒子 圆孔屋顶

项目概况

该项目建造主要用作———奥林匹克体育场-2020年奥运会。这座体育馆综合体最大的特征是：仅用4个支点就支撑起整个大屋顶，覆盖了所有观众区和比赛场。在天蓬上开了很多三角形的小洞口，为巨大的体育场室内带来自然光线，这座天蓬上还设计了一个可以伸缩自如的光伏玻璃盖，打开后就会变为露天体育场。

建筑设计

该设计主要秉持了安全性和便捷性两大原则，项目创建了一个海拔34m的平台，用来连接新体育场和体育馆，并且该平台作为观众观看比赛的座位区，它包括了35 006个座位，可以最大程度地满足体育爱好者的需求。此外，在观众席的看台内，还穿插设计了一条玻璃"矩形盒子"，这是为VIP提供的，在这里有最好的视野，可以全方位的看到赛场上的每一个精彩的瞬间。

主体育场的屋顶有一个特别的设计：一个120m宽的很大的圆孔，通过借助悬挂在屋顶上双星的钢铁和织物光圆顶控制开关。当它打开时，它会看起来像罗马万神殿。孔的大小对应于足球场下边，并能使草地获得充足的自然光线。奥运会期间，奥运火炬通过张力电缆将会悬挂在孔的中心，而且它会漂浮在观众的头顶上方。这里的气氛将会成为所有奥运会历史上独一无二的，对于观众而言，它将会是全新的、令人兴奋的。当屋顶关闭时，在孔中心，通过织物它将获得一半透明的屋顶。屋顶也包括一个750m和100m的运动员训练场地，运动员将在屋顶进行训练。这将是奥林匹克体育场的一种全新的使用方式。

| 文体建筑方案集成 | CULTURAL AND SPORTS ARCHITECTURE PROGRAM INTEGRATION |

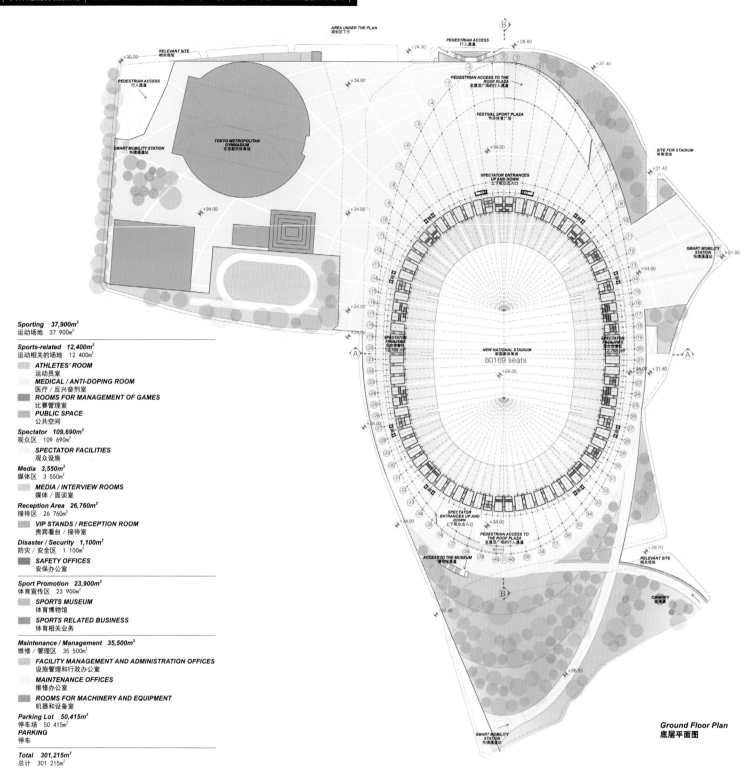

Sporting 37,900m²
运动场地 37 900m²

Sports-related 12,400m²
运动相关的场地 12 400m²
- ATHLETES' ROOM 运动员室
- MEDICAL / ANTI-DOPING ROOM 医疗／反兴奋剂室
- ROOMS FOR MANAGEMENT OF GAMES 比赛管理室
- PUBLIC SPACE 公共空间

Spectator 109,690m²
观众区 109 690m²
- SPECTATOR FACILITIES 观众设施

Media 3,550m²
媒体区 3 550m²
- MEDIA / INTERVIEW ROOMS 媒体／面谈室

Reception Area 26,760m²
接待区 26 760m²
- VIP STANDS / RECEPTION ROOM 贵宾看台／接待室

Disaster / Security 1,100m²
防灾／安全区 1 100m²
- SAFETY OFFICES 安保办公室

Sport Promotion 23,900m²
体育宣传区 23 900m²
- SPORTS MUSEUM 体育博物馆
- SPORTS RELATED BUSINESS 体育相关业务

Maintenance / Management 35,500m²
维修／管理区 35 500m²
- FACILITY MANAGEMENT AND ADMINISTRATION OFFICES 设施管理和行政办公室
- MAINTENANCE OFFICES 维修办公室
- ROOMS FOR MACHINERY AND EQUIPMENT 机器和设备室

Parking Lot 50,415m²
停车场 50 415m²
- PARKING 停车

Total 301,215m²
总计 301 215m²

Ground Floor Plan
底层平面图

Rectangular Box Circular Hole Roof

Concept
概念

1. **STADIUM ACCESS UP AND DOWN**
   1. 上下体育通道

2. **ADDITIONAL 10,600 SEATS FOR FOOTBALL**
   2. 足球场地附加 10 600 个座位

   FOOTBALL PLAYGROUND GOES DOWN 5M
   足球场地下降5m

   FESTIVAL SPORT PLAZA
   节庆体育广场

3. **COVERED OLYMPIC SQUARE**
   3. 包围奥林匹克广场

4. **PUBLIC SLOPES TO THE ROOF**
   4. 至屋顶的公共坡道

   ATHLETICS TRACK
   田径跑道

   CONCERTS / VOLLEYBALL / BASKETBALL / JUDO / SUMO...
   音乐会 / 排球 / 篮球 / 柔道 / 相扑

5. **PUBLIC ROOF SPORT TRAINING / EVENTS SPACE**
   5. 公共天台体育训练 / 活动空间

Project Overview

The main aim of the project is for Olympic Stadium—2020 Olympic Games. The most important feature of this Stadium Complex is that only four fulcrums can support the whole big roof, covering all the audience areas and playing field. Many small triangular holes opened on the canopy bring natural light into the huge stadium, and this canopy is also designed a retractable photovoltaic glass lid, which will become open stadium once it open.

Building Design

The design mainly upholds the two principles of safety and convenience to build a main platform of 34m above sea level connecting the new stadium with the gymnasium, and the platform as the spectators seating area, which includes 35,006 seats, can meet the needs of sports enthusiasts at the greatest degree. Moreover, in the audience stand, it is interspersed to design a glass "rectangular box", which is provided for the VIP, where there is the best view so that you can see the full range of every wonderful moment on the field.

The main roof of the stadium is characterized by a big circular hole 120m wide that can be closed or open by a light dome of steel and fabric sliding on the roof on binaries. When it's open, it will seem like the Pantheon in Rome. The size of the hole corresponds to the football field below and allows getting natural light enough for the grass. During the Olympic Games the Olympic torch will be suspended in the center of the hole by tension cables and it will be floating on the head of the spectators. The atmosphere will be unique in the history of all the Olympic Games and it will be very new and exciting for the spectators. When the roof is closed it will get a half transparent roof by fabric in the center. The roof includes also a training flied for the athletes—750m and 100m. The athletes will make training on the roof, visible by the public, and then will go down in the arena to play the game. It will be a completely new way for the Olympic stadium.

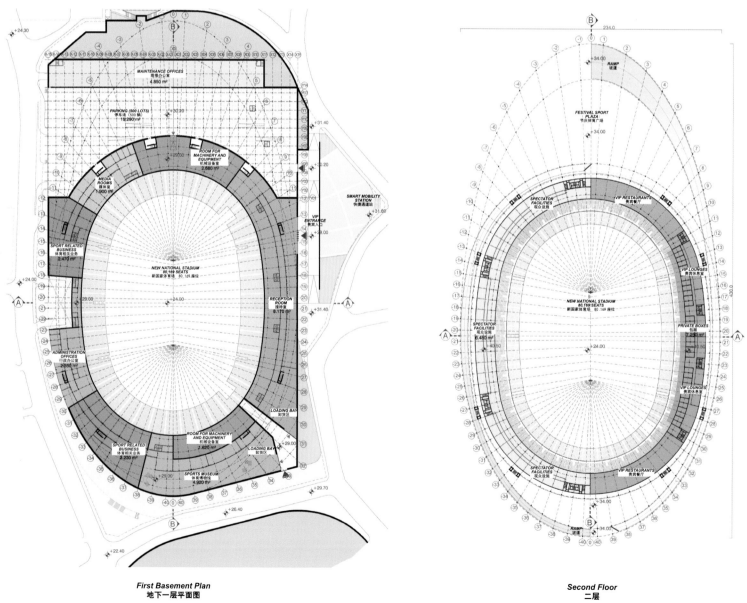

First Basement Plan
地下一层平面图

Second Floor
二层

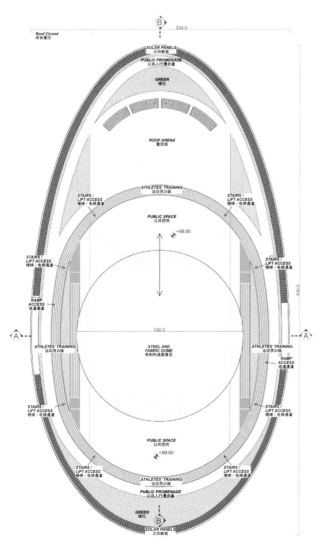

Roof Floor Plan-Closed
屋顶层平面图 – 闭合

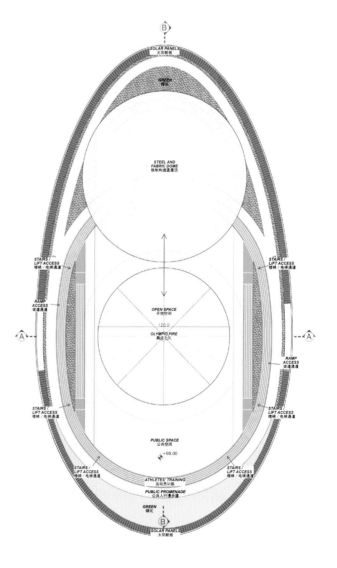

Roof Floor Plan-Open
屋顶层平面图 – 打开

| 24–25

新疆喀什市图书馆
Kashi Library

设计单位：深圳市瀚旅建筑设计顾问有限公司
委托方：喀什市人民政府
项目地址：新疆维吾尔族自治区喀什市
项目面积：83 398m²

Designed by: Shenzhen H&L architects and Engineers Co., Ltd.
Client: Kashi Municipal People's Government
Location: Kashi City, Xinjiang Uygur Autonomous Region
Area: 83,398m²

以人为本
可持续发展
地域文化与民族团结

项目概况

喀什市图书馆及附属设施配套建设项目现场地质状况良好，地形平整。项目选址在喀什市多来巴格尔乡，地处喀什东部新城区，位于深喀大道以南，紧邻未来喀什市政务中心。深喀大道连接直通喀什老城区的迎宾大道，地理位置优越，具备良好的自然生态条件。

建筑设计

项目致力于以人为本的人性化设计理念，秉承可持续发展原则，处理好建筑与周围环境的关系，实现所处区域的一体化，延续城市环境文脉的关系，提出最符合用地特征规律的整体结构，形成可持续发展的有机整体区域。

基于本项目用址的地形、地貌及周边环境，将图书馆主体建筑靠近用地北侧，预留出足够的后期发展空间。将馆区分为图书馆主体、特色广场、绿化区、停车区及预留区5个部分。各个区域既相对独立又紧密相连。

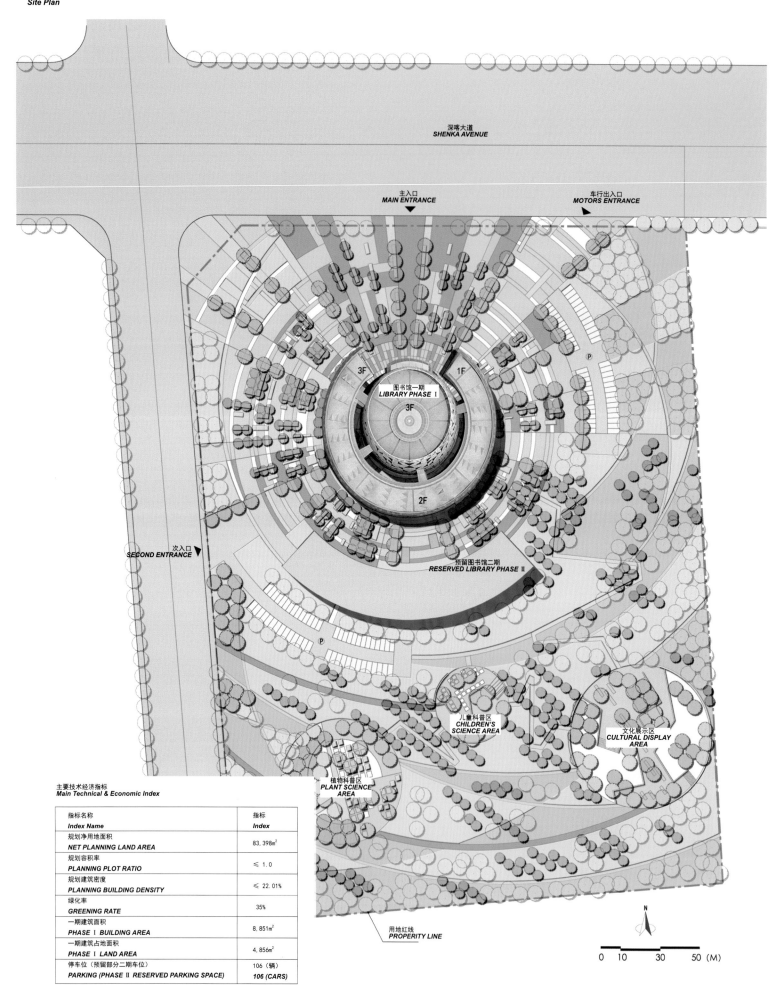

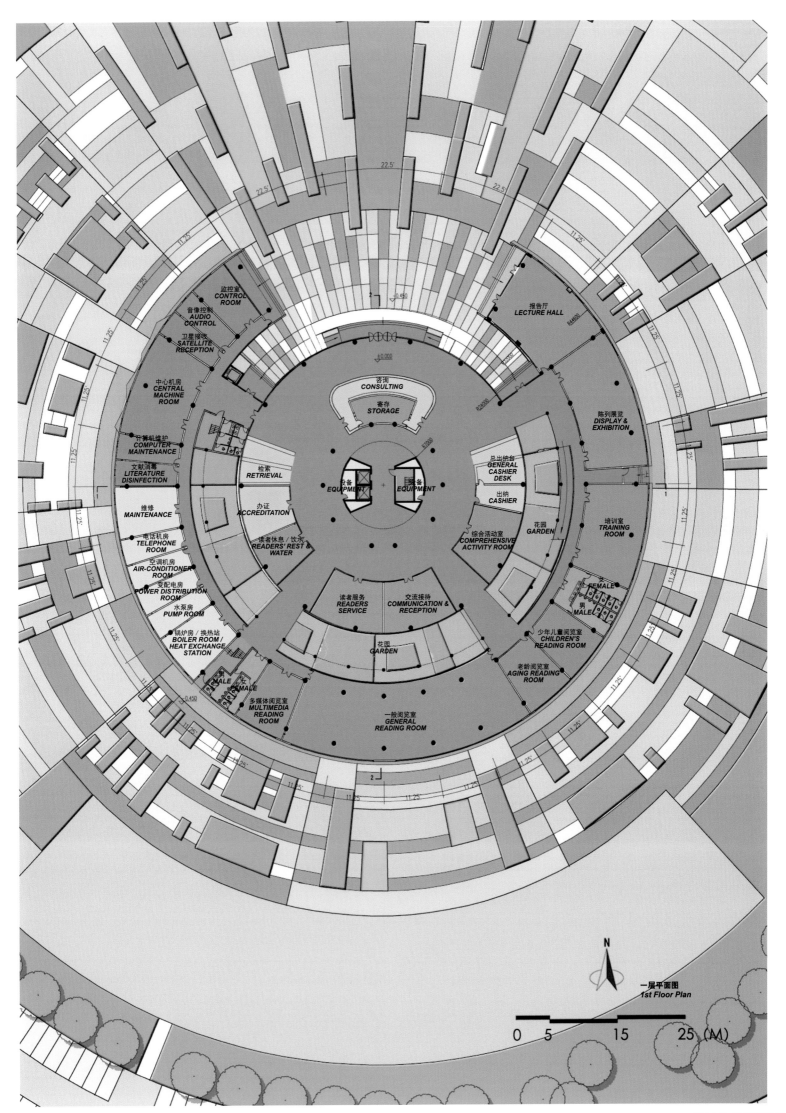

一层平面图
1st Floor Plan

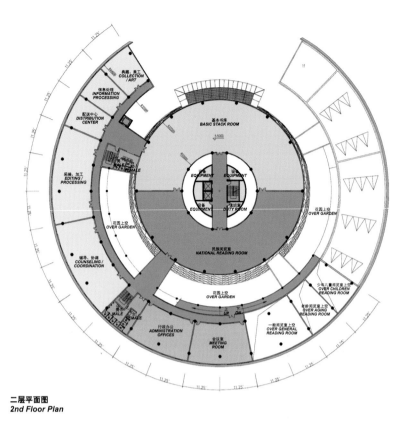

二层平面图
2nd Floor Plan

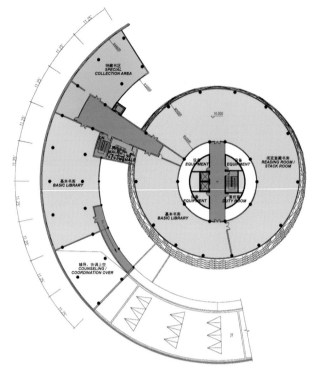

三层平面图
3rd Floor Plan

以图书馆为中心进行放射状的景观设计，主要由主入口广场、次入口广场、中心庭院景观、中心广场休闲区，三个特色主题庭院（文化展示庭院、儿童科普庭院、植物科普庭院）及深喀大道上的城市绿化景观带组成。主入口广场通过铺地、绿化带和园林小品对入馆人流进行指引和汇聚，既满足了功能要求，又从精神层面上体现出民族团结，共同发展的深层意义。

项目体现民族地域文化及民族大团结为创作的主要方向，形成具有时代文化特色的现代建筑风格。结合当地气候与建筑特点，强调虚与实、水平与垂直、明与暗的对比和融合，使建筑外观丰富多变。建筑外围实墙采用开窗面积较小的窗户组合，而内庭一侧则设有开敞的走廊，结合建筑西侧立面遮阳折板和屋顶的侧面采光天窗，既可以满足采光要求，又可以避免光线直射，还能降低能耗。毛面和光面材料的组合，使立面更加具有雕塑感。建筑造型既现代、活泼又不失稳重；既沉稳大方又不失时代气息。

东立面图
East Elevation

西立面图
West Elevation

南立面图
South Elevation

北立面图
North Elevation

屋顶平面图
Roof Plan

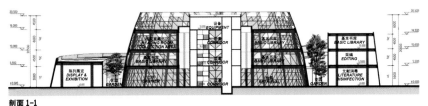

剖面 1-1
Section 1-1

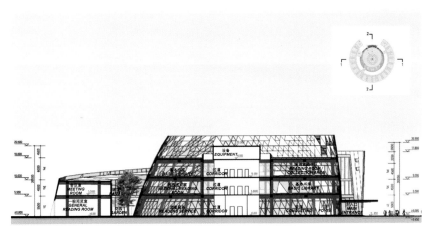

剖面 2-2
Section 2-2

People-oriented Sustainable Development Regional Culture and National Unity

Project Overview

The geology of project site of Kashi Library and ancillary facilities construction is in good condition with flat terrain. It is located in Bagger Town, Kashgar, and the eastern new city, Kashi, south of Shenka Avenue, and is near to the future Kashgar Government Center. Shenka Avenue is connected Welcome Avenue directly through the old city of Kashgar, so the location is superior with good natural ecological conditions.

Building Design

The project is committed to the people-centered humanized design philosophy and adhered to the principle of sustainable development to handle the relationship between buildings and the surrounding environment and achieve the regional integration, so that it will continue the relationship between the urban and environment context to propose the overall structure which is the most in line with the features of the land and to form an whole sustainable development organic area.

Based on the topography and the surrounding environment of the project with the site, the main building of the Library is placed to close to the north side of the land to set aside enough space for the later development. The library is divided into five parts, including the main library, featured square, green area, parking area and reserved area. Each of which is relatively independent but closely linked.

The radial landscape design with the library as the center is mainly composed by the main entrance plaza, second entrance plazâ, center courtyard, and lounge area in central plaza, three themed patio (cultural show yard, children's science garden, plant science yard) and urban green landscape belts on Shenka Avenue. The main entrance plaza through paving, green belts and garden ornaments guides the flows to enter or out of the library, which not only satisfied the functional requirements, but also reflects national unity on the spiritual level and the deep meaning of common development.

The project reflects the national regional culture and the national unity as the main direction of the creation to form modern architecture style with contemporary culture characteristics. The design is combined with the local climate and architectural features, and emphasized the contrasts and integrations between virtual and real, horizontal and vertical, light and dark to make the architectural appearance be rich and varied. The outer solid walls of the buildings are used smaller window combination, while the court has the open corridor on one side with the sun roof flaps on the west façade of the building and side skylights on the roof to meet the lighting requirements and avoid direct light, also reduce energy consumption. The combination of matte and glossy materials makes the façade with more sculptural feeling. The architectural style is modern, lively yet stable; and calm, generous, but contemporary.

辽宁大连公共图书馆
Dalian Public Library

设计单位：Architects Collective	Designed by: Architects Collective
委托方：大连市	Client: City of Dalian
项目地址：辽宁省大连市	Location: Dalian, Liaoning
建筑面积：50 000m²	Building Area: 50,000m²
设计团队：Andreas Frauscher　Patrick Herold　Richard Klinger　Martin Schorn　Kurt Sattler　Fei Tang	Design Team: Andreas Frauscher, Patrick Herold, Richard Klinger, Martin Schorn, Kurt Sattler, Fei Tang

地区地标
月季花
视景宽敞

项目概况

新建图书馆将被精心设计打造成为与海岸紧邻且关系密切的社区核心。建筑以地区地标的形象置身于公园景观之中，诠释打造成未来新普兰地区创新及友好的象征。它将成为大众汲取知识、沉思冥想和聚集往来的理想场所。

建筑设计

新图书馆的建筑造型创意来源于大连的市花——月季。新普兰是大连地区若干图书馆中的一支，因此图书馆建筑造型的每个分支代表一片花瓣，若干分支结合在一起犹如若干花瓣组成了完整的月季花朵。

建筑被园林式的绿色景观环境所环抱，园内种植多种当地的草植、灌木和花簇。植株栽种线沿弧形景观元素展开，使人们联想到海滨城市的"绿色海浪"——由图书馆庭院翻涌而起并充填入周边的城市空间中。位于环形建筑中心的是被打造为具备中国园林景观特色的图书馆庭院，建筑自身的形体屏蔽了来自道路及高速公路的噪声影响，以此将中心庭院打造成为供人们赏读、慢踱、休憩、沉思的"水榭桥台"园林景观。

建筑北翼的标高设计大大高于建筑南翼的标高，使得建筑坐北面海，以保证图书馆的高处区域享有宽敞辽阔的普兰湾视景。建筑被划分为北面的公共区域和南面的非公共区域。公共区域包括书籍借阅区、信息服务区、公共互动区、园林庭院以及对外停车场。非公共区域包括书籍存储区、后勤服务区、行政管理及商务区。此外，建筑包含有四个交通核，有两个位于南翼用于公共交通；另外两个位于北翼，其中一个为公共使用，另外一个为非公共使用。一层和二层通过自动扶梯相连。所有楼层均由展览坡道相连，其引导所有路径抵达顶层。

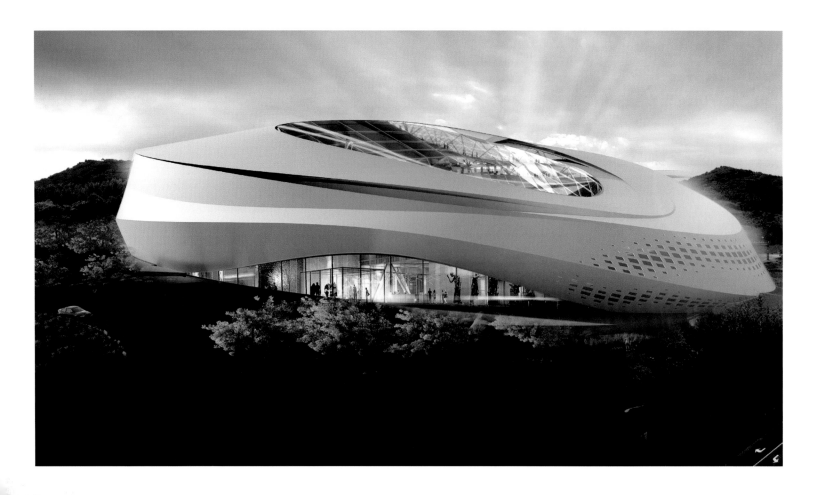

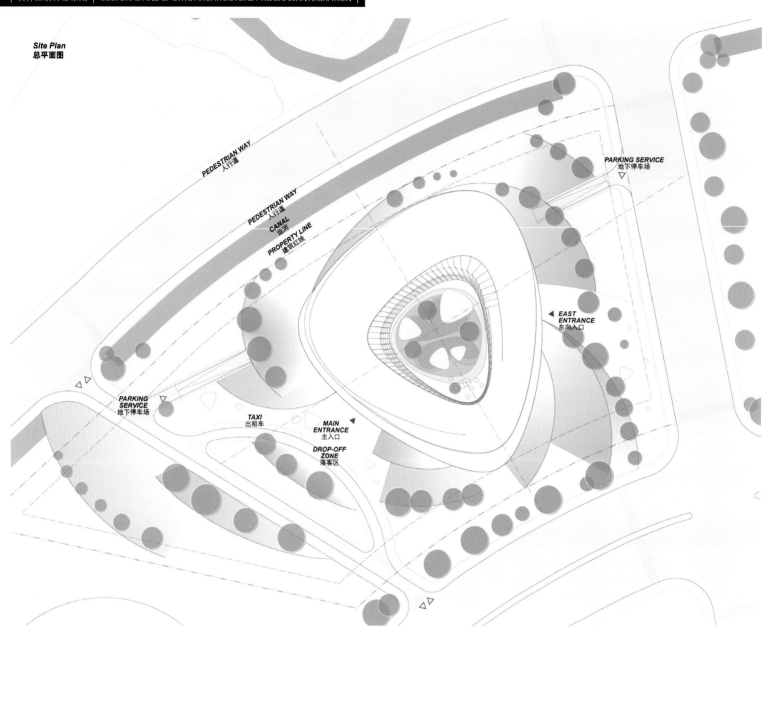

Site Plan
总平面图

Mezz 1
阁层 1

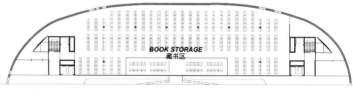

Mezz 3
阁层 3

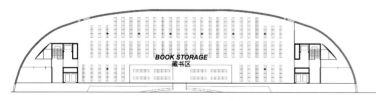

Mezz 2
阁层 2

Mezz 4
阁层 4

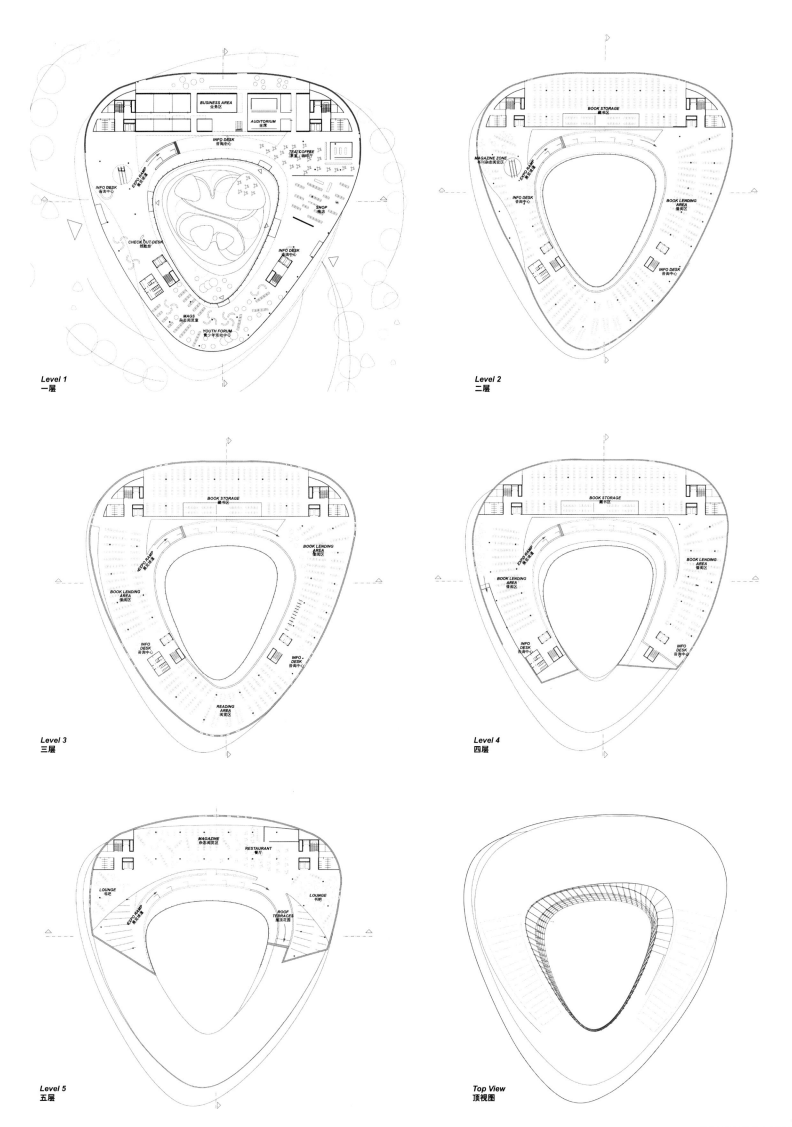

Area Landmark
Monthly Rose
Spacious Scene

Relation to Garden and Landscape
关于园林景观

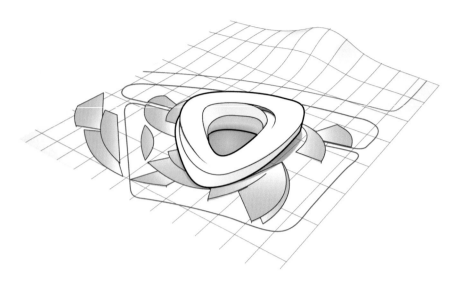

Project Overview

The new library is designed to become the center and heart for the local community with a strong relationship to the ocean and the bay. The building is placed in a park setting and aims to be a landmark for locals and visitors and a symbol for the creative and environmentally friendly future of New Pulan. It is a place for the public to read, contemplate and come together.

Building Design

The basic architectural form for the new library is the rose petal, since the Monthly Rose is the city flower of Dalian. The new Pulan branch is one of several library branches in the Dalian region, and with each branch, it is symbolizing a petals and forming a rose flower.

The building is surrounded by a park-like green space with a variety of local grasses, bushes and groups of flowers. Lines of trees are arranged along the edges of curved landscape elements which evoke water waves that start from the library courtyard sending out waves across the city. At the center of the ring-shaped building is the library courtyard that features elements of a Chinese landscape design. The building form shields the courtyard from loud noise of the streets and the highway and creates a quite interior with a pond with bridge and terraced platforms for reading, walking and sitting down and contemplating.

The north wing of the building is substantially higher than the south wing and the buildings toward the waterfront. The upper levels of the library feature wide-angle views of Pulan Bay. The building is divided in a Public Zone in the North and a Non-Public Zone in the South. The Public Zones include the Book Lending Area, Information & Service Areas, Public Activities, Courtyard and Public Parking. The Non-Public Areas include the Book Storage, Logistic Support, Administration and Business Area.

The building has four circulation cores with two cores in the south that are public and two in the north that are divided by a public and a non-public core. The first floor and the second floor are connected by an escalator. All floors are connected by an exhibition ramp that leads all the way to the top floor.

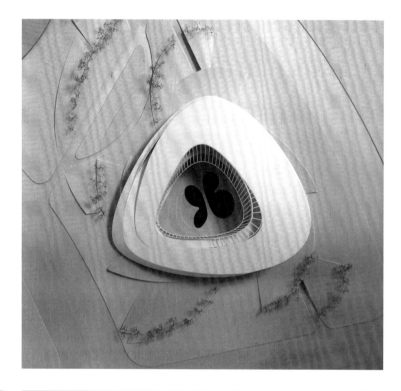

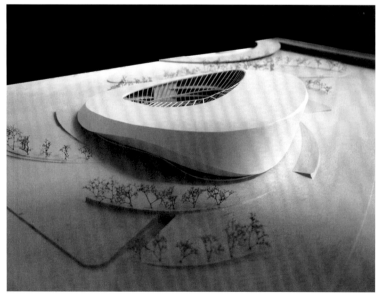

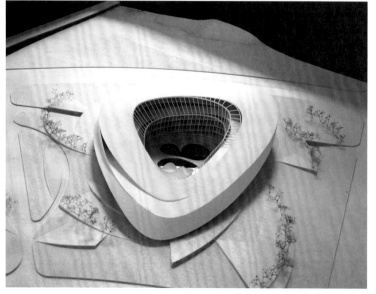

Section
剖面图

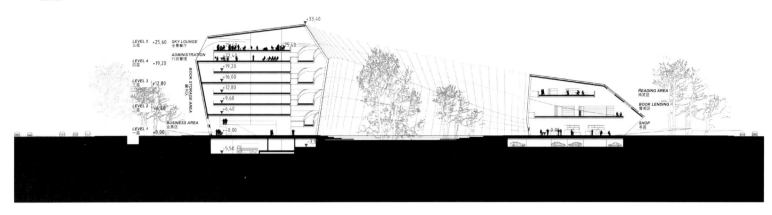

Section
剖面图

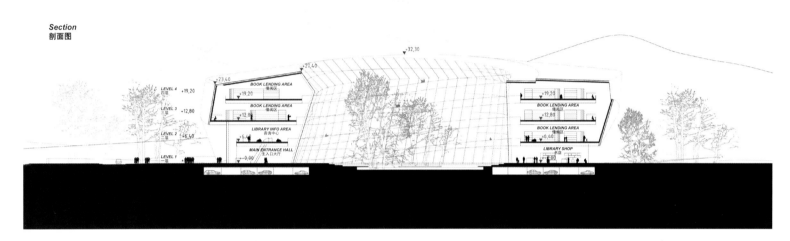

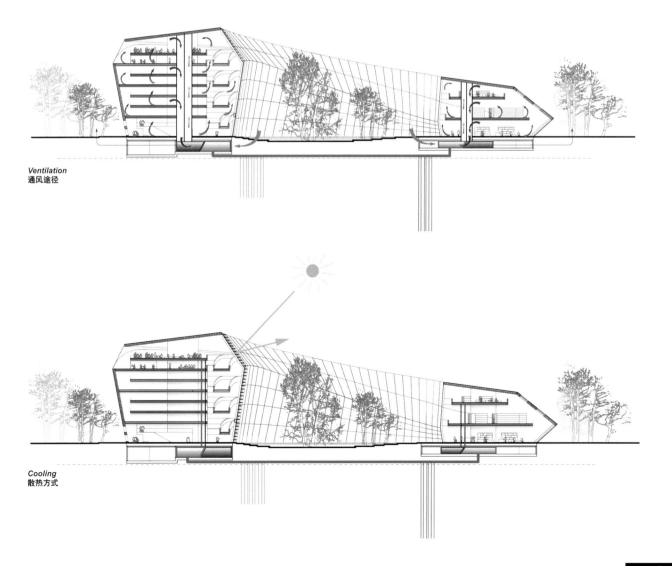

Ventilation
通风途径

Cooling
散热方式

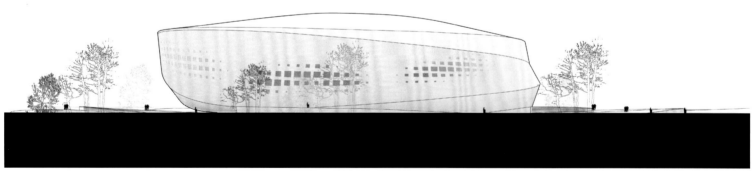

North Elevation
北立面图

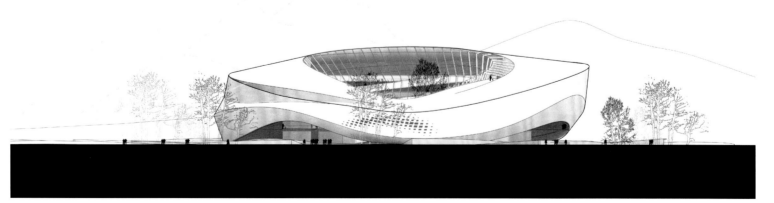

South Elevation
南立面图

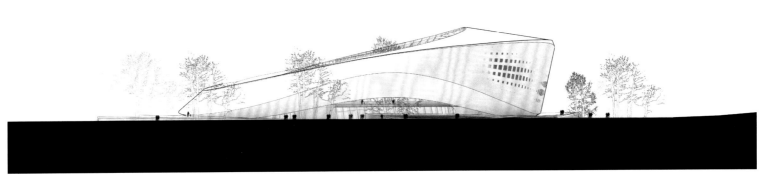

East Elevation
东立面图

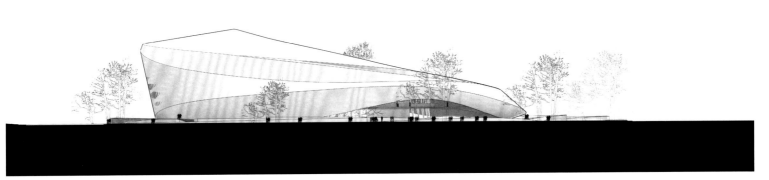

West Elevation
西立面图

土耳其伊斯坦布尔恰姆勒贾清真寺
Camlica Mosque

设计单位： TUNCER CAKMAKLI ARCHITECTS
项目地址： 土耳其伊斯坦布尔
占地面积： 58 000m²
建筑面积： 65 000m²

Designed by: Tuncer Calmakli Architects
Location: Istanbul, Turkey
Land Area: 58,000m²
Building Area: 65,000m²

深山藏古寺
壁龛墙
伊甸园

项目概况

"深山藏古寺"的思想被合理地运用在伊斯坦布尔的恰姆勒贾清真寺的设计中，被安置在大自然中心的清真寺不仅作为一个祈祷的空间功能结构，而且还作为一个穆斯林聚集的地方，是一种宗教信仰、宗教精神的外显。

建筑设计

建筑设计综合了奥斯曼帝国时期所有社会综合体设计的精华，在该项目的设计中，清真寺的墙壁变成了固定的壁龛墙，并辅助设计了一个环绕院墙的伊甸园花园。

在清真寺的内部和外部装饰了许多不同颜色的花卉，如玫瑰、茉莉等，设计试图将其与院墙一起为伊甸园打造一种宁静、和谐、温馨的氛围和感觉。坐落在清真寺朝拜方向的两个尖塔将作为竖井引入新鲜空气，并为祈祷室提供自然通风。

Ancient Temple under Deep Mountain
Mihrab Wall
Garden of Eden

Project Overview

The idea of "ancient temple under deep mountain" is reasonably used in Istanbul Camlica Mosque's design. The Mosque is located in the heart of the nature which is not only as a structure with the function of a prayer space, but as a place for Muslims gathering and a kind of explicit for religious belief and spirit.

Building Design

The building design combines all the essence of the social complex design in Ottoman Empire period. In the design of the project, the walls of the Mosque became into the mihrab wall, and aided design to build a Garden of Eden around the courtyard wall. The Mosque, with courtyard walls decorated with roses, jasmines and other flowers in many different colors on the inner and outer surfaces, which will try to create a quiet, harmonious and warm ambiance and sensation for Garden of Eden. Two minarets located on the sides of the mosque in Qibla direction will act as shafts aspirating fresh air and provide natural ventilation for the prayer room.

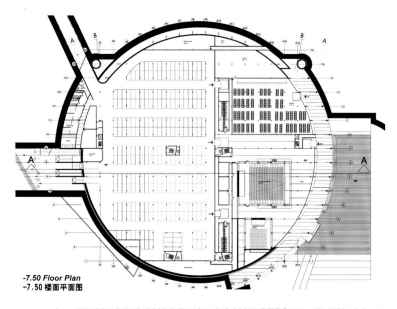

-7.50 Floor Plan
−7.50 楼面平面图

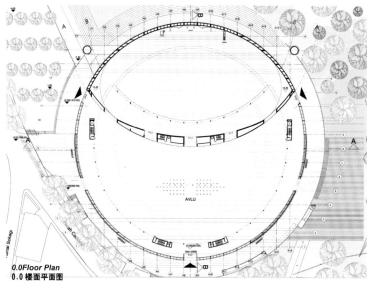

0.0 Floor Plan
0.0 楼面平面图

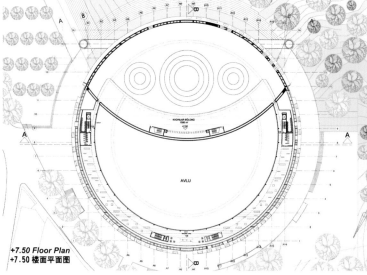

+7.50 Floor Plan
+7.50 楼面平面图

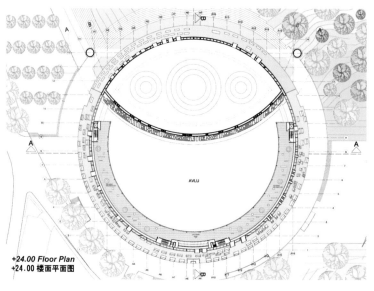

+24.00 Floor Plan
+24.00 楼面平面图

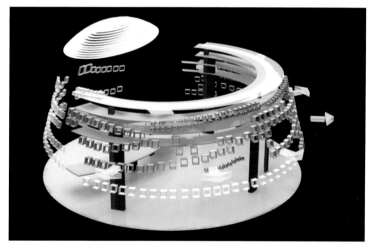

Mosque Plan Development
清真寺计划发展

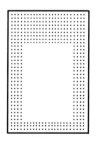

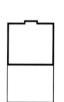

SAMARRA MOSQUE (9, CENTURY)　　ISABEY MOSQUE (14, CENTURY)　　SELIMIYE MOSQUE (16, CENTURY)　　CAMLICA MOSQUE
萨马拉清真寺（9世纪）　　　　　　伊萨贝清真寺（14世纪）　　　　　塞利米耶清真寺（16世纪）　　　　　恰姆勒贾清真寺

Camlica Mosque Symbolic and Spatial Organization
恰姆勒贾清真寺象征性和空间性的组织

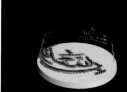

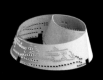

革新与极致造型

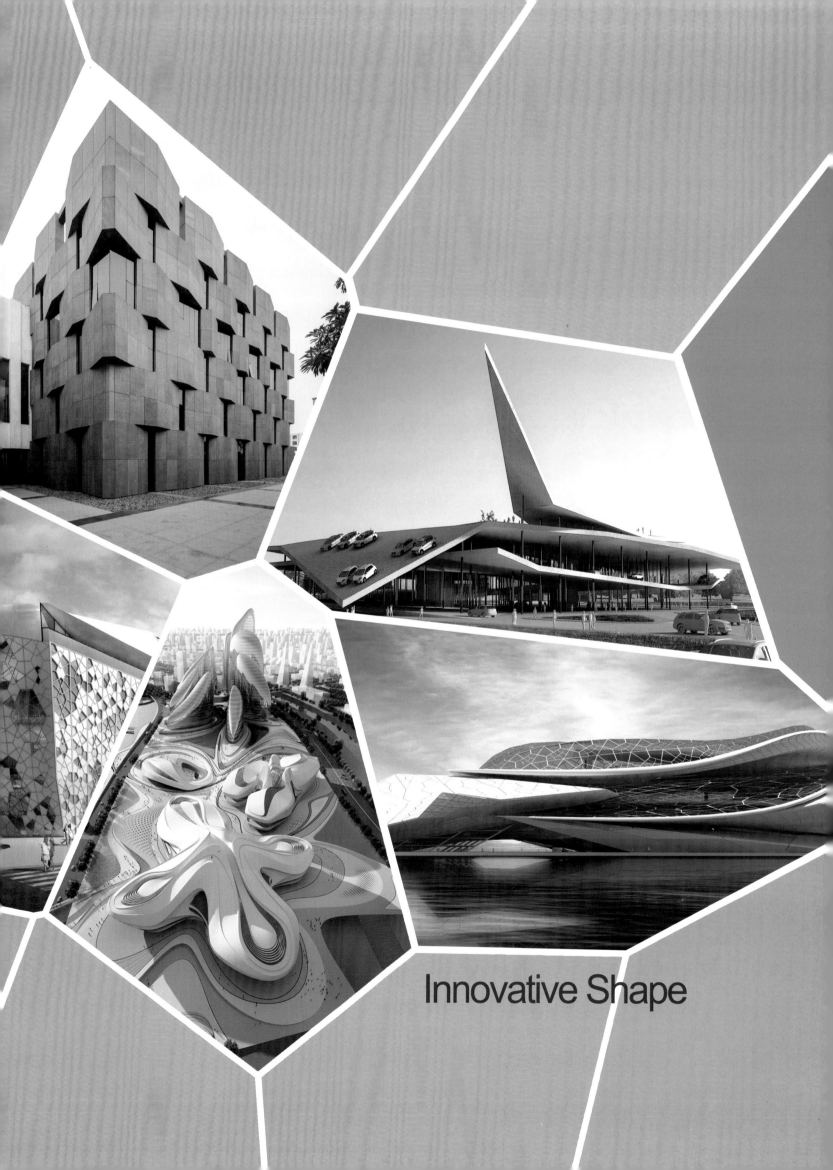

Innovative Shape

| 文体建筑方案集成 | CULTURAL AND SPORTS ARCHITECTURE PROGRAM INTEGRATION |

阿根廷布宜诺斯艾利斯艺术博物馆
Buenos Aires Contemporary Museum of Art

设计单位：MA2
委托方：AC-CA
项目地址：阿根廷布宜诺斯艾利斯
项目面积：8 500m²

Designed by: MA2
Client: AC-CA
Location: Buenos Aires, Argentina
Area: 8,500m²

视觉冲击
几何结晶
相交轨迹

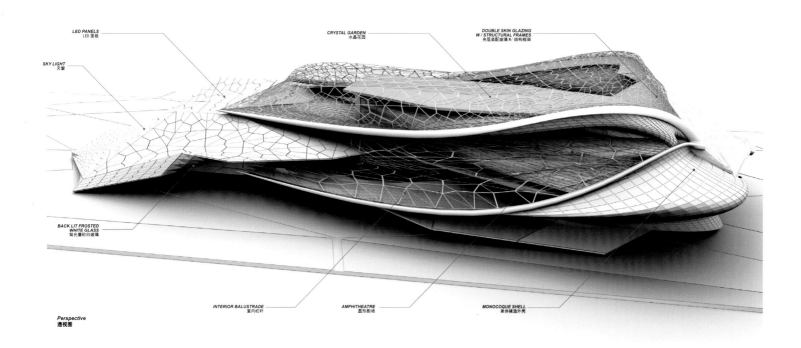

Perspective
透视图

项目概况

项目位于阿根廷首都布宜诺斯艾利斯，是多层次建筑，由辐射性的体量和片块构成，在构造性的操作点、结构和感观阵列中汇聚。多样化的液体和结晶成分被完美配置的一个动态结构，勇敢地扮演观众感官体验和视觉冲击作用的角色。

建筑设计

博物馆是在制作的构造和空间"流动视角协同"的工作领域中形成的。流体形式和协作线的合作互动领域也产生了一系列所需的图案，类似于经典线性工作的"连接环"、或者说建立"连接环效应"。

外部包层是许多弯曲和结晶形式的几何形状，由与轨迹和相交表面的组织行为模式执行的数组。创建的几何形状的灵感来源于参数笛卡尔方程的扭索图案。利用基本的参数方程综合表面的装饰和结构，它允许了一系列的模式可以转化为一种有效的工具用于不同的拼板和晶格结构。除了进一步调查参数的公式，类似内旋轮线、外旋轮线和摆线，对这一系列的连接环形成是有益的。

设计采用不同大小的抛光和亚光玻璃石或白色的 Neoparies，以展示建筑光滑的表面和外部包层。博物馆共有 5 个等级的展览空间，出入口是由户外广场隔开来创建一个循环人流的流动空间，以体验整个博物馆。

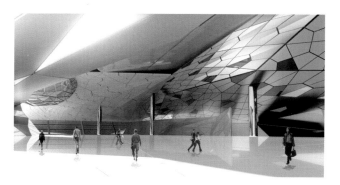

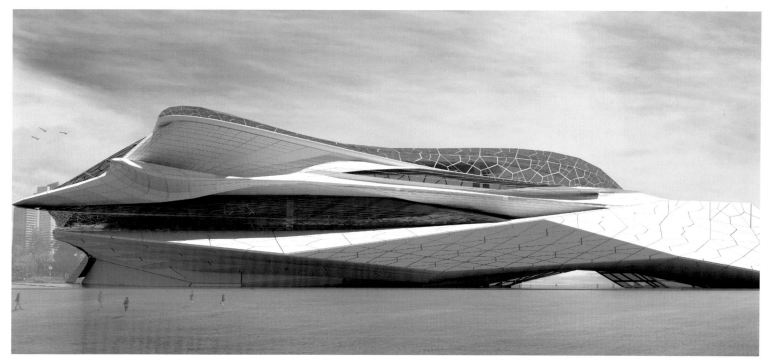

1. BRIDGE (CALATRAVA) 桥（卡拉特拉瓦）
2. PLAZA 广场
3. BOARDWALK 木板路

Visual Impact
Geometric Crystals
Intersected Trajectories

Project Overview

The project is located in Buenos Aires, the capital of Argentina. The museum is multi-layered and composed of radiant volumes and pieces which converge in a poly-operational, structural, and sensuous array of tectonics. The effect of multi-generative forms grouped and working together as a performative whole is a diverse set of fluid and crystallized components strategically placed for an outcome of a dynamic structure, which valiancy plays a role in the visual and experiential impact of the viewer.

Building Design

A museum is produced in which tectonics and space is operating in a field of "Fluxion Synergies". These fields of cooperative interactions of fluid forms and synergized lines which also produce a series of desired patterns like that of the classical line work of the "Guilloche", or creating "The Guilloche Effect".

The external cladding is an array of curved and crystallized formal geometries by which they perform with an organizational behavior pattern of trajectories and intersecting surfaces. The geometries created are resulted from guilloche patterns that have been derived from parametric Cartesian equations. By utilizing basic parametric equations for comprehensive surface decor and structure, it has allowed for a series of patterns which can be transformed into an effective tool for varied penalization and lattice structures. In addition to further investigate parametric formulas similar to the hypotrochoid, epitrochoid and hypocycloid are useful in this series of guilloche rosette formation.

The design is used varied sizes of polished and honed Glass Stone or White Neoparies to show the sleek surfaces and external cladding. The Museum has five levels of exhibition space; in addition the entry and exit are separated by the outdoor plaza to create a flow spaces that circulate up and around to experience the entire museum.

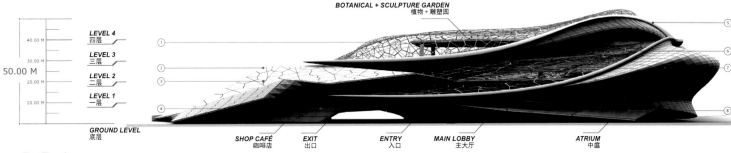

East Elevation
东立面

LEVEL 4 / 四层
LEVEL 3 / 三层
LEVEL 2 / 二层
LEVEL 1 / 一层
GROUND LEVEL / 底层

BOTANICAL + SCULPTURE GARDEN / 植物＋雕塑园
SHOP CAFÉ / 咖啡店
EXIT / 出口
ENTRY / 入口
MAIN LOBBY / 主大厅
ATRIUM / 中庭

1. CRYSTALLIZED STRUCTURAL FRAME W / DOUBLE GLAZING
1. 晶体结构构架 W / 双层玻璃结构
2. ULTRA WHITE _ CRYSTALLIZED NANO STONE PANELS _ POLISHED
2. 超白 _ 纳米结晶石面板 _ 抛光砖
3. CHROME PLATED REVELS W / LED STRIP INSERTS
3. 镀铬陶 W/ LED 条形刀片
4. EXIT / ENTRANCE 2 _ GLAZING
4. 出口 / 入口 2 _ 装配玻璃
5. STRUCTURAL STEEL TUBING _ PAINTED W / WHITE CAR PAINT
5. 结构钢管 _ 涂料 W / 白色车漆
6. CURVED GLAZING W / HAIRLINE STAINLESS STEEL FRAMES
6. 弧形玻璃 W/ 极细不锈钢框架
7. MONOCOQUE SHELL W / ULTRA WHITE STONE GLASS TILES
7. 单体横造梁 W/ 超白石玻璃瓦板
8. WHITE WASHED CONCRETE EXTERIOR WALL
8. 粉饰现浇混凝土外墙

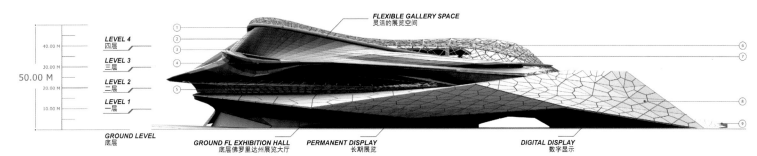

West Elevation
西立面

FLEXIBLE GALLERY SPACE / 灵活的展览空间
GROUND FL EXHIBITION HALL / 底层佛罗里达州展览大厅
PERMANENT DISPLAY / 长期展览
DIGITAL DISPLAY / 数字显示

1. STRUCTURAL STEEL TUBING _ PAINTED W / WHITE CAR PAINT
1. 结构钢管 _ 涂料 W / 白色车漆
2. ULTRA WHITE STONE GLASS TILES _ POLISHED
2. 超白石玻璃瓦板 _ 抛光砖
3. WHITE TINTED BLACK GLAZING
3. 白色着黑色玻璃砖
4. SEMI-MONOCOQUE SHELL W / CAST ALUMINIUM PANELS _ PAINTED WHITE
4. 半单体横造外壳 W/ 铸铝面板 _ 白色涂漆
5. CURVED GLAZING
5. 弧形装配玻璃
6. GLASS ATRIUM
6. 玻璃中庭
7. CURVED GLAZING W / HAIRLINE STAINLESS STEEL FRAMES
7. 弧形玻璃 W/ 极细不锈钢框架
8. POLISHED _ ULTRA WHITE NANO STONE PANELS
8. 抛光 _ 超白纳米石面板
9. LED LIGHT STRIPS
9. LED 灯条

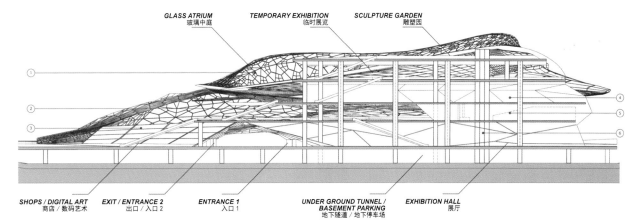

Section
剖面

GLASS ATRIUM / 玻璃中庭
TEMPORARY EXHIBITION / 临时展览
SCULPTURE GARDEN / 雕塑园
SHOPS / DIGITAL ART / 商店 / 数码艺术
EXIT / ENTRANCE 2 / 出口 / 入口 2
ENTRANCE 1 / 入口 1
UNDER GROUND TUNNEL / BASEMENT PARKING / 地下隧道 / 地下停车场
EXHIBITION HALL / 展厅

1. CRYSTALLIZED STRUCTURAL FRAME W / DOUBLE GLAZING
1. 晶体结构构架 W / 双层玻璃结构
2. ESCALATORS IN BLACK STAINLESS STEEL _ MIRROR FINISH
2. 黑色不锈钢的自动扶梯 _ 镜面抛光
3. BLACK HAIRLINE STAINLESS STEEL _ INTERIOR PANELS
3. 极细黑色不锈钢 _ 内饰板
4. GALLERY DISPLAY WALL IN ACRYLIC PANELS
4. 丙烯酸面板的画廊显示墙
5. AMPHITHEATRE SEATING
5. 圆形剧场座位
6. GLASS ELEVATORS
6. 玻璃电梯

江苏南京汽车博物馆
Nanjing Automobile Museum

设计单位：3GATTI Architecture Studio
委托方：海德投资集团有限公司
项目地址：江苏省南京市
建筑面积：15 000m²

Designed by: 3GATTI Architecture Studio
Client: Jiangsu Head Investment Group Co., Ltd.
Location: Nanjing, Jiangsu
Building Area: 15,000m²

汽车体验 "折纸" 概念

项目概况

南京汽车博物馆位于南京高新科技区，这个博物馆不仅以一种不同寻常的方式来展示汽车，还可以让人们通过汽车来感受这个建筑。设计师以"汽车体验"为理念，在这个过程中，车成为了探索、学习、展示的对象。该项目设计旨在呈现出一种不同于通常所见的开放式空间的一种新的构造。

建筑设计

设计整体以车辆为主角，因此，建筑结构在自然结合车辆、人类个体以及有机世界这三者以外，尤其突出对车辆这个主题的关注。

建筑以"折纸"为概念，打造了仿佛会飞翔的博物馆。建筑放弃了连接楼层的楼梯、墙壁、电梯等以人类为主要思考对象的衍生产物，取而代之的是旋转上升，一路绵延的大型坡道，可以让参展的车子直接开进去，具有很好的空间流动感。

建筑的主要结构是一条由玻璃划分为内外区域的坡道，内部区域设为人行区域，因此坡度较小；外部设为车行区域，坡度较大。从宏观来说，整个建筑的层面实际与路面相似，类似于城市高架的行车道，观众可以驾驶自己的汽车进入展馆，在外围坡道的曲折起伏之中，体验驾驶的极致乐趣。

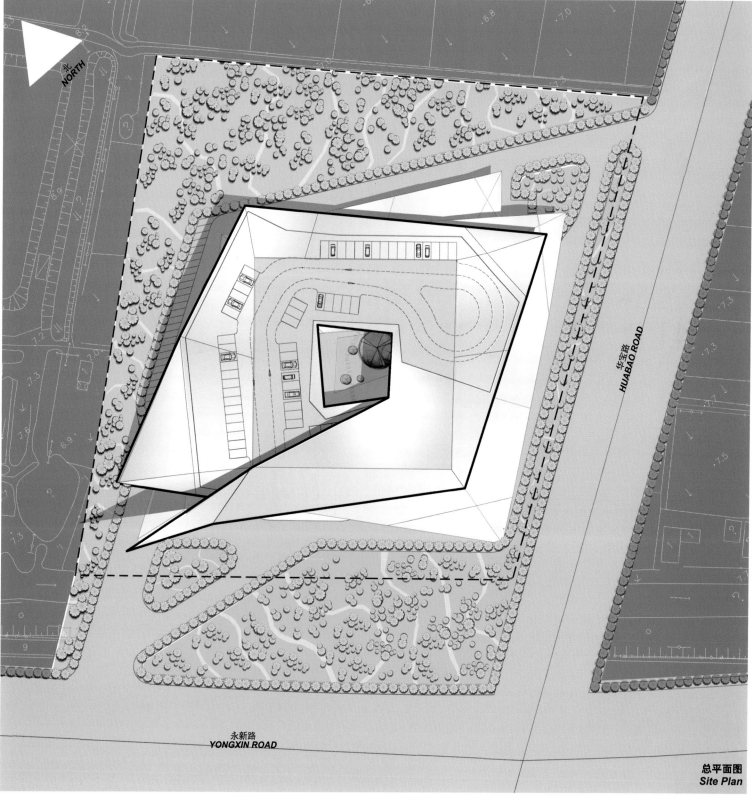

总平面图
Site Plan

| 文体建筑方案集成 | CULTURAL AND SPORTS ARCHITECTURE PROGRAM INTEGRATION |

底层平面图
Ground Floor Plan

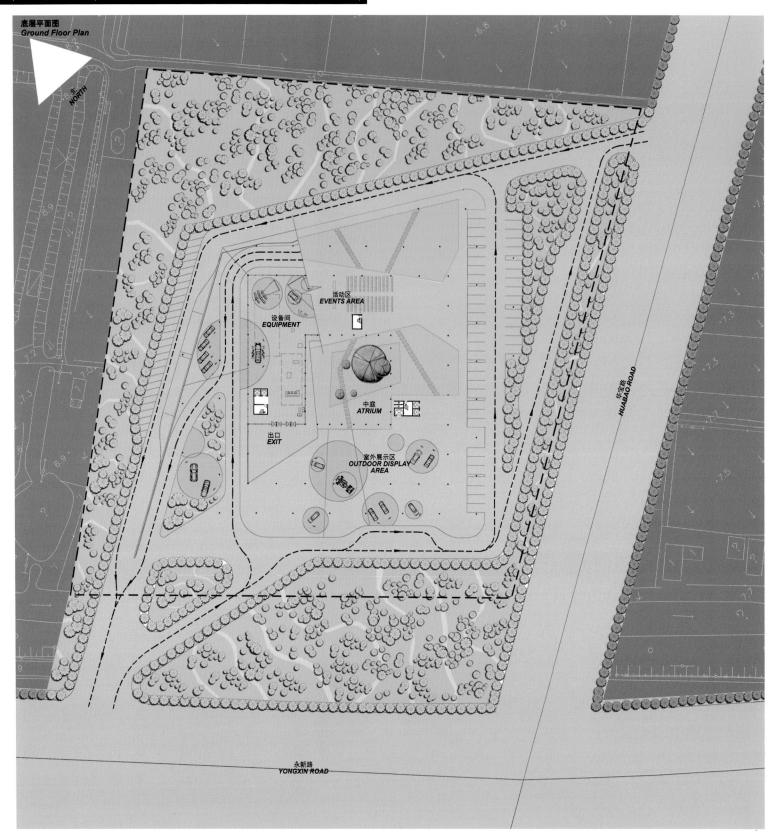

视线分析图
View Sight Analysis Chart

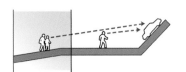

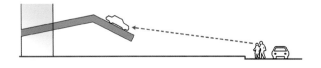

外部坡道展览观光图解
EXTERNAL RAMP EXHIBITION SIGHT SCHEME

会议室视线看展示
CONFERENCE ROOM VIEW TO SEE DISPLAYER

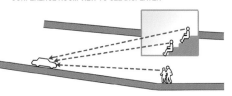

室外视线看展示
URBAN VIEW TO SEE DISPLAYER

室内视线看展示
INTERIOR VIEW TO SEE DISPLAYER

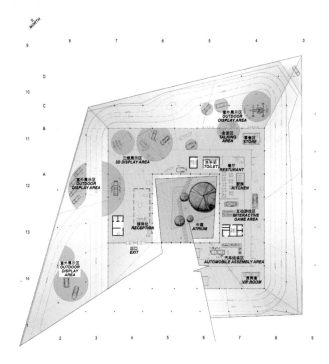

一层平面图
First Floor Plan

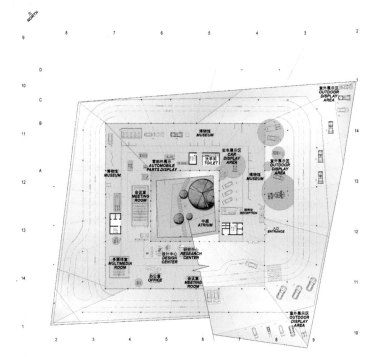

二层平面图
Second Floor Plan

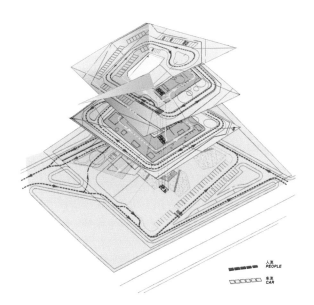

立体流线分析图
3D Flow Line Analysis Chart

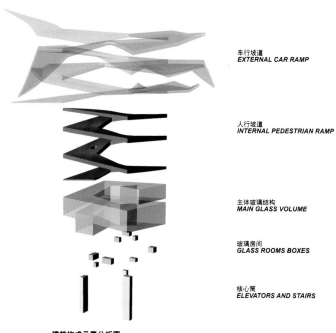

建筑构成元素分析图
Explode Model Scheme Analysis Chart

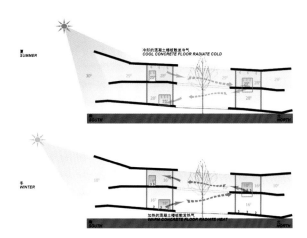

空调分析图
Air Conditioner Analysis Chart

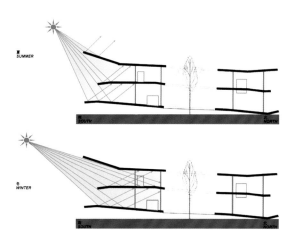

日光分析图
Sunlight Analysis Chart

| 52-53 |

| 文体建筑方案集成 | CULTURAL AND SPORTS ARCHITECTURE PROGRAM INTEGRATION |

A-A 剖面图
Section A-A

B-B 剖面图
Section B-B

C-C 剖面图
Section C-C

D-D 剖面图
Section D-D

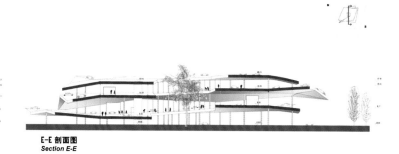
E-E 剖面图
Section E-E

Car Experience "Origami" Concept

Project Overview

Nanjing Automobile Museum is located in Nanjing High-tech Zone, both with an unusual way to show the car and allow people to feel this building by car. The designers with the "Car Experience" for the concept make the car become an object to explore, a technology to study, an article to display. Here the world of the automobile intersects with the human and organic world creating a new tectonic structure with methods differing from the usual flat open spaces.

Building Design

The whole design is with the car for the point of reference. Therefore, the building structure is naturally besides combined with car, human individual as well as organic world, especially prominent the concern for the subject of "Car".

The buildings with "Origami" for the concept would seem to create a flying museum. Here one will not find stairs to different floors, walls and elevators, but ramps which wind sinuously upwards creating a fluid conception of space and where the flux of cars can move freely and reach the different levels of the edifice.

The principal structure of the building is a spiral ramp with a glass partition dividing the exterior from the interior. In the internal part, reserved for pedestrians, the incline is more gradual, whereas the exterior and steeper side is for the transit of cars. On an overall scale the area tectonically resembles a road, with a structure similar to that of an elevated motorway or a car park, but on a more human scale, the structure is as complex, ergonomic and sophisticated as the interior of a car.

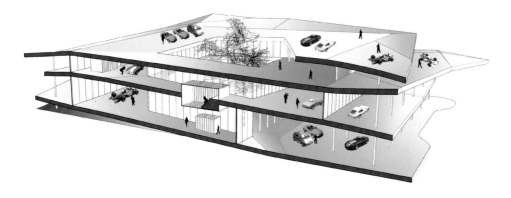

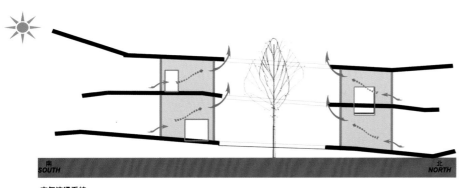

空气流通系统
Ventilation

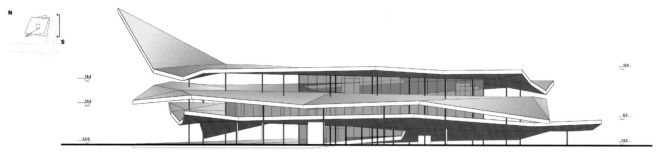

东南立面图
Southeast Elevation

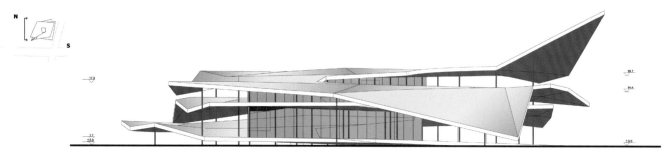

西北立面图
Northwest Elevation

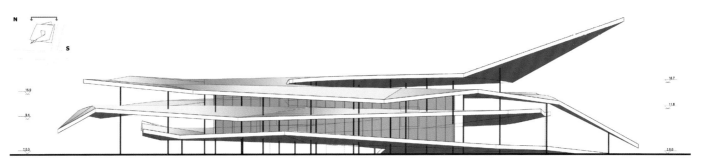

东北立面图
Northeast Elevation

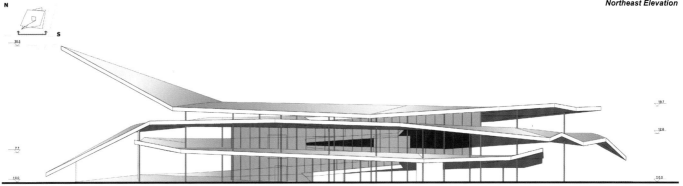

西南立面图
Southwest Elevation

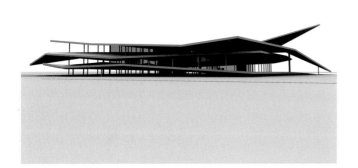
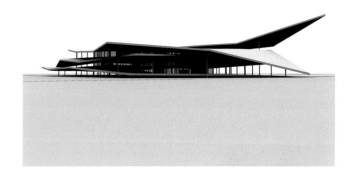
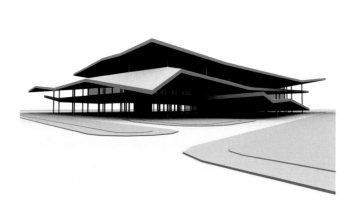

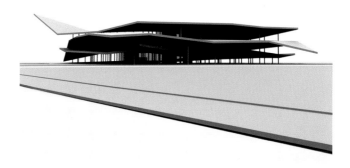

阿联酋迪拜中东现代艺术博物馆
The Museum of Middle Eastern Modern Art, Khor Dubai, UAE

设计单位：UNStudio
委托方：迪拜地产
项目地址：阿联酋迪拜
建筑面积：41 200m²

Designed by: UNStudio
Client: Dubai Properties
Location: Dubai, UAE
Building Area: 41,200m²

独桅帆船
露天剧场
精品酒店

项目概况

中东现代艺术博物馆位于文化村艺术之岛上，它汇集了海的元素和迪拜的航海传统。它作为迪拜新文化中心的一部分，是由阿联酋迪拜的副总统和总理共同推出的。借助该项目，我们有机会来认识一个全新类型的博物馆：它由一个充满活力的城市中心组成，那里有专业人士、收藏家和相互会面的公众；此外，中东现代艺术博物馆还将成为该市内的社区建设机构，并成为中和外来游客和当地居民相互交流的"调色板"。

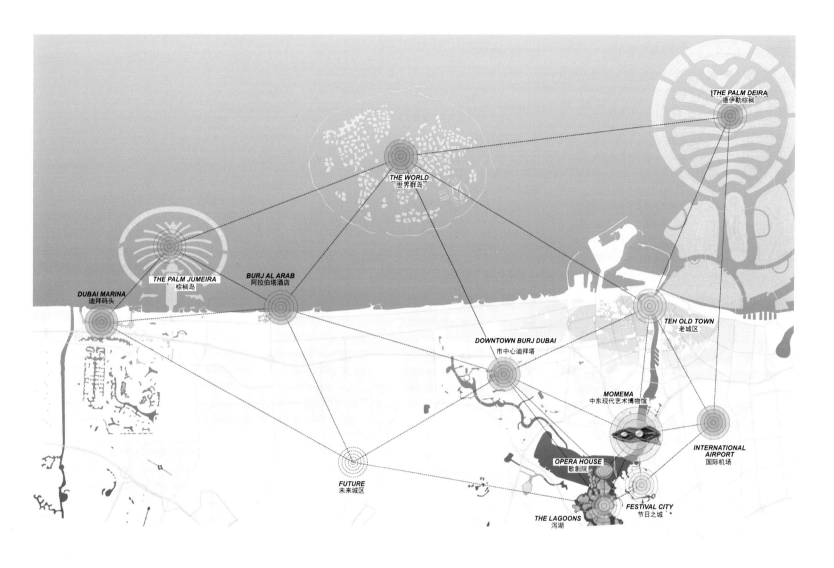

建筑设计

博物馆的设计初衷是：使阿联酋成为多元文化中心的战略枢纽。该建筑的定位是充分利用了在文化村突出的位置，凭借其像船头升起的独桅帆船，使该建筑可欣赏到周边地区的全景。

该博物馆包括一个可用于现场表演的露天剧场，在这里将举行各种各样的空间展示，例如文化展览、艺术画廊等；一个用于展示本地及国际艺术的展览大厅和博物馆陈列室、一个作为传统单桅帆船建设的造船厂以及一些商业和零售区。

除此之外，中东现代艺术博物馆提供了一个拥有 60 个房间的精品酒店和一个精品零售长廊，以及一个可 360° 全视角欣赏到迪拜河的顶级高端特色餐厅。

在内部，这个新馆突破传统的强制受限光学领域，并在有限的空间里激发了无限的沉思，使参观者感受到博物馆内部的动态和可感知性。在博物馆内，公众、事件、艺术与商业互相碰撞，并相互以此为生，介质和时间可以毫不费力地排列在一起，并可进行重新安排，进一步展现了该建筑设计的灵活性与机动性。

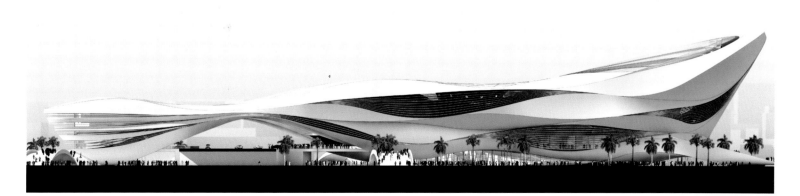

Dhow
Outdoor Theatre
Boutique Hotel

Project Overview

MOMEMA is situated on Art Island part of Culture Village, and it brings together elements of the sea and Dubai's tradition of seafaring. As a part of a new cultural hub in Dubai, it was launched by the Vice President and Prime Minister of the Dubai, UAE. With the help of the project, we have a chance to recognize an entirely new type of museum, which consists of a vibrant urban center, where professionals, collectors and public meet each other. In addition, MOMEMA will be a community-building institution within the city, and offer to both visitors and residents a continuously "changing palette" of experiences and events.

Building Design

The Museum is designed to make the UAE as a strategic hub of the multicultural center. The building is positioned to take full advantage of the prominent location in the Culture Village. With its Dhow-like prow rising up, the building offers panoramic views to the surroundings, and vice versa.

The Museum includes an outdoor theatre for live performances, which will hold a variety of spaces to exhibit Arts and Culture such as exhibitions and art galleries; a showcase of hall and museum gallery for local and international art exhibition; and a shipyard as a traditional dhow builder as well as some commercial and retail districts. In addition, MOMEMA offers a boutique hotel with 60 keys and a boutique retail promenade on the active Culture Village waterfront, as well as a high end signature restaurant on the top level, with 360 degree views of Dubai Creek.

Inside, the new Museum broke the restricted optical field and stimulated the infinite contemplation in the limited space to make the visitors feel the dynamic and perception inside the Museum. The public, event, art and business inside the Museum meet each other and feed on each other. Formats, mediums, and times can be effortlessly arranged together and rearranged. It further demonstrates the flexibility and mobility of architectural design.

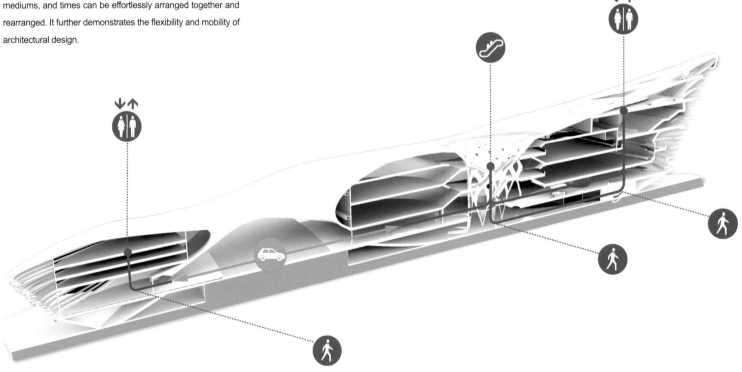

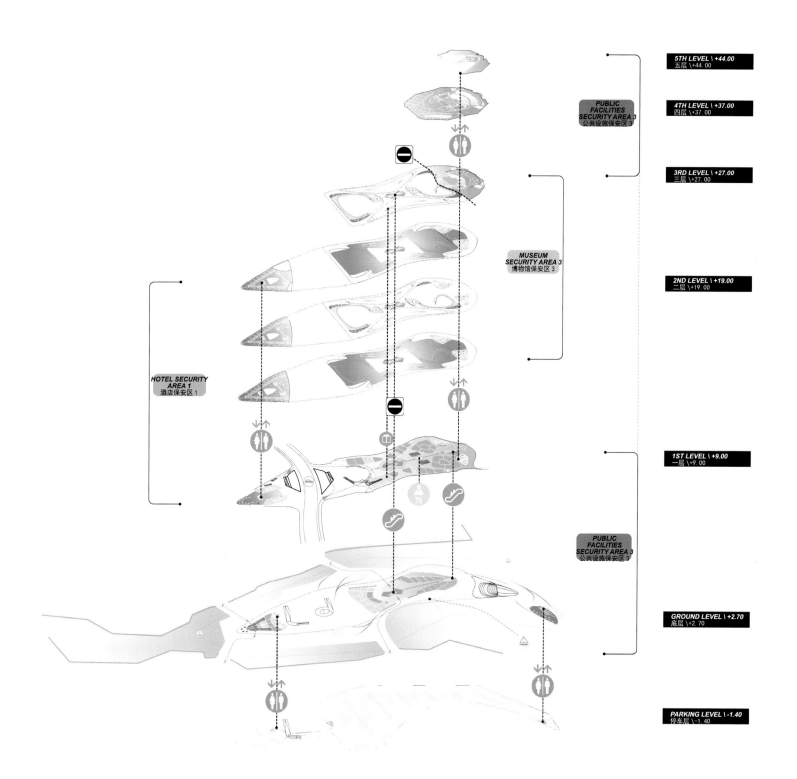

立陶宛维尔纽斯博物馆
Museum in Vilnius

设计单位：Zaha Hadid Architects
委托方：立陶宛维尔纽斯市
项目地址：立陶宛维尔纽斯市
项目面积：98 500m²
渲染图纸：Zaha Hadid Architects

Designed by: Zaha Hadid Architects
Client: City of Vilnius, Lithuania
Location: Vilnius, Lithuania
Area: 98,500m²
Renders and Drawings: Zaha Hadid Architects

文化元素
加速曲线
悬浮视觉

项目概况

2009年，维尔纽斯成为欧洲文化之都。正因为有了对艺术的浓厚兴趣和不懈追求，维尔纽斯继而发展成为一个文化中心，而艺术文化和公共生活之间的相互连接是至关重要的。本案博物馆的拟建将成为一个可以尝试将空间的复杂性和运动性实验性地结合在一起。

建筑设计

项目的规划建立是由所罗门·R·古根海姆基金会和国家冬宫博物馆共同开展的可行性研究的一部分，它将是这座城市文化生活的一个关键元素。

对于未来的建筑语言，其设计要点是扎哈·哈迪德建筑师创新研究轨迹的一部分：揉合最新的数字化设计技术和制造方法，使建筑的特性加速曲线和雕刻表面调制从图纸开始实现无缝转移。

博物馆的雕塑体块用流动性、速度和亮度的概念术语来进行设计。该建筑看起来就像一个神秘的浮动物体，看似没有重力。曲线线条与建筑的轮廓拉长相呼应，体现了一个神秘的存在，与维尔纽斯商业区的垂直天际线形成对比，这也是城市新文化意义的体现。

| 文体建筑方案集成 | CULTURAL AND SPORTS ARCHITECTURE PROGRAM INTEGRATION |

Site Plan
总平面图

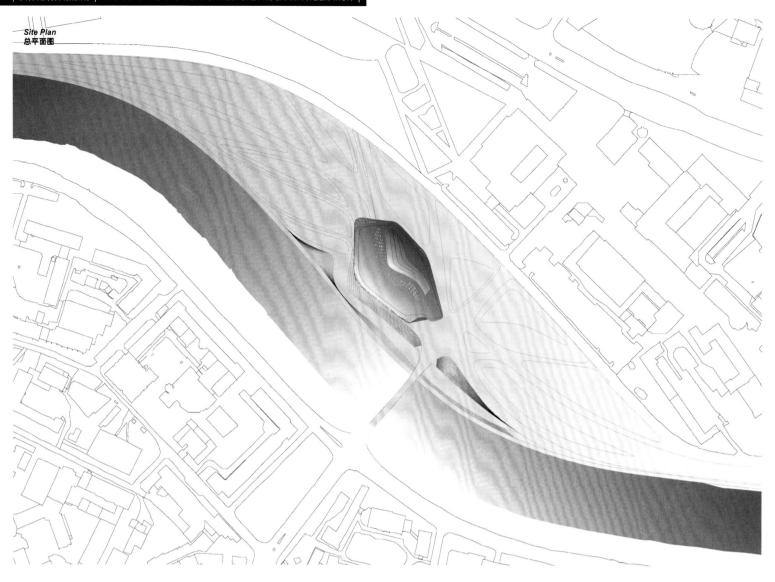

Cultural Elements Acceleration Curve Suspended Vision

Project Overview

Vilnius will be the European Capital of Culture in 2009 and has a long history of art patronage. With such an interest in the arts, Vilnius will continue to develop as a cultural centre where the connection between culture and public life is critical. This museum will be a place where you can experiment with the idea of galleries, spatial complexity and movement.

Building Design

The creation of the new centre of contemporary and media art in Vilnius is part of a feasibility study undertaken by the Solomon R. Guggenheim Foundation and The State Hermitage Museum. It will be a critical element of the cultural life of the city.

The design points towards a future architectural language that is part of an innovative research trajectory within Zaha Hadid Architects; embracing the latest digital design technology and fabrication methods to enables a seamless transfer of Zaha Hadid Architects' characteristic acceleration curves and sculpted surface modulations from drawing board to realization.

The museum's sculptural volume is designed with the conceptual terms of fluidity, velocity and lightness. The building appears like a mystical floating object that seemingly defies gravity. Curvilinear lines echo the elongated contours of the building, offering an enigmatic presence that contrasts with the vertical skyline of Vilnius' business district. It is a manifestation of the city's new cultural significance.

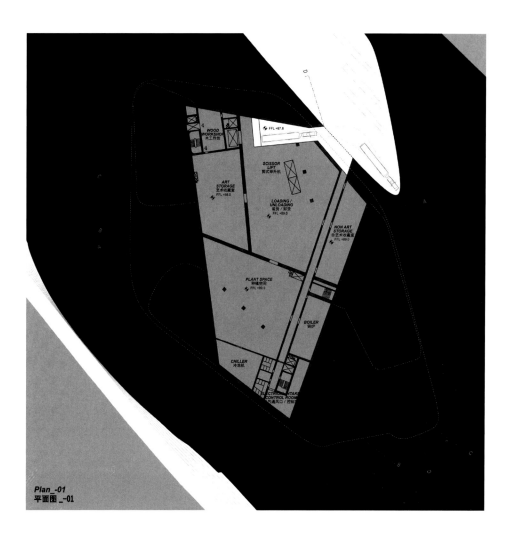

Plan_-01
平面图_-01

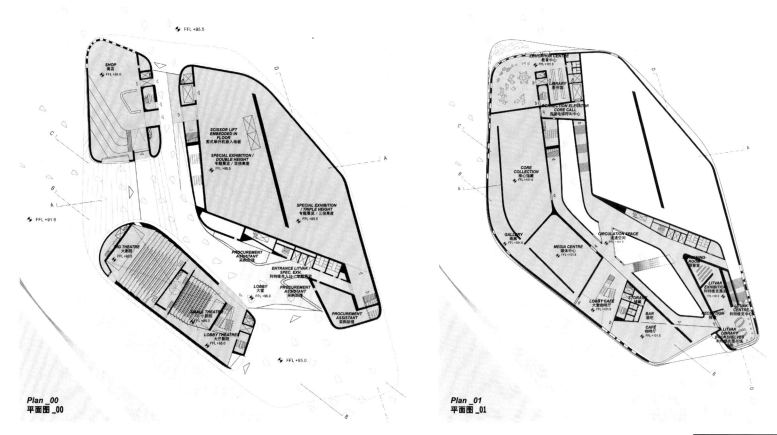

Plan_00
平面图_00

Plan_01
平面图_01

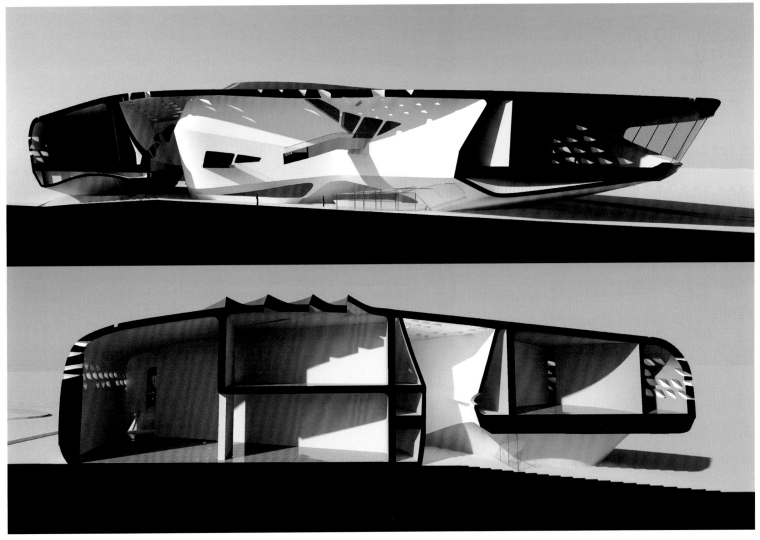

辽宁大连城市规划博物馆
Dalian Urban Planning Museum

设计单位：Architects Collective ZT-GmbH
委托方：大连市
项目地址：辽宁省大连市
总建筑面积：25 000m²
设计团队：Andreas Frauscher　Patrick Herold　Richard Klinger
　　　　　Kurt Sattler　　　　Matthew Tam

Designed by: Architects Collective ZT-GmbH
Client: City of Dalian
Location: Dalian, Liaoning
Gross Floor Area: 25,000m²
Design Team: Andreas Frauscher, Patrick Herold, Richard Klinger,
　　　　　　Kurt Sattler, Matthew Tam

**街道枢纽
三角面
低能耗**

项目概况
大连城市规划博物馆是大连市中心的一道门户，它将几条繁忙的街道重新引向主干道。这座耗资2 000万欧元的建筑尽量减少能耗，创造舒适和稳定的室内环境。内外部空间是连续的，形成与城市脉络之间的物理联系。内部设置了休息室、阳台、步行桥和平台，参观者可以方便的在其间行走，并营造了与城市之间的视觉连接。

建筑设计
该建筑的肤色由透明和不透明的三角形面闭合组成，这个简单而正规的规划使建筑物避免夏季过热，并在冬天有足够的自然光线。该建筑通过使用低科技和高科技的方法来使能源消耗最小化，并创造舒适和稳定的室内环境。主要的建筑目标是与连续内部或外部空间创建一个非常的公共建筑，从而营造城市环境的物理关系。

在25 000m²的建筑外部，上面的三层扭转并与三个盒子结构的边角相连，形成连续的向上盘旋。闭合与打开的三角面的安排使闭合和揭示内部空间与城市环境之间相互影响。

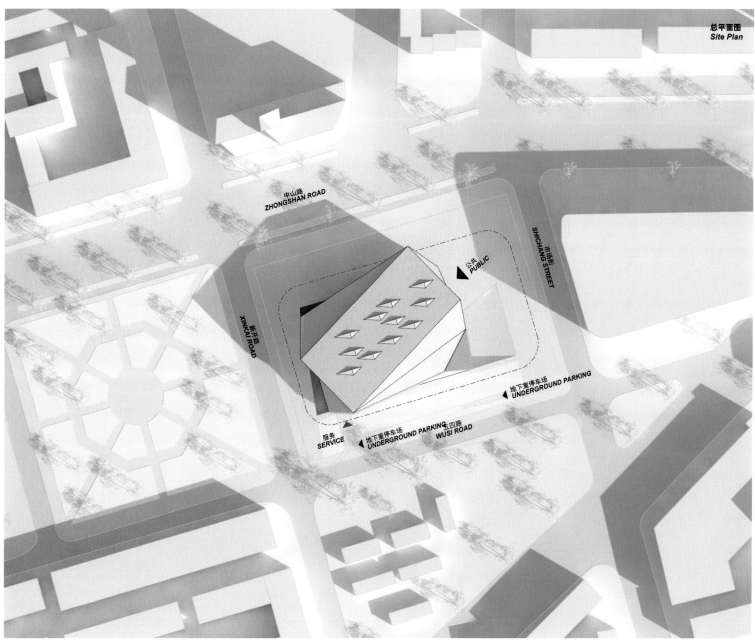

总平面图
Site Plan

| 文体建筑方案集成 | CULTURAL AND SPORTS ARCHITECTURE PROGRAM INTEGRATION |

流通图
Circulation

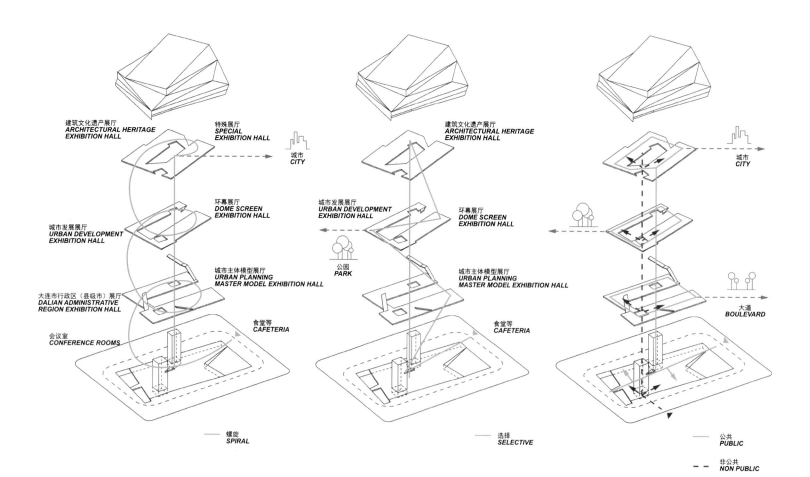

Street Hub
Triangle
Low energy consumption

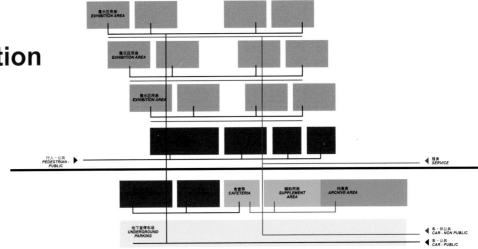

Project Overview

Dalian Urban Planning Museum in Dalian is a gateway to the city center, and a few busy streets will be redirected to the main roads. This building spent $20 million Euros to minimize the energy consumption, creating a comfortable and stable indoor environment. Internal and external space is continuous to form the physical contact among the urban contexts. The interior space sets up lounge, balcony, pedestrian bridge and platform, and the visitors can easily walk among them, which create a visual connection to the city.

Building Design

The building skin consists of a careful arrangement of transparent and opaque triangular surfaces closing and reveals the interior of the museum to the urban surrounding. This simple formal strategy keeps the building from overheating in summer and allows enough natural light in winter. The building used low-tech and high-tech methods to minimize energy consumption and created comfortable and stable interior environments. The primary architectural ambition is to create a very public building with continuous inside / outside spaces creating physical relationships to the urban context. In the 25,000 square meters of building exterior, the upper three rotating levels connect to the corners of three boxes so that it forms a continuous upwards spiral. The arrangement of closed and opened triangular surfaces creates an inter-play of closing and revealing interior spaces and the urban environment.

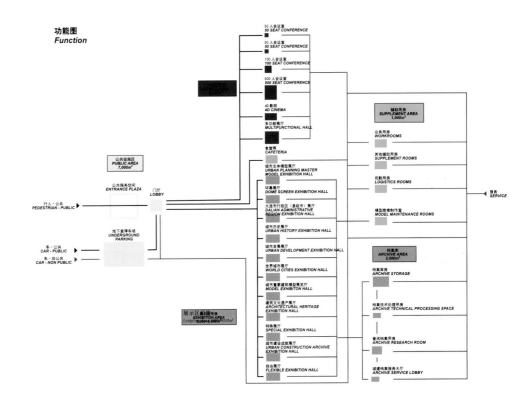

底层
Ground Floor

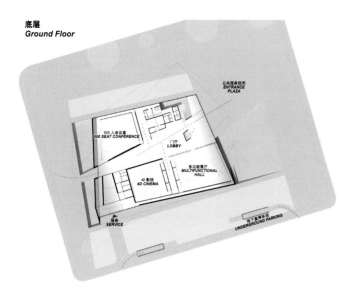

一层平面图
1st Floor Plan

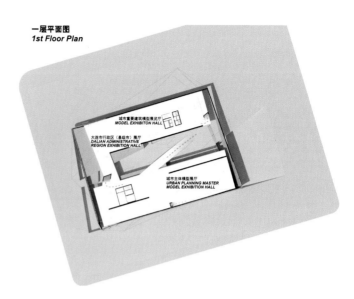

二层平面图
2nd Floor Plan

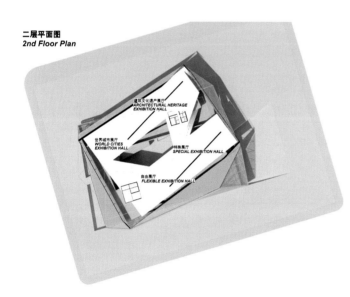

三层平面图
3rd Floor Plan

形态
Morphology

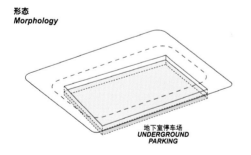
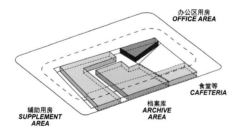
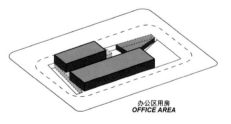

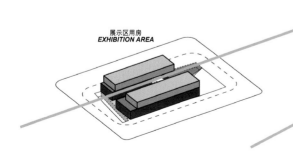
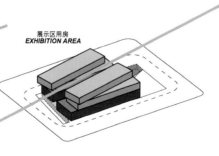
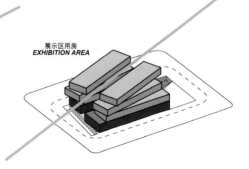

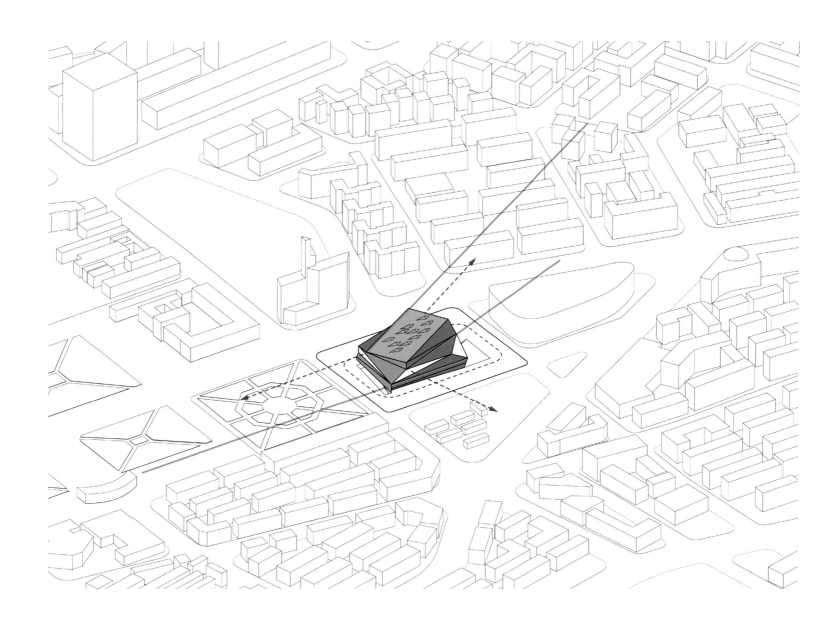

东立面
East Elevation

北立面
North Elevation

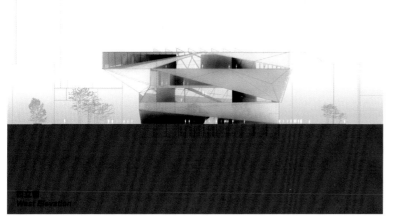

西立面
West Elevation

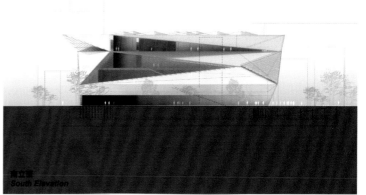

南立面
South Elevation

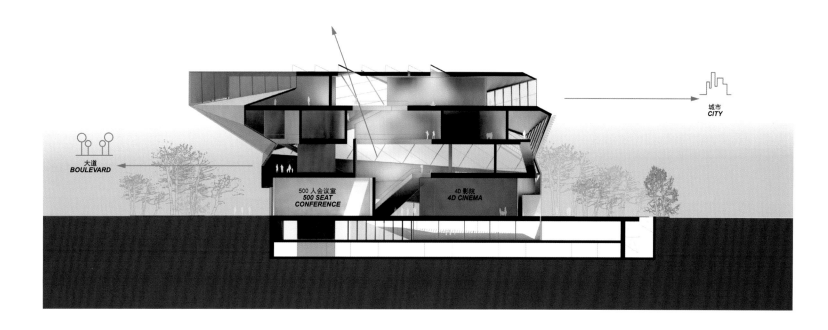
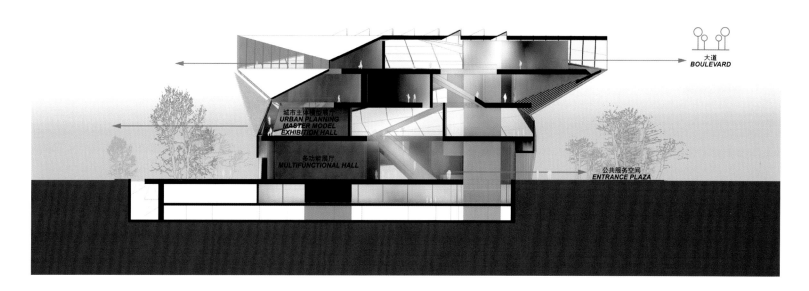

江苏武进第八届中国花卉博览会科技馆 & 特色馆
Pavilion of Art & Pavilion of Science, The 8th China Flower Expo

设计单位： LAB ARCHITECTURE STUDIO
委托方： 江苏花博投资发展有限公司
项目地址： 江苏省武进市
建筑面积： 8 000m²

Designed by: Lab Architecture Studio
Client: Jiangsu Huabo Investment and Development Co.,Ltd.
Location: Wujin, Jiangsu
Building Area: 8,000m²

水和花瓣
有机外壳
孪生设计

项目概况

该项目位于西太湖上，设计主要探讨了水和花瓣之间的关系美学——整个花瓣状的设计使本项目从视觉和构造上都呼应了环境，创造出独特的有强大感染力的场所，成为了第八届中国花博会的一大亮点。此外，项目同时也对自然景观与标志性文化建筑的张力关系进行了新的尝试。两个相互关联的展馆（艺术馆，科学馆）在设计上利用了自然地域特点并使用了生态建筑的设计原则，在提高艺术、建筑和环境之间关系的同时，也体现了花博会推广可持续性以及自然生态的理念。

建筑设计

作为文化类型的建筑，它具有社会、文化方面的象征意义。整个建筑设计对于能源、资源的合理利用，环境污染的减少，建筑环境质量的提高以及生态设计理念的综合运用等各方面进行了全面的考量，旨在满足建筑生态要求的同时，设计出具有创意的形式和空间。

建筑体量中的两个展馆拥有 140m 以上的延伸有机外壳形体，该形体由三维网格钢构件构造，并由 8 000 多个控制点和高度透明的 ETFE 防雨膜组成。建筑顶棚采用成熟的膜结构以及网架，建筑立面为玻璃幕墙系统，有自动开启功能，使整个建筑达到全自然通风采光。这些举措大大减低了能源的消耗，达到了低碳绿色环保的标准。

艺术馆和科技馆被视为孪生设计，它们在不同的基地位置上共同拥有类似的表达形式及基因元素，有利于不同的展览流线与建筑语言的组织。艺术馆展示的是规模较小的展品，所以建筑空间具有较强的封闭性和集中性。科技馆设有大型投影和动画，需要更大的、能让参观者互动的空间，其建筑形式也就更为开放，使人们在展览之间有一个呼吸的空间。两个展馆的曲线式流线组织，使得大规模展会人流的出入交通流线得以最优化。

| 文体建筑方案集成 | CULTURAL AND SPORTS ARCHITECTURE PROGRAM INTEGRATION |

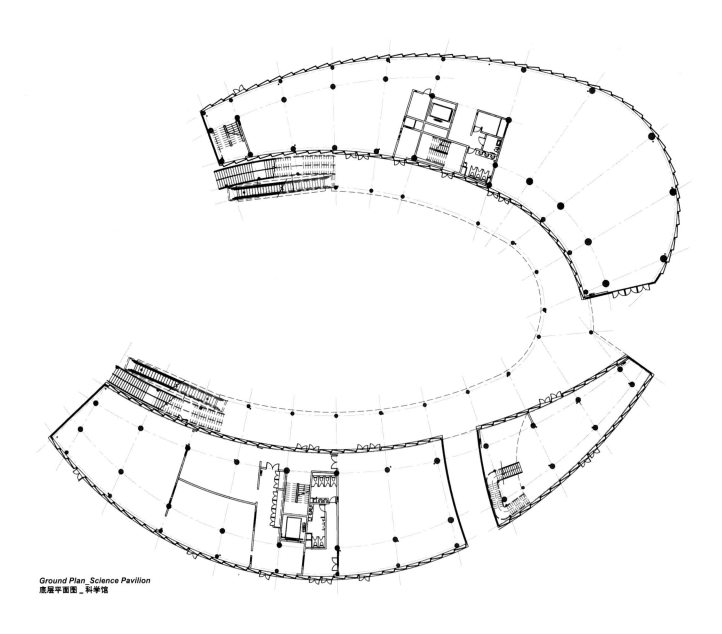

Ground Plan_Science Pavilion
底层平面图_科学馆

Water and Petals Organic Shell Twins Design

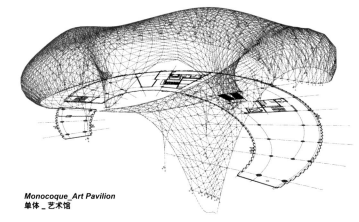

Monocoque_Science Pavilion
单体_科学馆

Monocoque_Art Pavilion
单体_艺术馆

Project Overview

The project is located on the West Tai-hu Lake, and the design mainly explores the aesthetics of the relationship between the water and the petals—the entire petal-like composition of the project responds visually and tectonically to its environment to create a distinct and powerful sense of place as well as a breathtaking backdrop for the 8th Chinese Flower Expo. Also it makes an experiment the tension between the Natural landscape and the iconic culture body. The designs of the two related pavilions (Art Exhibition Pavilion, Science Exhibition Pavilion) draw upon the natural geographical features of the area and use ecological building design principles to improve the relationship between art, architecture, and the environment while furthering the Flower Expo building society's view to promote sustainability and the natural ecology.

Building Design

The building as a cultural type of architecture is with the symbolic significance in the aspects of society and culture. The whole building design has been made an comprehensive consideration for all aspects like the rational use of energy and resources, reducing environmental pollution, improving the quality of the built environment and the integrated use of eco-design concepts, to meet the ecological requirements of the building, while to design creative form and space.

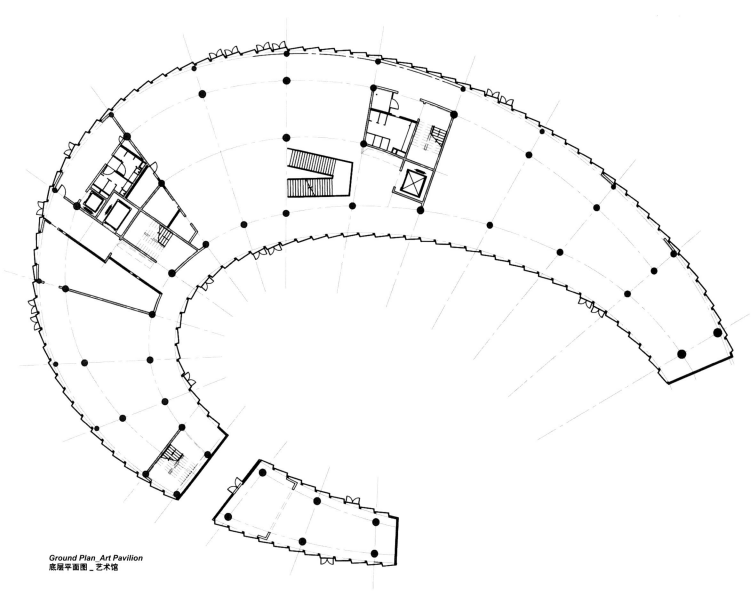

Ground Plan_Art Pavilion
底层平面图_艺术馆

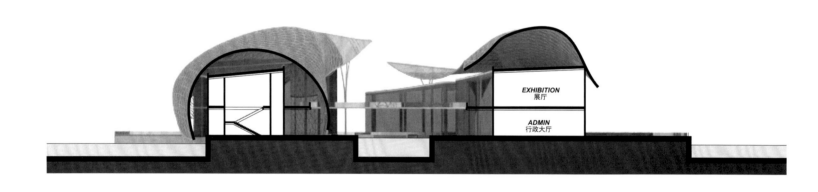

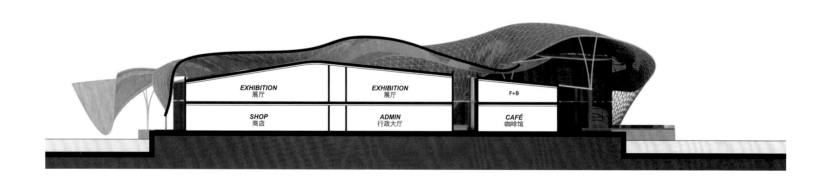

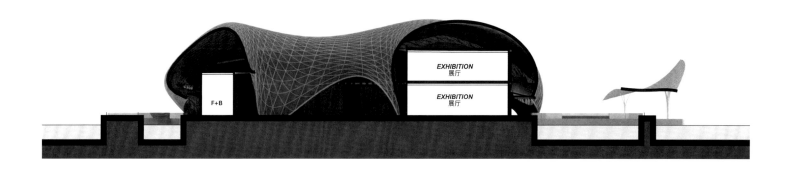
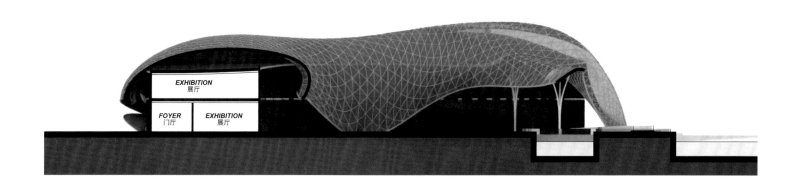

Both pavilions have a 140+ meter expanse of sweeping, curvilinear forms constructed of tri grid steel members; with more than 8000 control points, in association with highly translucent ETFE weatherproof membrane. The building roof adopts the mature membrane structure and grid, and the building façades is for glass curtain wall system with the function of automatic opening so that it makes the whole building be natural ventilation in a day and light. These measures greatly reduce the energy consumption and achieve the standard of low-carbon green environmental protection.

The pavilions, both the Art Exhibition Pavilion and Science Exhibition Pavilion, operate as twinned, but separate sites, following similar formal expressions and DNA, as befits their different exhibition sequences and architectural language. The Art Exhibition Pavilion is more enclosed and intimately focused, with smaller displays and objects for viewing and study. The Science Exhibition Pavilion displays projected films and animations, which is more outbound arrangement. It allows breathing space for visitors between the visual exhibitions. The exhibition relies on large audience interactive with the forums and spaces. Two pavilions curvilinear arrangements to deal with large-scale expo of passengers entering and exiting traffic and exhibitions circulation optimization results.

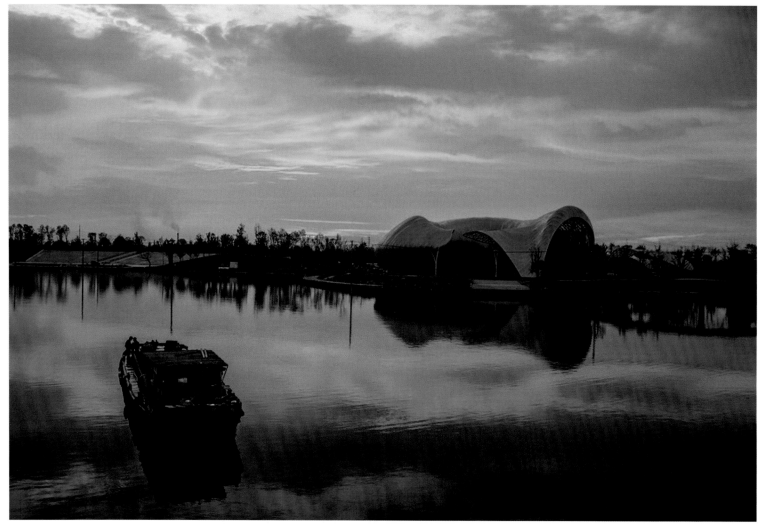

宁夏银川规划展览馆
Yinchuan Urban Planning Exhibition Hall

设计单位：绿舍都会
委托方：银川市规划局
项目地址：宁夏回族自治区银川市
项目面积：25 302m²

Designed by: SURE Architecture
Client: Yinchuan Planning Bureau
Location: Yinchuan, Ningxia
Area: 25,302 m²

贺兰石
穆斯林文化
城市沙盘

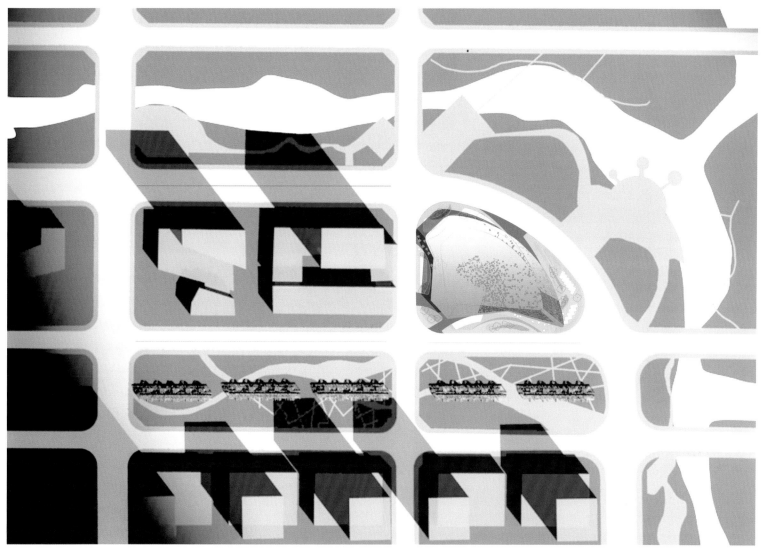

项目概况

银川规划展览馆位于银川市德馨公园西南角，占地面积 8 415m²。馆内设有城市沙盘模型、历史文化各城屋、展廊、小展厅等共 16 个展区。通过现代科技手段，图文资料、多媒体影像等多元手段，全面反映了银川 2 000 多年来，特别是建国和改革开放以来的城市变迁及发展脉络，以及银川城乡规划建设取得的辉煌成就。

建筑设计

贺兰石是宁夏五宝之一，素有"一端二歙三贺兰"之说，该项目的设计灵感就是来自于此。此外，银川也是穆斯林文化的重要集中地之一，设计师们综合两种文化背景，利用宁夏的贺兰石，再加上一些中国手艺，把这些贺兰石打造成穆斯林图案，将银川的历史、中国艺术石的传统与穆斯林文化都融合在该规划展览馆当中。

展览馆最引人注目的是一幅巨大的现代城市沙盘，站在二楼向下望去，银川环城高速路内，高楼林立、路网纵横、湖泊连片，湖城银川的城市风貌尽收眼底。

银川规划展览馆的成功开馆，不仅证实了该设计在展览展示行业中的坚强实力，更为银川城市对外宣传打造了新名片、城市的会客厅，它将对银川发展脉络的展现及城乡建设规划起到不可磨灭的作用。

建成后的银川规划展览馆必将成为见证历史的标志性建筑，作为中国和阿拉伯轴线的起点，银川也将承载历史使命，成为中国与阿拉伯世界之间的桥梁。

1st Floor Plan
一层平面图

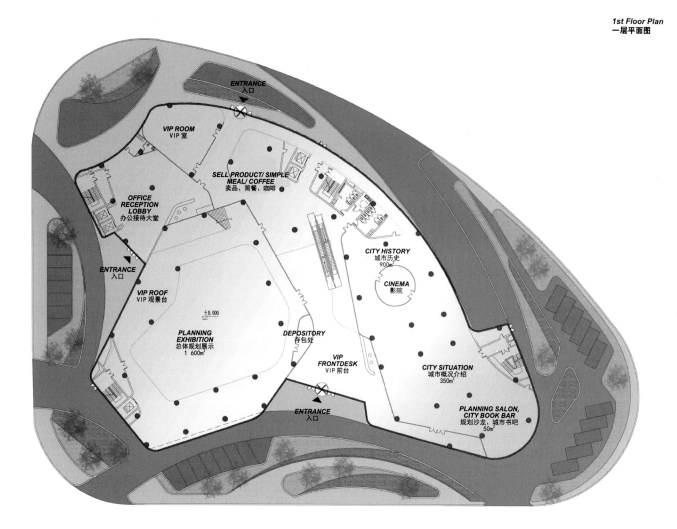

2nd Floor Plan
二层平面图

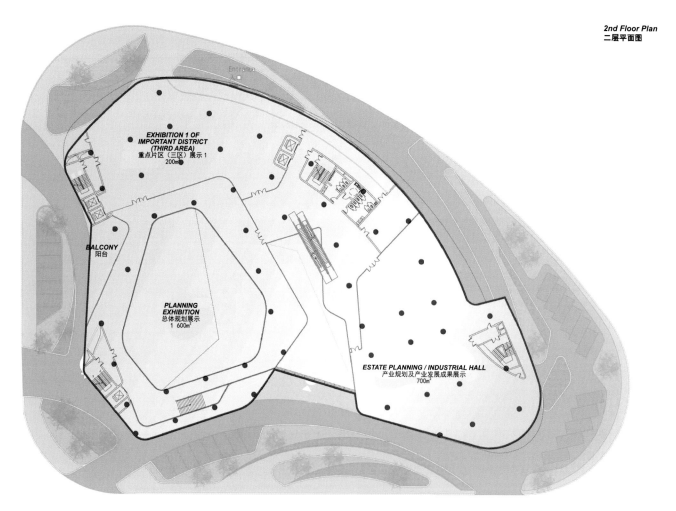

Helan Stone
Muslim Culture
City Sandbox

Project Overview

Yinchuan Planning Exhibition Hall is located in the southwest corner of Dexin Park, covering an area of 8,415m^2. The Hall houses a city sandbox model, famous historical and cultural houses, galleries, small exhibition halls and so on with a total of 16 exhibition areas. Through multiple means of modern technology, graphic data and multimedia video, it fully reflects the change and development context of the city as well as the brilliant achievements of Yinchuan urban and rural planning and construction of the city over 2,000 years, especially, since the founding and the reform and opening up of the country.

Building Design

Helan stone is the head of the five treasures of Ningxia; Helan Yan is "one end of the Xi three Helan". And the inspiration is from this. At the same time Yinchuan city is one place of full of Muslim Culture. The architects combine the two cultural backgrounds and use the West Xia Helan stone, and as some Chinese techniques we start to work in the stone to create the Islamic pattern. The history of Yinchuan, the tradition of the Chinese Art Stone and the Muslim Culture were fusion in The Urban Planning Exhibition Centre.

The most striking in the Exhibition is a huge modern city sandbox. Looking down from the second floor is skyscraper everywhere, road network, lakes contiguous in the Yinchuan ring highway, that is to say, Yinchuan, the Lake City, with the panoramic view is all in your eyes.

The successful opening of Yinchuan Planning Exhibition Hall not only confirms our strong strength in the exhibition industry, but more promotes for Yinchuan city to create a new card and the drawing room of the city, so it will play an indelible role on the development of urban context and urban and rural construction planning.

After the completion of Yinchuan Planning Exhibition Hall, it will become a landmark of the history as evidence and will be the starting point of the China and Arab Axis; furthermore, Yinchuan will bear the historical mission of China's bridge to the Arab world.

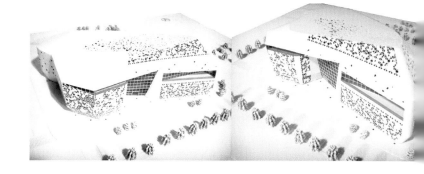

3rd Floor Plan
三层平面图

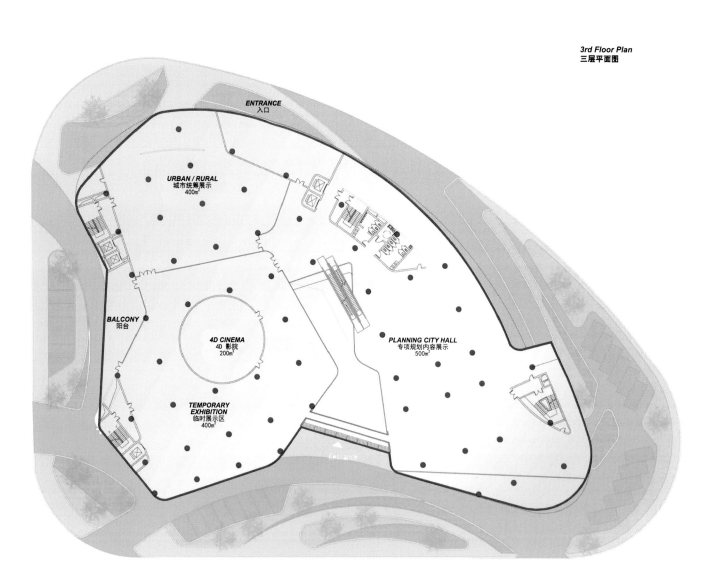

4th Floor Plan
四层平面图

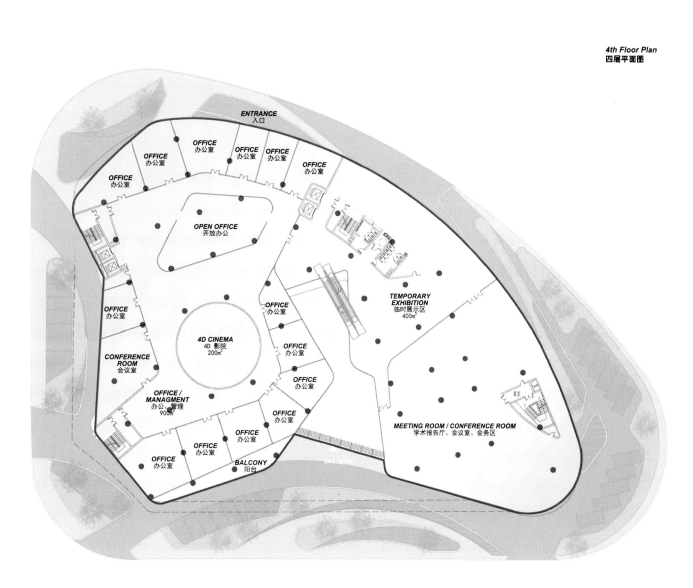

| 文体建筑方案集成 | CULTURAL AND SPORTS ARCHITECTURE PROGRAM INTEGRATION |

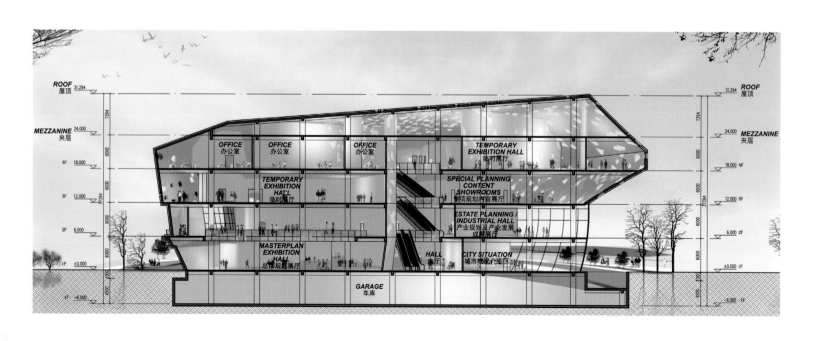

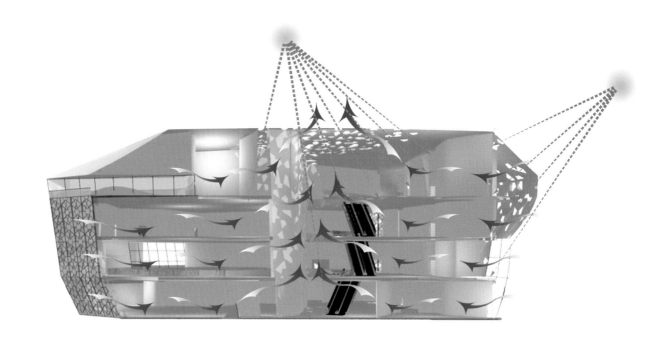

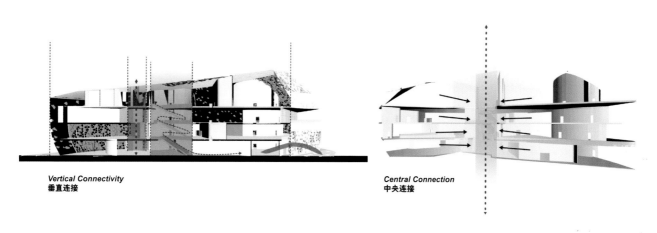

Vertical Connectivity
垂直连接

Central Connection
中央连接

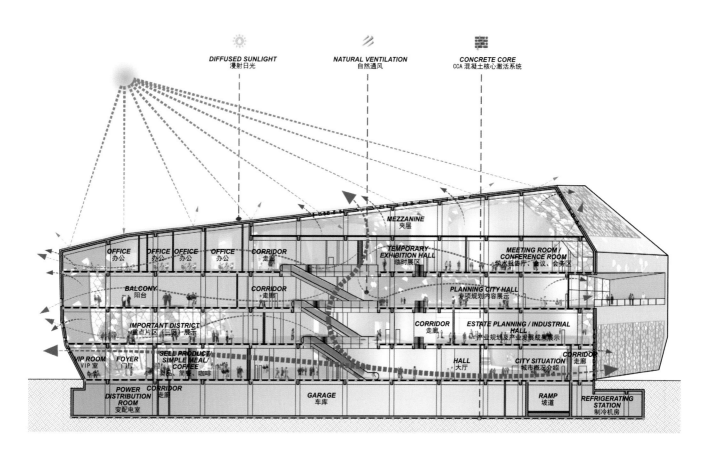

DIFFUSED SUNLIGHT
漫射日光

NATURAL VENTILATION
自然通风

CONCRETE CORE
CCA混凝土核心激活系统

Exhibition Flow
展览流线

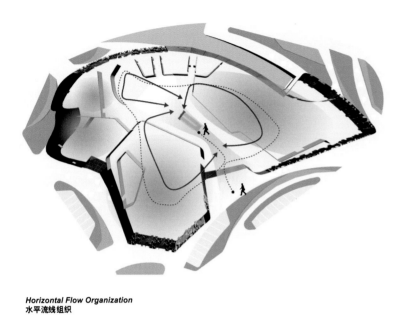

Horizontal Flow Organization
水平流线组织

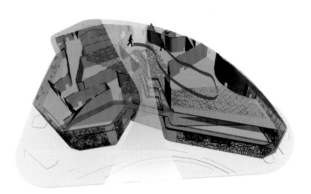

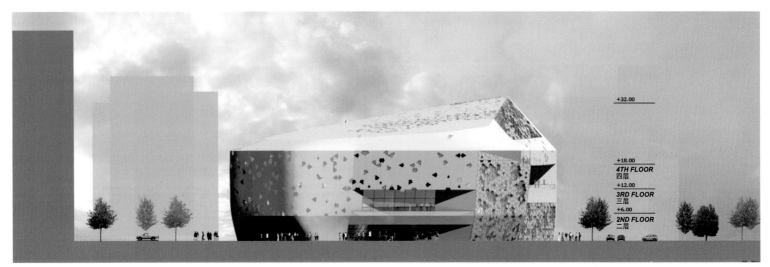

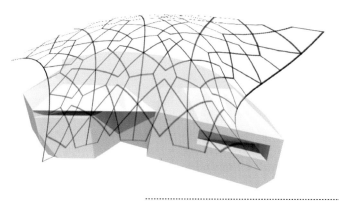
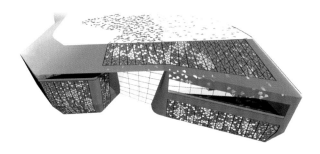

Building
建筑

Building + Muslim Pattern
建筑 + 穆斯林图案

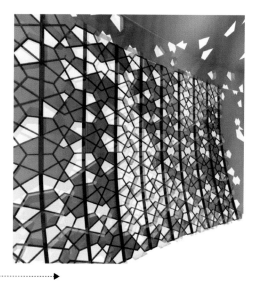

Muslim Pattern
穆斯林图案

Building + Muslim Pattern
建筑 + 穆斯林图案

韩国丽水世博会展览馆
Expo Pavillon in Yeosu, South Korea

设计单位：ATELIER BRÜCKNER
委托方：Peopleworks Promotion Co. Ltd. (PW), Seoul, Korea (for GS Caltex).
项目地址：韩国丽水
项目面积：2 174m²
图片版权：Nils Clauss
　　　　　Michel Casertano

Designed by: ATELIER BRÜCKNER
Client: Peopleworks Promotion Co. Ltd. (PW), Seoul, Korea (for GS Caltex).
Location: Yeosu, South Korea
Area: 2,174m²
Photography Copyright: Nils Clauss
　　　　　　　　　　　Michel Casertano

项目概况

韩国丽水世博会展览馆体现的是一种能源的象征，是由ATELIER BRÜCKNER 设计，它提供了富于梦想的空间，为能源与自然的和谐理念提供了一个三维立体的表达。

建筑设计

展览馆的建筑以一个动态整体呈现，乍一看时不禁让人想起一片宽广的稻田。高达 18m 的"叶片"如稻草般摇曳在风中，他们连续的运动象征着自然界永不停息的能量流动。

当夜幕降临时，这 380 片彩色"叶片"在黑夜里闪闪发光，并以波浪的形式展开。每一位游客都能独自探索这一片场地。隐身于其中的是一个星型的展览馆，如镜子般的外立面使整个场地看起来像是可以无限延展。通过星形建筑凸起的转角，游客可以进入底层同样如镜子般的入口区，棱镜的折射使每位游客拥有一个共同的空间体验，能进一步丰富彼此的交流。

展览馆的中心在上层楼面，一个全景投射的 7m 圆形空间。诗意的黑白美学图片符合世博会"生机勃勃的海洋与海岸"的口号，这一叙述的主角是珍珠潜水员和鲸鱼。他们象征着海洋世界和大陆、自然和人类影响的连接。游客同时也能够亲身参与到其中：在影片的最后一个章节，他们投射的影子成为了背投影的表面。这样，可持续发展的信息便直接传递给了游客个人。

玻璃叶片
星型建筑
可持续发展

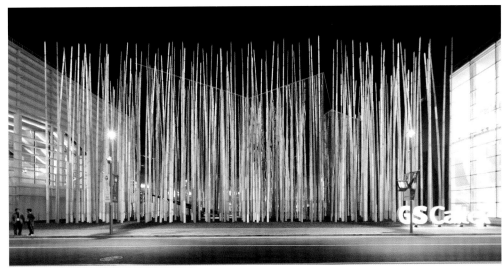

Glass Leaves
Star Building
Sustainable Development

Project Overview

The Expo reflects a symbol of energy designed by ATELIER BRÜCKNER. It offers visionary spaces that give three-dimensional expression to the idea of energy in harmony with nature.

Building Design

The pavilion architecture is presented as a dynamic ensemble which, at first glance, is reminiscent of an outsized rice field. 18 meters high, so-called blades sway like grass in the wind, whereby their continuous motion symbolizes the never-ending flow of energy in nature. When darkness falls, the 380 colored blades shine brightly into the night. Touching activates individual sensitive blades and initiates pulses that spread out in the shape of waves over the entire "energy field".

Each visitor can individually explore the site, which is around 2,000 meters in size and contains a centrally located star-shaped pavilion building that is optically withdrawn in its entirety. Its mirrored façades make the energy field appear to stretch into infinity. Via raised corners of the star, the visitor can gain access to the also mirrored entrance area on the ground floor. Prismatic refractions encourage a collective spatial experience of social networking to further enrich each other's exchanges.

The centre of the pavilion is on the upper floor; a seven-meter-high round room with panoramic projection. Poetic images in a reduced black-and-white aesthetic are in line with the Expo's slogan "Living Ocean and Coast". The protagonists of the narration are a pearl diver and a whale. They symbolize the connection between marine world and mainland, between nature and human influence. The visitor also becomes involved: in the last chapter of the film, the shadow he/she throws becomes a surface for a back-projection. In this way, the message of sustainability is directly aimed at the individual.

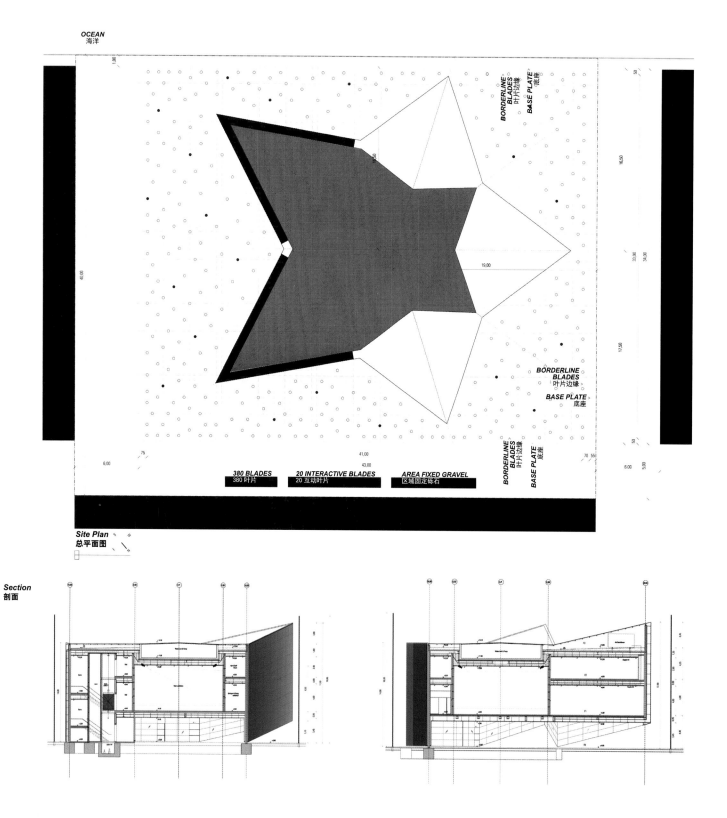

Site Plan
总平面图

Section
剖面

Floor Plan F0_Option 1
底层平面图

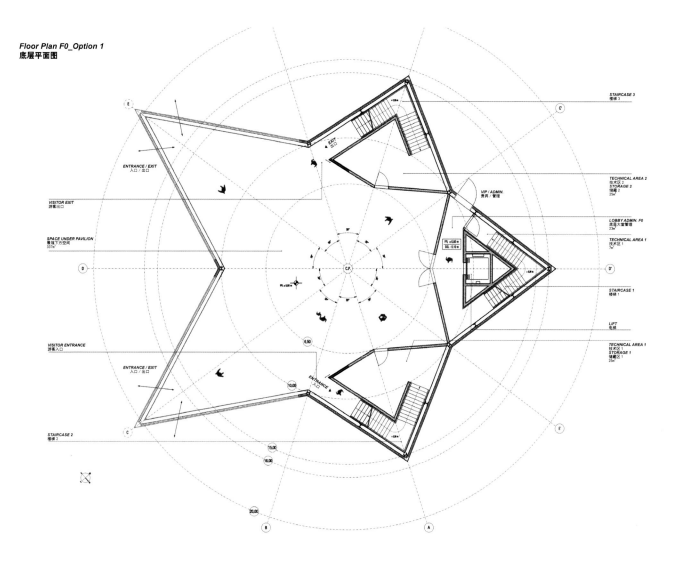

Floor Plan F1_Option 1
一层平面图 _ 选择 1

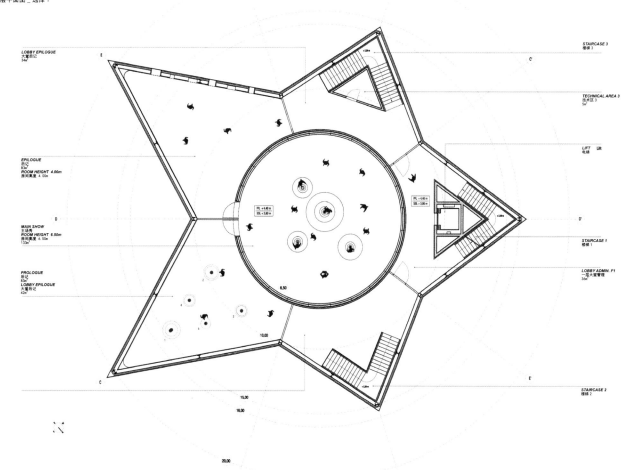

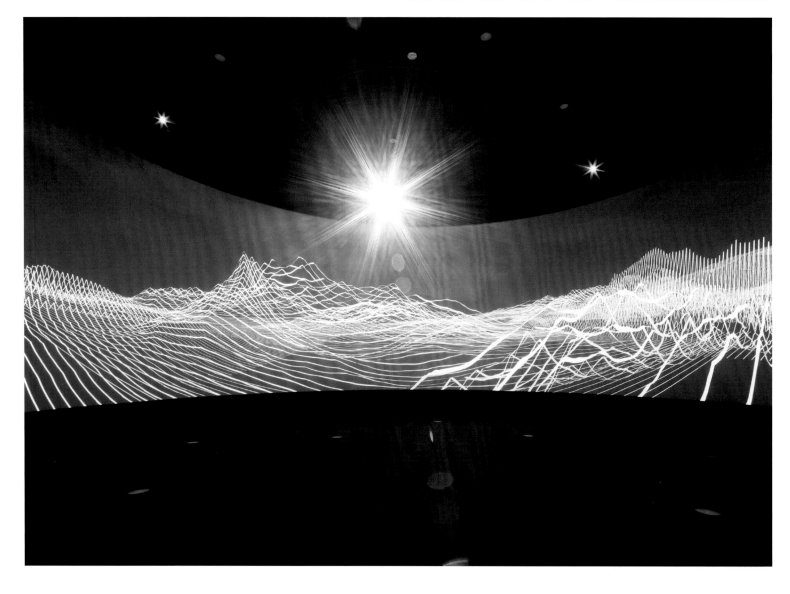

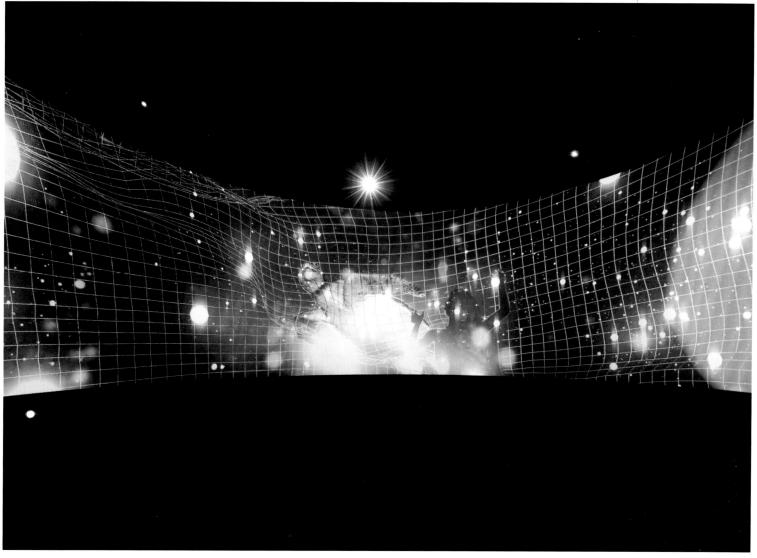

| 文体建筑方案集成 | CULTURAL AND SPORTS ARCHITECTURE PROGRAM INTEGRATION

福建泉州鼎立雕刻馆
Fujian Quanzhou Dingli Sculpture Art Museum

委托方：福建鼎立雕刻有限公司
项目地址：福建省泉州市
建筑面积：3 600m²
建筑设计：王彦
　　　　　雕刻馆立面设计
　　　　　接待中心建筑设计
建筑摄影：吕恒中

Client: Fujian Dingli Sculpture Co., Ltd
Location: Quanzhou, Fujian
Building Area: 3,600m²
Building Designer: Wang Yan
　　Sculpture Museum Façade Design
　　Building Design in Reception Center
Photography: Lv Hengzhong

项目概况

崇武古城始建于1387年，是中国现存比较完好的明代石头城。这里距福建泉州大约半小时车程，素有"中国石雕之乡"的美誉，而鼎立雕刻馆就位于直达古城的惠崇国道旁。

建筑设计

鼎立雕刻馆坐北面南，处于广场中轴位置。东西两侧分别是保留下来的原有办公楼和新建的接待中心。它们与艺术馆形成U字型广场，环抱中间水池。

雕刻馆外观像是错位堆叠起来的巨石堆，方正大气，暗示着石雕馆功能特征，同时给人以质朴拙然的视觉感受。众多折面的石材幕墙单元在日光下分出光影，视觉层次丰富；转角处钝角转折处理更增加了浑厚有力的建筑感。雕刻馆内部空间体现天圆地方的主题，中心圆形内庭空间统摄全局，四隅展厅分别设置石雕作品。顶层露台设置休憩空间，俯瞰南侧广场。

雕刻馆立面选用了当地易采的普通花岗岩，崇武人称它为"G654"，是经常被用作石雕的建筑材料。崇武古城是座明代石头城，而且传统石房子历来都是就地取材，使得本项目用石块垒叠砌筑，独具特色。雕刻馆立面巨石垒叠的设计概念不禁让人与崇武当地的"出砖入石"建筑传统联系起来。而简洁纯粹的垒叠形象极富现代感，也抽象地表达了建筑与当地人文历史的联系。

错位堆叠
花岗岩立面
出砖入石

| 文体建筑方案集成 | CULTURAL AND SPORTS ARCHITECTURE PROGRAM INTEGRATION |

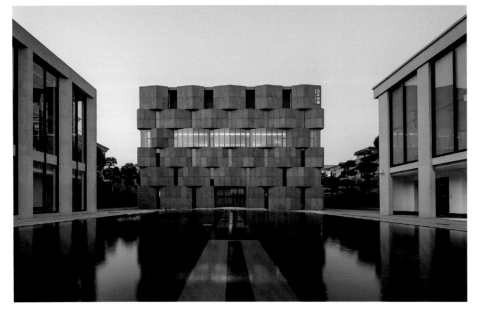

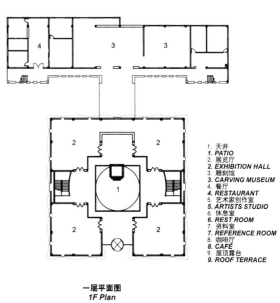

1. 天井 **1. PATIO**
2. 展览厅 **2. EXHIBITION HALL**
3. 雕刻馆 **3. CARVING MUSEUM**
4. 餐厅 **4. RESTAURANT**
5. 艺术家创作室 **5. ARTISTS STUDIO**
6. 休息室 **6. REST ROOM**
7. 资料室 **7. REFERENCE ROOM**
8. 咖啡厅 **8. CAFÉ**
9. 屋顶露台 **9. ROOF TERRACE**

一层平面图
1F Plan

Offset Stacking
Granite Façade
Brick into a Stone

Project Overview

Chongwu ancient town was built in 1387, the best preserved ancient stone city in China. It is located about 30 minutes to drive away from Quanzhou city of Fujian province. City of Chongwu has long history of fame as "Town of Stone Sculpture" in China. And Dingli Stone Carving Museum lies beside the Chonghui National Road leading to the ancient town.

Building Design

The museum is facing to the south and located on the center axis of the plaza. The retaining original office and the new reception center were respectively built to the east and the west. They form a U-shaped plaza with the art galleries, surrounding the middle of the pool.

The façade of the museum looks like many huge stones stacking over the other, silent and generous. The façade implies the functions of the building; meanwhile it gives a visual feeling out a kind of nature and simplicity. The folding surfaces of stone wall create vivid shadow and the obtuse angle stone makes the corner look more firm and powerful. The gallery interior space layout is symmetry, there is a round patio in the center, which surrounded by four exhibition rooms on each floor level. From the terrace on the top, people can enjoy the leisure and overlook the plaza on the south.

The elevations of the museum are chosen the common local granite which is easy to mine as the material. This kind of stone is locally called "G654" which is widely used as stone curving material. Chongwu ancient town is a stone town in Ming Dynasty, and many traditional houses are made of local materials, stacking by the stone with unique feature. Thus the museum design concept of stacking stones would arise the thinking of relationship between Dingli Museum and local traditional architectures of "brick into stone". The pure form of stacking stone has a strong modern sense and abstractly expresses the connection between the building and the local humanity history.

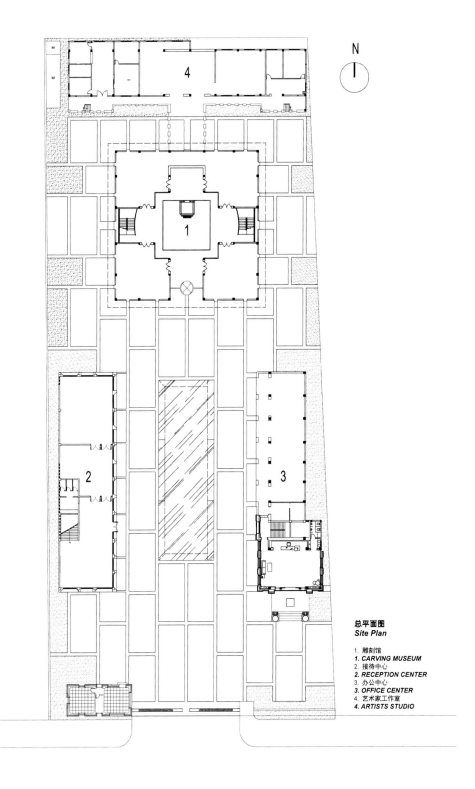

总平面图
Site Plan

1. 雕刻馆
   1. CARVING MUSEUM
2. 接待中心
   2. RECEPTION CENTER
3. 办公中心
   3. OFFICE CENTER
4. 艺术家工作室
   4. ARTISTS STUDIO

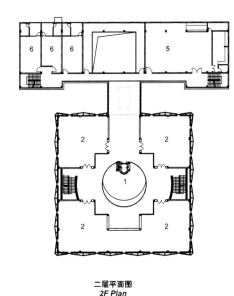

二层平面图
2F Plan

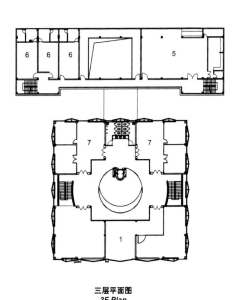

三层平面图
3F Plan

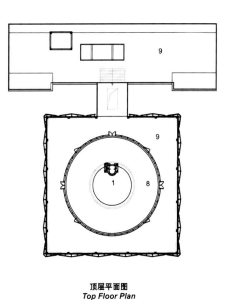

顶层平面图
Top Floor Plan

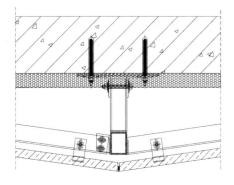
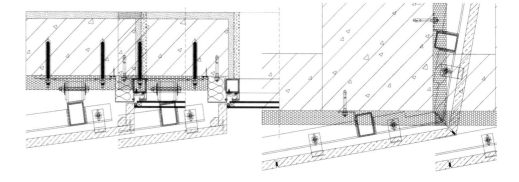

石材节点
STONE MATERIAL NODE

石材与窗交接处横剖节点
STONE MATERIAL AND WINDOW JUNCTION CROSS-SECTIONAL NODE

建筑转角处节点 1
NODE 1 IN BUILDING CORNER

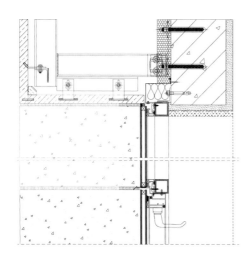
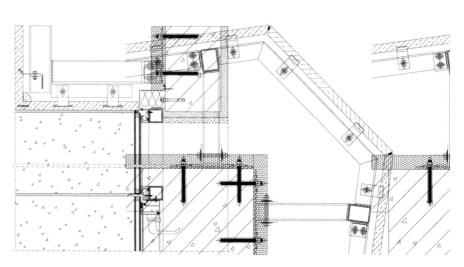

石材与窗交接处纵剖节点
STONE MATERIAL AND WINDOW JUNCTION CROSS-LONGITUDINAL NODE

建筑转角处节点 2
NODE 2 IN BUILDING CORNER

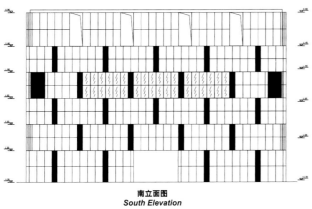

南立面图
South Elevation

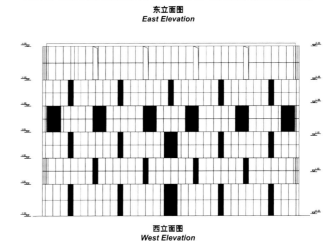

东立面图
East Elevation

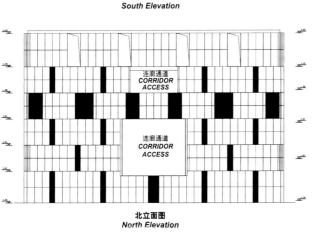

北立面图
North Elevation

西立面图
West Elevation

| 文体建筑方案集成 | CULTURAL AND SPORTS ARCHITECTURE PROGRAM INTEGRATION |

台湾新北市立美术馆
New Taipei City Museum

设计单位：Oporto Office for Design and Architecture (OODA)
委托方：台湾新北市政府
项目地址：台湾省台北市
建筑面积：20 000m²

Designed by: Oporto Office for Design and Architecture (OODA)
Client: New Taipei City Government, Taiwan (R.O.C.)
Location: Taipei, Taiwan
Area: 20,000m²

超立方体
菱形表皮
标志性建筑

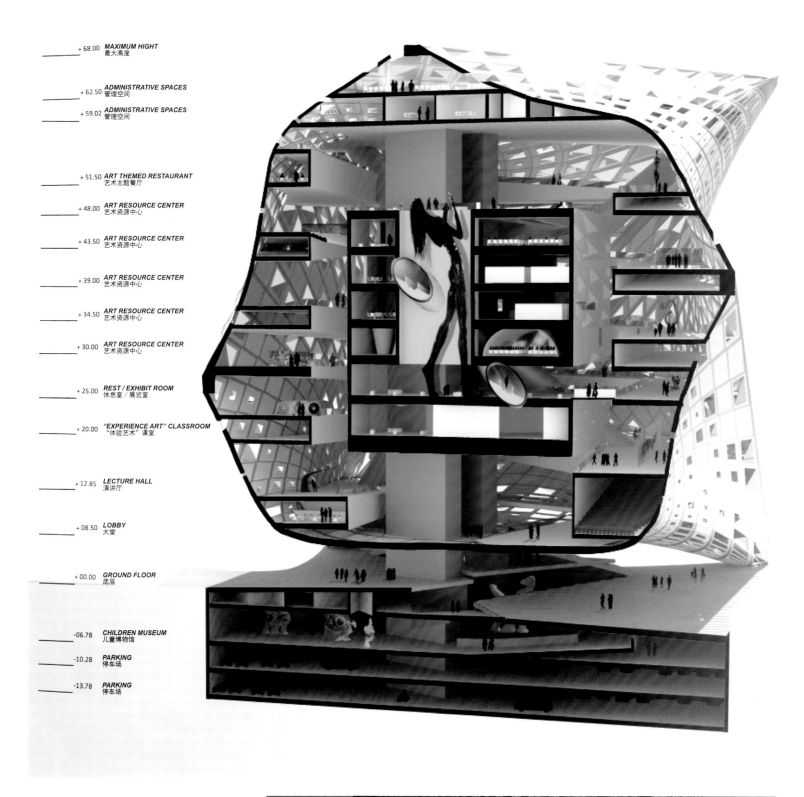

标高	英文	中文
+68.00	MAXIMUM HIGHT	最大高度
+62.50	ADMINISTRATIVE SPACES	管理空间
+59.02	ADMINISTRATIVE SPACES	管理空间
+51.50	ART THEMED RESTAURANT	艺术主题餐厅
+48.00	ART RESOURCE CENTER	艺术资源中心
+43.50	ART RESOURCE CENTER	艺术资源中心
+39.00	ART RESOURCE CENTER	艺术资源中心
+34.50	ART RESOURCE CENTER	艺术资源中心
+30.00	ART RESOURCE CENTER	艺术资源中心
+25.00	REST / EXHIBIT ROOM	休息室／展览室
+20.00	"EXPERIENCE ART" CLASSROOM	"体验艺术"课室
+12.85	LECTURE HALL	演讲厅
+08.50	LOBBY	大堂
+00.00	GROUND FLOOR	底层
−06.78	CHILDREN MUSEUM	儿童博物馆
−10.28	PARKING	停车场
−13.78	PARKING	停车场

项目概况

项目位于台湾新北市（现台北市），总建筑面积 51 045m²，景观面积达 12 000m²，设计期望用开拓创新的设计理念，将台北的新精神用划时代的象征性手法展现出来。

| 文体建筑方案集成 | CULTURAL AND SPORTS ARCHITECTURE PROGRAM INTEGRATION |

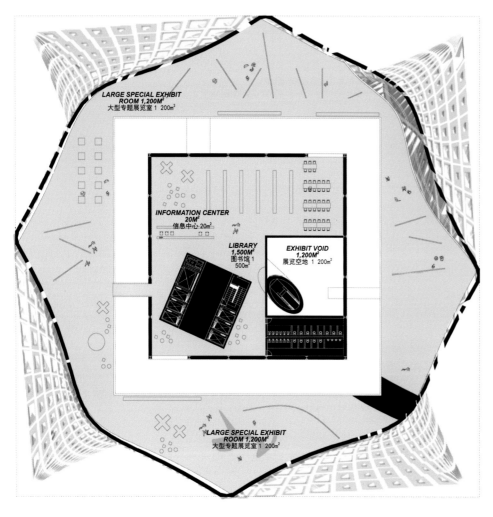

Contemporary Museum of Art-Exhibit Space (+35.00 Plan)-1:500 Scale
艺术展览空间的现代博物馆（+35.00 平面图）——1：500 比例

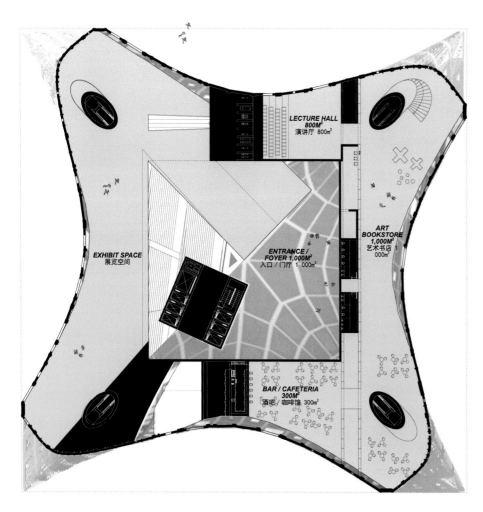

Contemporary Museum of Art-Lobby (+12.00 Plan)-1:500 Scale
艺术大厅的现代博物馆（+12.00 平面图）——1：500 比例

建筑设计

　　这是一个运用数理手段实现的超立方体概念，大的扭曲结构外形之中包含小的核心立方体，外形自然而然的形成底部开放的拱状广场。主馆位于广场上方，儿童艺术博物馆在由主雕塑作品创造的庇护广场之下。艺术博物馆的展览空间就流动在超酷外表皮与中心核体之间。整个建筑被菱形表皮所覆盖，可容纳建筑集成太阳能电池板，并允许自然采光扩散。使博物馆既融于环境，又成为一座标志性建筑。

　　设计通过扭转创建弧度，从而使建筑就像一个碗来收集雨水。它还确保建筑随着太阳在天空中移动，接收全天的太阳能。通过在体块内插入中庭空间来实现建筑内部的最佳通风。设计将 750 到 1 250 棵树木种植在场地的周边，以抵消将被释放的碳，从而达到有益于环保的目的。

STEEL STRUCTURE 4M WIDE SECTION	STEEL STRUCTURE 1.5M WIDE SECTION	REINFORCED CONCRETE VERTICAL ACCESSES CORE	STEEL STRUCTURE MESH 0.5M SECTION
钢结构 4m 宽截面	钢结构 1.5m 宽截面	钢筋混凝土垂直通道核心	钢结构网格 0.5m 截面

   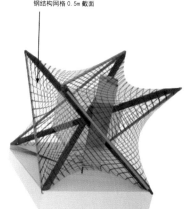

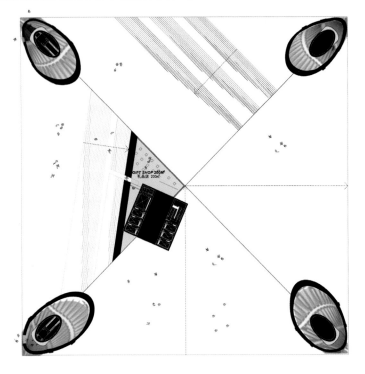

Ground Floor Plaza (+05.00 Plan)-1:500 Scale
底层广场（+05.00平面图）-1:500 比例

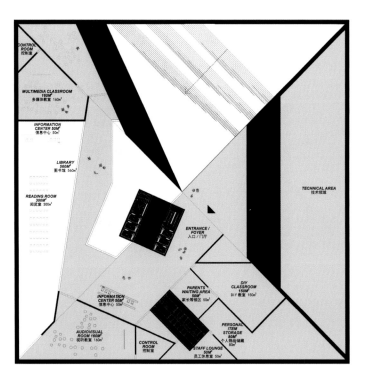

Children Museum of Art-Entrance (-06.00 Plan)-1:500 Scale
艺术入口的儿童博物馆（-06.00平面图）-1:500 比例

Big Volumetric Cube Diamond-shaped Skin Landmark Building

Project Overview

The project is located in the new Taipei City with total floor are of 51,045m² and landscape area of 12,000m². The competition intention was to create a pioneering and innovative design concept which will stand as a new-age landmark and a symbolic voice to the world of Taipei City new spirit.

Building Design

The concept form proposed emerges from a big volumetric cube in confrontation with a smaller inner structure cube-hypercube-in which is applied a mathematical form that relates both. Then the big outer bisectrices clamp (as structural elements) and sucked the in between surfaces to a central point-hypercube as core. The main museum is positioned above and the Children Museum of Art is located below the sheltered square that is created under the main sculptural piece. The Museum of Art itself flows on a continuous spiral ramp between the outer skin and the hyper core cube inside all the way to the top. The whole building is wrapped in a diamond-shaped skin that can accommodate building integrated solar panels and permits diffused natural lighting so that the building is both integrated into environment and become a landmark building. The curvature created by the twists acts like a bowl that captures rainwater. It also ensures that the building receives the optimum amount of solar energy throughout the day as the sun travels across the sky. Optimal ventilation is achieved by inserting atrium spaces within the volume. And finally, the design team has proposed that 750-1250 trees be planted on the property's perimeter in order to offset the carbon that will be released while realizing this project.

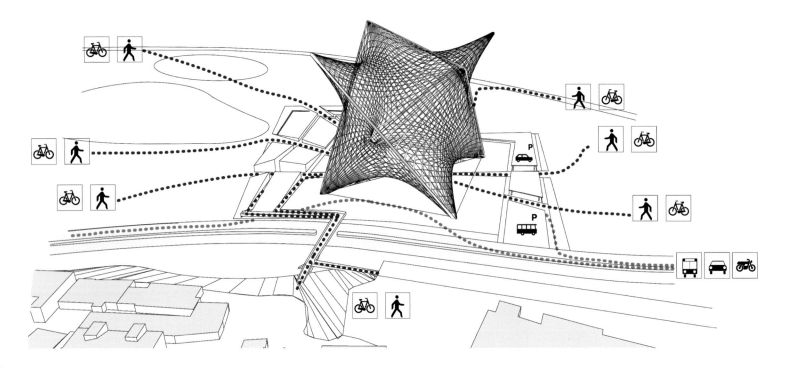

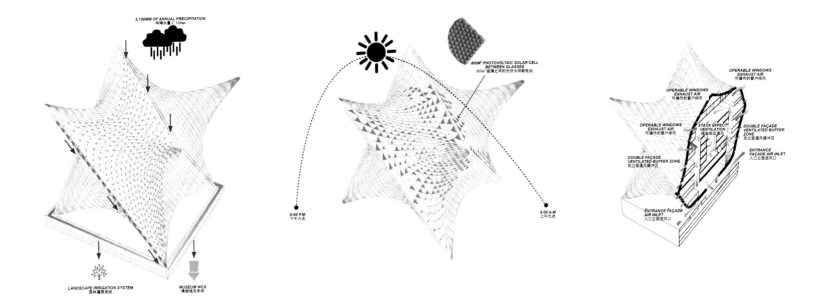
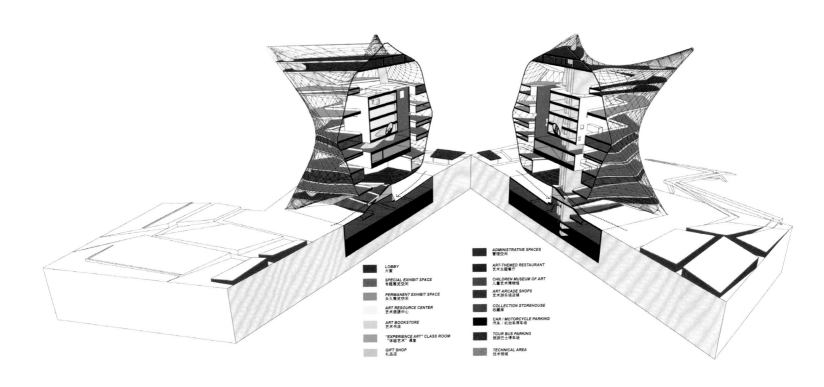

台湾高雄爱的小海湾
Love Cove

设计单位：Maxwan architects + urbanists	Designed by: Maxwan architects + urbanists
委托方：高雄市	Client: City of Kaohsiung
项目地址：台湾省高雄市	Location: Kaohsiung, Taiwan
建筑面积：60 000m²	Building Area: 60,000m²

饮食城
城市广场
户外剧院

项目概况

项目的设计构想是"海上流行音乐活动的可视化展示"。作为海上流行文化产业一个真正的图标，该空间不仅仅是简单的、形式上的象征，它更是从功能上展示海上音乐文化的实际运作和活动。

一般而言，特别是在台湾高雄这样人口密集、面积狭小的城市中，空地是作为一件奢侈品而存在的。所以，利用海湾搭建建设平台，是为了将可用空地最大化。区域内的建设项目包括：城市"海滩"、游艇码头、儿童水上游乐区、室外水族馆和世界级客运港口等，在这些空间内所有年龄段的游客都可以进行不同体验。

建筑设计

该场址包含三个区域：东面的饮食城、与饮食城相毗邻的宏伟的城市广场，以及西面的户外圆形剧场。该建筑是由东到西的连接结构，创造一个可眺望到海湾的充满活力的零售桥梁。该建筑被赋予一个单薄的轮廓，为了应对外观挤压活动，给予公众更大的视觉接触，同时还提供面向海港不变的风景。把所有用途聚集到一个建筑实现了两个目标：一是它允许功能区域的结合，从而使它们最小化；二是它同时允许分享半公共空间，将它最大化并增加它们的宏伟性。

建筑的顶端是一个空中公园。通过自动扶梯直接连接地面，给公众一个浮动的绿色空间与整个城市的广阔景致。海上中心悬浮在爱河上，创造了海上活动和对象的一个标志性通道。演出大厅顶部作为户外露台的两角，为高架餐馆提供俯瞰海湾的场地。

公共空间功能布局简约，有三个停车场区域，从邻近的街道可直接通达。这些区域也允许装载的货和物流直线上升到该建筑。这三个区域的遮盖物是一个起伏的公共空间。此外，它们构造三个主要的人行入口到达该建筑。每一个空间涉及每个使用的入口处：流行音乐入口被最新的流行音乐器具厂商所包围，主要表演场地入口处有与之相邻的大跳劲爆热舞游戏块，以及海上中心入口包含了航运产业的产业遗迹。

Diet City
City Square
Outdoor Theatre

Project Overview

The design concept of the building is that "a visual showcase of maritime and pop music activity".

As a true icon of the maritime and pop culture industries, the space should be more than simply and formal symbol, it should function as a visual display of bustle of the actual workings and events of pop and maritime culture.

In Taiwan in general, and in Kaohsiung specifically with the city of density population and small size, open space is a precious commodity. So, to make use of bay to build construction platform is made to maximize the available open space on site and the building projects in the area include that: an urban "beach", yacht marina, a children's water play area, an outdoor aquarium labyrinth and the world class passenger harbor. All of these spaces permit all ages of visitors to interact with water for a variety of experiences.

Building Design

The site includes three areas: the diet city in the east side, the grand city plaza next to the diet city and the outdoor amphitheater in the west side. The building is a connecting structure from east to west, creating a vibrant retail bridge with views out over the cove. The building is given a thin profile, in order to press the activities against the façade, giving the public greater visual access while also providing those in the building constant views out toward the harbor. Putting all uses into one building accomplishes two goals: one is that it allows a combining of functional areas, thus minimizing them; the second is it simultaneously allows for semi public spaces to be shared, maximizing them and increasing their grandeur.

Topping the building is a sky park. Linked directly to the ground via an escalator, the public is given a floating green space with expansive views over the city. The maritime center is suspended over the Love River, creating an iconic gateway of maritime activity and objects. The top of the performance hall doubles as an outdoor terrace for the elevated restaurants to look out over the cove.

The public space is laid out for simple functionality. There are three zones for parking, directly accessed from the adjacent streets. These zones also allow for loading and logistics straight up into the building. Covering these three areas is an undulating public space. Additionally, they frame the three primary pedestrian entrances to the building. Each space relates to the entrance for each use: the pop music entrance is surrounded by vendors of the latest pop music paraphernalia, the primary performance venue entrance has the DDR game blocks adjacent to it, and the marine center entrance contains industrial relics of the shipping industry.

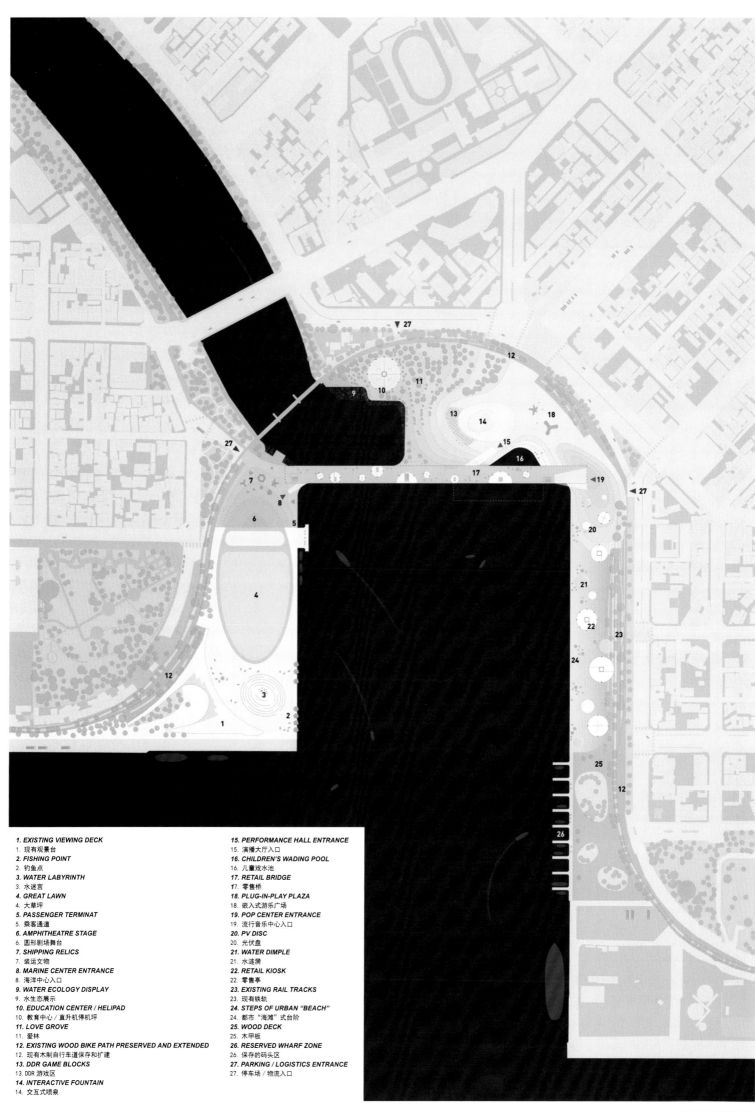

1. **EXISTING VIEWING DECK**
1. 现有观景台
2. **FISHING POINT**
2. 钓鱼点
3. **WATER LABYRINTH**
3. 水迷宫
4. **GREAT LAWN**
4. 大草坪
5. **PASSENGER TERMINAT**
5. 乘客通道
6. **AMPHITHEATRE STAGE**
6. 圆形剧场舞台
7. **SHIPPING RELICS**
7. 装运文物
8. **MARINE CENTER ENTRANCE**
8. 海洋中心入口
9. **WATER ECOLOGY DISPLAY**
9. 水生态展示
10. **EDUCATION CENTER / HELIPAD**
10. 教育中心 / 直升机停机坪
11. **LOVE GROVE**
11. 爱林
12. **EXISTING WOOD BIKE PATH PRESERVED AND EXTENDED**
12. 现有木制自行车道保存和扩建
13. **DDR GAME BLOCKS**
13. DDR 游戏区
14. **INTERACTIVE FOUNTAIN**
14. 交互式喷泉
15. **PERFORMANCE HALL ENTRANCE**
15. 演播大厅入口
16. **CHILDREN'S WADING POOL**
16. 儿童戏水池
17. **RETAIL BRIDGE**
17. 零售桥
18. **PLUG-IN-PLAY PLAZA**
18. 嵌入式游乐广场
19. **POP CENTER ENTRANCE**
19. 流行音乐中心入口
20. **PV DISC**
20. 光伏盘
21. **WATER DIMPLE**
21. 水涟漪
22. **RETAIL KIOSK**
22. 零售亭
23. **EXISTING RAIL TRACKS**
23. 现有铁轨
24. **STEPS OF URBAN "BEACH"**
24. 都市 "海滩" 式台阶
25. **WOOD DECK**
25. 木甲板
26. **RESERVED WHARF ZONE**
26. 保存的码头区
27. **PARKING / LOGISTICS ENTRANCE**
27. 停车场 / 物流入口

| 116–117

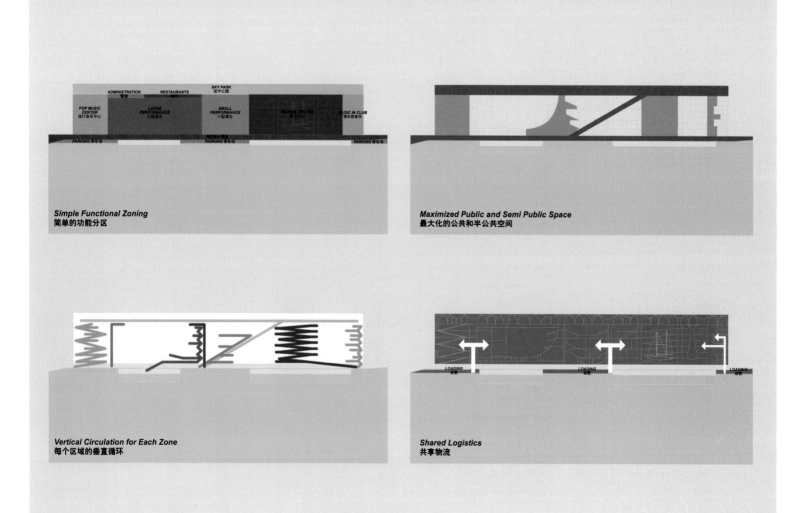

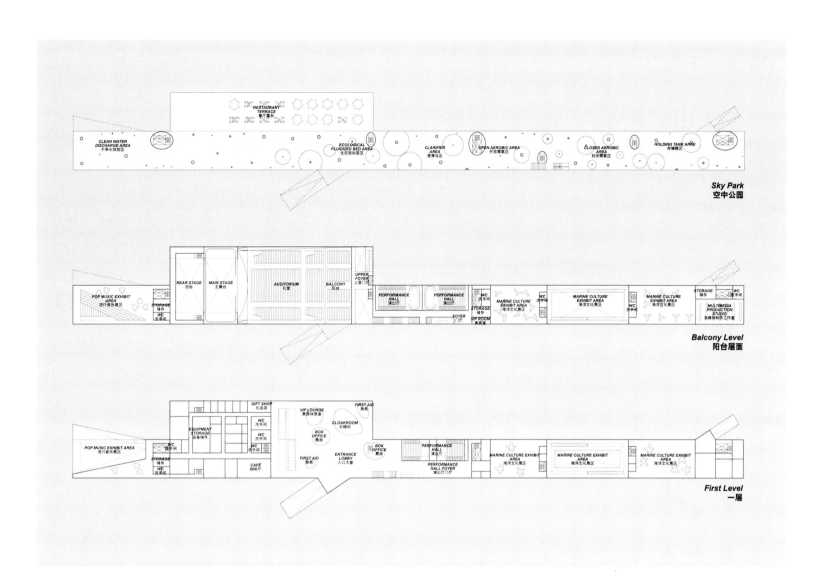

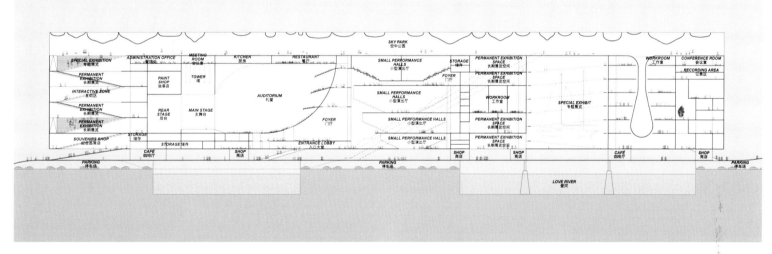

荷兰斯贝克尼萨剧院
Netherlands Theatre Spijkenisse

设计单位: UNStudio
委托方: 斯贝克尼萨市物价局
项目地址: 荷兰斯贝克尼萨
建筑面积: 5 800m²

Designed by: UNStudio
Client: Municipality of Spijkenisse
Location: Spijkenisse, Netherlands
Building Area: 5,800m²

交通流
透明幕墙
水元素

项目概况

斯贝克尼萨剧院的设计专注于项目在该城市位置的定点和定位，同时兼顾到用地面积和公共空间拓展等功能诉求。

该建筑体量呈圆形，这是为了尽可能地减少对附近工厂的干扰；对建筑材料和建筑技术的审慎考量，既确保了对环境的保护，也降低了外加的维护费用。

剧院在整体设计上秉持着可持续发展的理念，并将这种可持续的观念纳入到智能楼宇的建造中，在这里的建筑目标不是妥协，而是保持设计的重点。

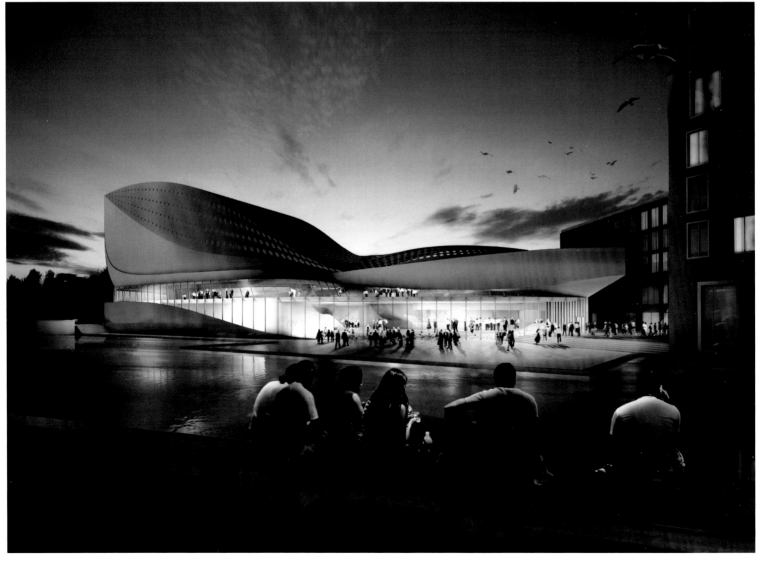

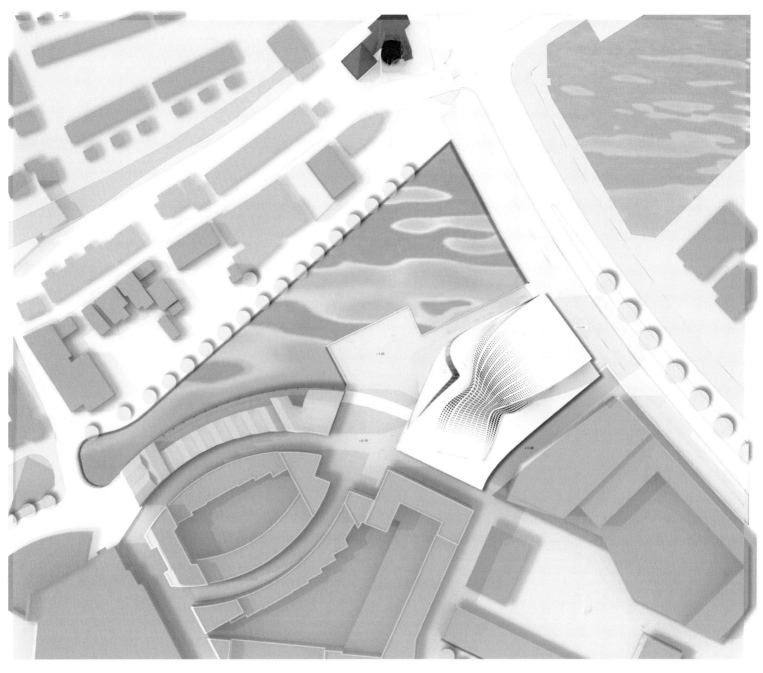

建筑设计

交通流在该建筑物的组织中发挥着重要作用：通往剧院的路线形成城市必不可少的动脉，除了为剧院的游客提供识别外，在整体上将被用作名片为城市形象建设发挥作用。

透明幕墙的大厅增大了两个前庭的可见度，两个悬垂的门廊设计增加了空间的可体验性和戏剧效果。色彩的运用，再加上门厅的几何形状，为戏剧的社会功能创造了一个关键点。门厅向透明的屋顶垂直开辟，通过这个可以看到夜空。白天，太阳透过外立面的空隙产生不断变化的光，照进门厅及通过内饰颜色的煽动元素照射到广场外。

独特的地理位置是其与水和视图到附近的风车"Nooitgedacht"的连接。该视图的周围是通过透明幕墙的方式来进一步照亮，而剧院咖啡厅毗邻水，形成了门厅和公共广场之间的连接元素。以一个圆形剧场的形式，该剧院咖啡厅被设计成第三剧院。

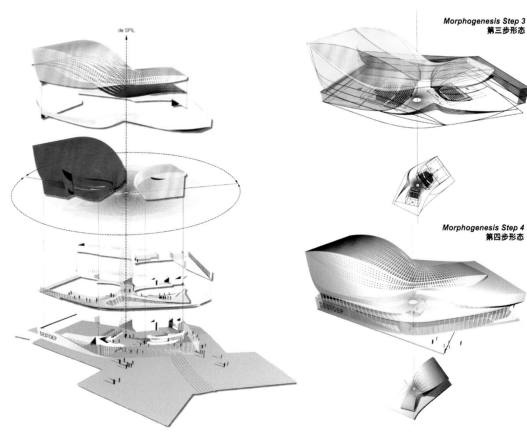

Morphogenesis Step 3
第三步形态

Morphogenesis Step 4
第四步形态

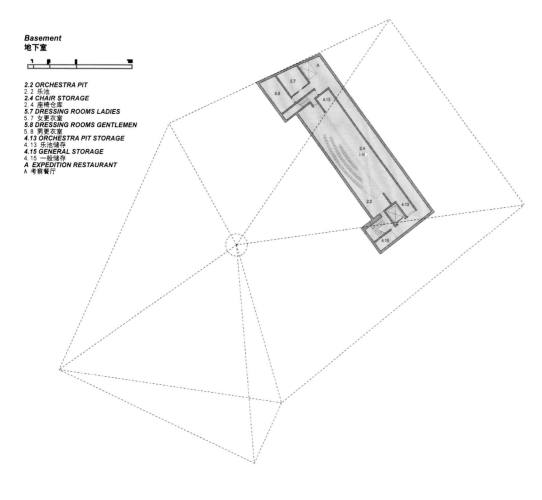

Traffic Flows
Transparent Walls
Water Element

Project Overview

The design for the Theatre Spijkenisse focuses on the placement and orientation of the building in the urban location, while taking into account the land area and public space development and other functional demands.

The building volumes are rounded, so as to ensure as little wind disturbance to the nearby mill; the choice of materials and use of the newest technologies ensures that the building has minimal effect on the environment, coupled with low maintenance costs.

The whole design for the Theatre Spijkenisse adheres to the concept of the sustainable development, where sustainability is incorporated intelligently into the building, but where architectural aims are not compromised and remain the focal point in the design.

Building Design

Traffic flows play an important role in the organization of the building. Access routes to the theatre form an essential artery for the city, with the theatre providing a point of recognition for returning visitors and functioning as a visiting card for the city as a whole.

The draping of the transparent façade over the two atria allows for only glimpses of the warm colors of the interior from the outside. Inside the foyer the visibility of the two atria increases and visitors can experience the full theatrical effect of the wall treatments. The use of color, coupled with the geometry of the foyer, creates a pivotal point in the social functioning of the theatre. Vertically the foyer opens up towards a transparent roof, through which night skies can be seen. During day time, the sun creates an ever changing play of light through the perforations in the façade, both into the foyer and by fanning elements of the interior color on to the square outside.

Unique to the location is its connection with the water and the view to the nearby windmill, 'Nooitgedacht'. The view to the surroundings is further illuminated by means of the transparent façade, while the theatre café is located adjacent to the water, forming the linking element between the foyer and the public square. The theatre café is designed as a third theatre in the form of an amphitheatre.

Ground Floor
底层

1.1 ENTRANCE
1.1 入口
1.2 TICKET SALES/BOX OFFICE
1.2 售票 / 票房
1.3 FOYER
1.3 门厅
1.4 BUFFETS
1.4 餐柜
1.5 CLOAK ROOM
1.5 衣帽间
1.6 TOILETS
1.6 洗手间
2.1 LARGE THEATRE
2.1 大剧院
2.3 ORCHESTRA PIT
2.3 乐池
2.5 FRONT STAGE
2.5 前台
2.6 STAGE TOWER
2.6 舞台塔
2.7 SIDE STAGE
2.7 侧舞台
2.8 BACK STAGE PASSWAY
2.8 后台通道
3.1 SMALL THEATRE
3.1 小剧院
4.1 ARTIST AND STAFF ENTRANCE
4.1 艺术家和员工入口
4.2 PORTER'S LODGE
4.2 门卫室
4.3 TRUCK GARAGE
4.3 卡车车库
4.4 LOADING DOCKS
4.4 装卸码头
4.6 LIGHTS AND FILTER STORAGE
4.6 灯和滤器仓库
4.14 STAGE STORAGE
4.14 舞台存放
5.1 SMALL KITCHEN
5.1 小厨房
5.2 BEVERAGE STORAGE
5.2 饮料储存
5.3 KITCHEN STORAGE
5.3 厨具储存
5.4 PACKAGING
5.4 包装
5.5 FRIDGE
5.5 冰箱
5.6 TOILETS
5.6 洗手间
5.9 KITCHEN THEATRE CAFÉ
5.9 厨房剧院咖啡厅
5.10 PREPARATION KITCHEN
5.10 备用厨房
7.3 GARBAGE STORAGE
7.3 垃圾储放
8.6 TRANSFORMER ROOM
8.6 变压器室
8.7 ELECTRICITY CONTROL ROOM
8.7 电控室
A EXPEDITION RESTAURANT
A 考察餐厅
B THEATRE SHOP
B 剧院商店

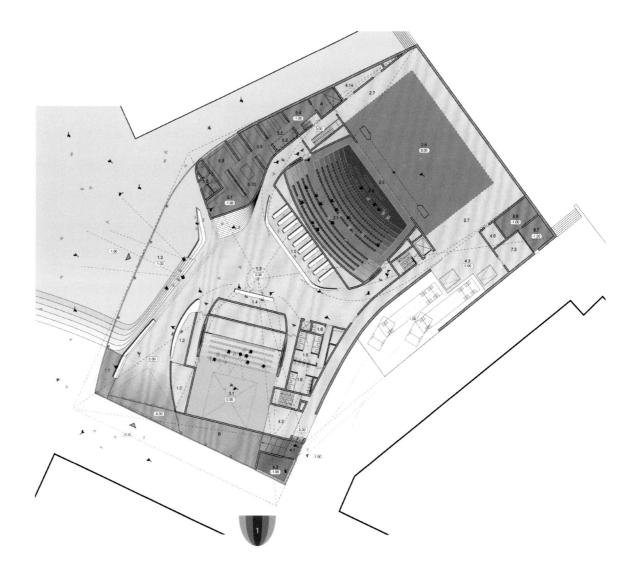

Second Floor
二层

2.1 LARGE THEATRE BALCONY
2.1 大剧院阳台
4.7 DRESSING ROOMS ARTISTS
4.7 艺术家更衣室
4.9 TOILETS / SHOWERS
4.9 洗手间 / 淋浴
4.10 ARTIST BAR
4.10 艺术家酒吧
4.11 KITCHEN ARTIST BAR
4.11 艺术家酒吧厨房
4.12 STORAGE ARTIST BAR
4.12 艺术家酒吧储存
4.15 GENERAL STORAGE
4.15 一般储存
6.1 OFFICE
6.1 办公室
6.2 DIRECTOR'S OFFICE
6.2 主任办公室
6.3 OFFICE TECHNICAL STAFF
6.3 办公室技术人员
6.4 OFFICE MANAGEMENT
6.4 办公室管理
6.5 OFFICE STAGE MANAGER
6.5 办公室舞台监督
6.6 OFFICE RESTAURANT MANAGER
6.6 办公室餐厅经理
6.7 TOILETS
6.7 洗手间
6.8 ARCHIVE / STORAGE
6.8 档案 / 储存
6.9 SERVER ROOM
6.9 服务器机房
7.1 GENERAL STORAGE
7.1 一般储存

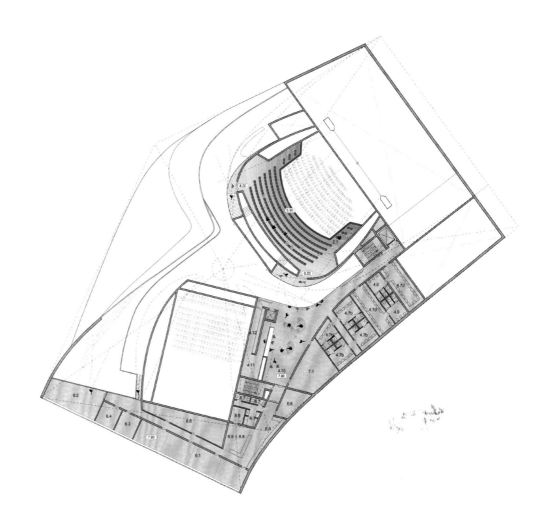

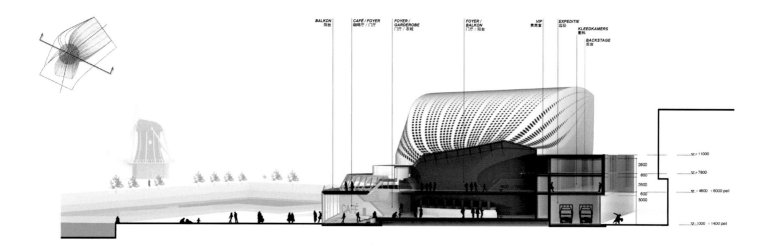

Section
剖面

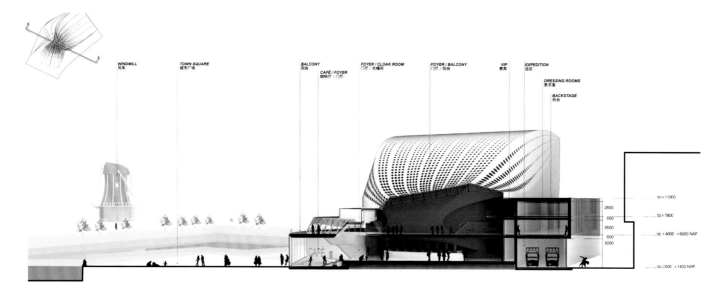

Cross Section
横剖面

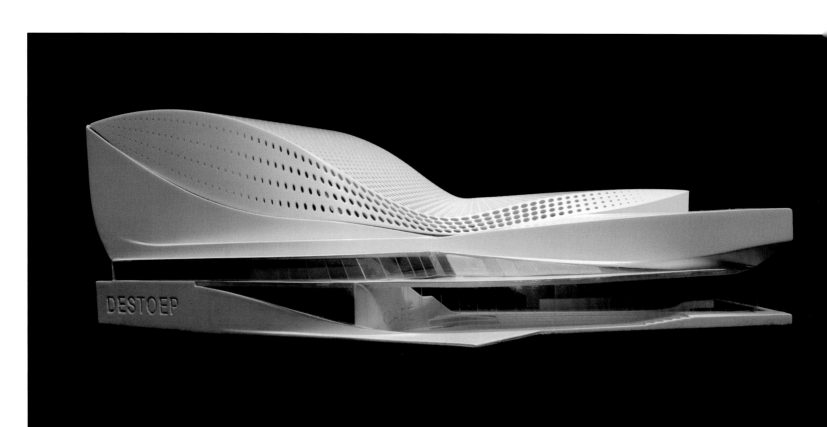

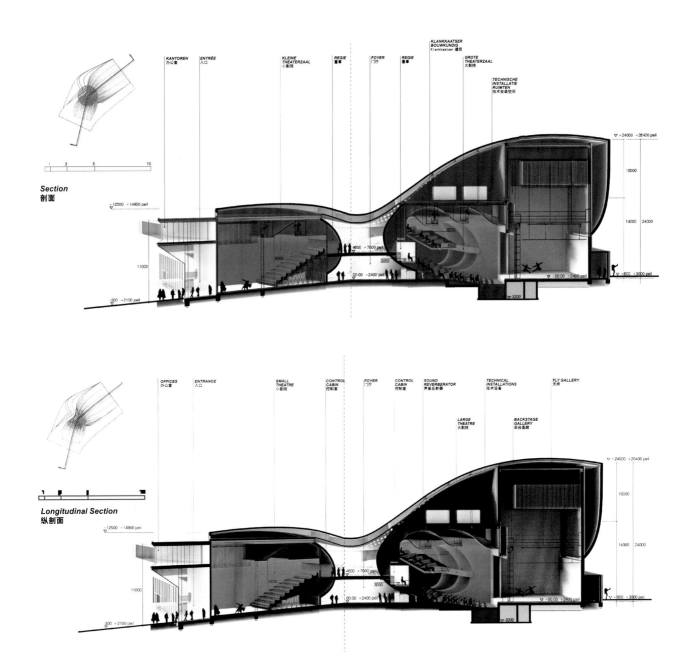

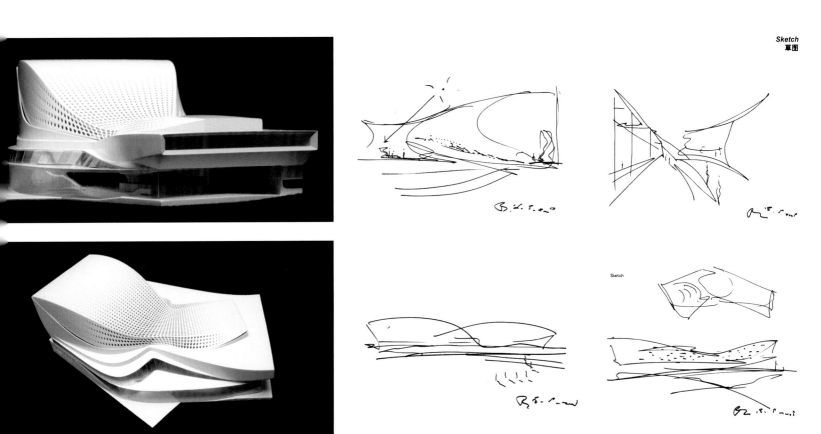

广东广州增城大剧院
Zengcheng Opera House

设计单位：eLANDSCRIPT（译地）设计事务所 　　　　香港大学可持续发展设计团队（合作） 　　　　广州大学建筑设计研究院（合作） 委托方：广州市增城区政府 项目地址：广东省广州市 项目面积：54 129m²	Designed by: eLANDSCRIPT LLC. 　　　　HKU Sustainable Design Team (Collaborator) 　　　　Guangzhou University Architecture Design and Research Institute (Collaborator) Client: Guangzhou Zengcheng District Government Location: Guangzhou, Guangdong Area: 54,129m²

城市名片
尊重自然
生态环保

项目概况

增城大剧院位于广州市增城区，设计将当地的人文风貌和自然美景置于方案之中。大剧院旨在与市民的日常生活相融合，使其在成为增城城市名片的同时，也成为市民日常活动的中心，从而激活整个社区，体现公共建筑的价值。

建筑设计

基地东面缓缓升起的连续景观带，既保留了基地的原始小山包形态，又给予道路对面居住区最小的空间压迫和景观改变。同时，考虑到未来交通繁忙的新城大道为整个区片带来更多人流的同时亦带来噪声的困扰，这一设计手法可为建筑内部及西边中心公园，形成一条绿色的噪声屏障，创造出被山丘保护的感觉。

建筑整体向西南面展开，如雕塑般形成门户，优雅地落在抬升一层的景观带上，以亲切的姿态迎接与引导着来自公园与绿道的访客。商业空间沿其设置，便利了访客，同时也提升了建筑自身的整体商业价值。体量较大的综合剧场位于基地的北边，成为主干道上的标识，形体呼应城市的风格轴线，被切割成相对硬朗的原石形态。相对小巧的多功能剧场安排在基地南边，呼应周边自然景观，如圆润的宝石般静坐在围挂绿湖意向延伸的水景之上。

生态环保设计——综合剧场的外表面采用了双层表皮的设计，极大地减少了外墙内表皮受到的热辐射量，从而降低室内空调制冷的能耗。果肉般晶莹的小剧场采用双曲面雾化玻璃作为外立面，小剧场汲取环绕四周的水景，不仅创造了宜人的视觉效果，也大大降低了玻璃表面的热辐射，从而降低室内空调能耗；大剧院屋顶架设的可自动调节角度的太阳能光伏板，充分利用可再生能源。景观设计利用了基地原有的覆土和挖建地下室的土方，不仅造就了自然有机的城市景观，也让大剧院的施工在土方运输方面减少60%的碳排放量；天然的雨水渗透系统，形成水资源的良性循环。

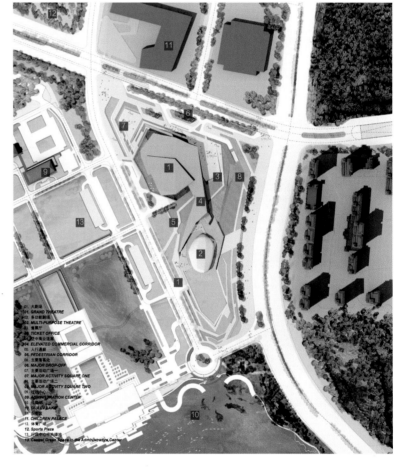

01. 大剧场 GRAND THEATRE
02. 多功能剧场 MULTI-PURPOSE THEATRE
03. 售票厅 TICKET OFFICE
04. 架空商业街 ELEVATED COMMERCIAL CORRIDOR
05. 人行通道 PEDESTRIAN CORRIDOR
06. 主要落客区 MAJOR DROP-OFF
07. 主要活动广场一 MAJOR ACTIVITY SQUARE ONE
08. 主要活动广场二 MAJOR ACTIVITY SQUARE TWO
09. 行政中心 ADMINISTRATION CENTER
10. 儿童馆 CHILDREN PALACE
11. 体育广场 SPORTS PLAZA
12. 行政中心中央绿地 Central Green Space in the Administrative Center

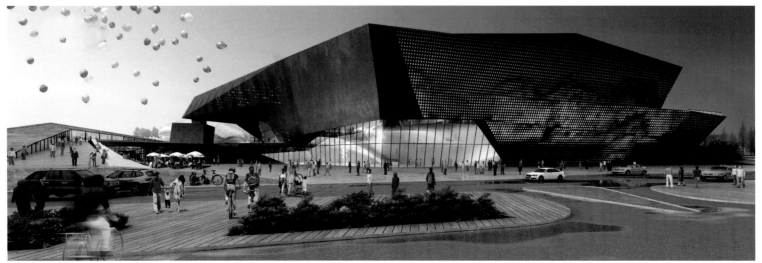

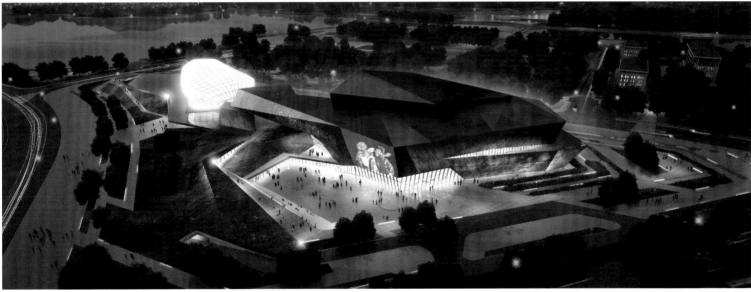

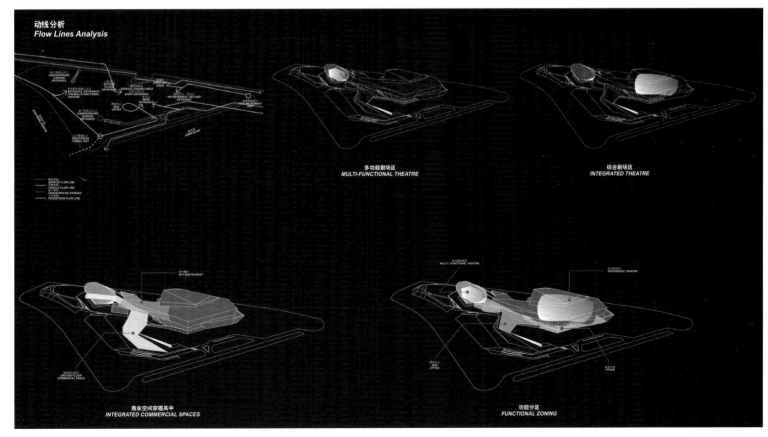

动线分析
Flow Lines Analysis

多功能剧场区
MULTI-FUNCTIONAL THEATRE

综合剧场区
INTEGRATED THEATRE

商业空间穿插其中
INTEGRATED COMMERCIAL SPACES

功能分区
FUNCTIONAL ZONING

| 文体建筑方案集成 | CULTURAL AND SPORTS ARCHITECTURE PROGRAM INTEGRATION |

A-A' 剖面

B-B' 剖面

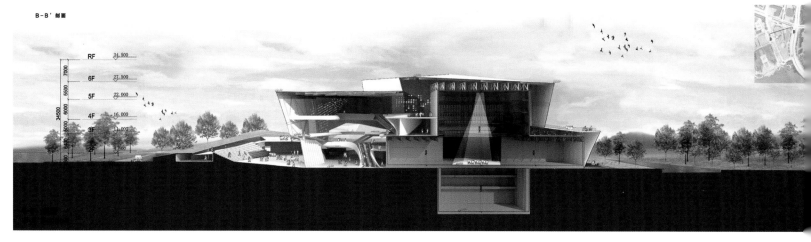

Urban Namecard
Respect for Nature
Ecological Environmental Protection

Project Overview

The project is located in Zengcheng City, Guangdong Province. The 1500 seats Integrated Theatre and the 500 seats Multi-form Theatre are included in the program of this project. The design is intend to integrate the opera house into people's daily life, to become Zengcheng business cards, the center of people's daily activities, thereby activating the entire community to better reflect the value of public buildings.

Building Design

The continuous landscape belt slowly rising from the east of the base not only retains the original hillock shapes, but also produces a minimum space oppression and landscape change for the residential areas across the road. Meanwhile, the future new avenue with the heavy traffic will bring more flow for the entire region and noise problems, so this design approach can form a green noise barrier for the building inside and the west side of Central Park to create the feeling of being protected by the hills.

The whole building is expanded to the southwest like a sculpture to form the portal, elegantly falling in landscape belt uplifted layer, with a cordial attitude to meet and guide the visitors from park and greenways. The business space is setting along it to facilitate the visitors and enhance the overall business value of the building itself. The integrated theater in larger volume is located at the north of the base to become the mark on the main road, and the shape echoed the style axis of the city has been cut into relatively tough original form. While the relative small multi-function theater is arranged at the south of the base to echo with the surrounding natural landscape, just as the rounded jewel is statically sitting on the waterscape. Eco-friendly design—the outer surface of the integrated theater with a double skin design, greatly reduces the heat radiated amount on exterior epidermis of the interior wall, thus reducing indoor air conditioning refrigeration energy consumption; glistening small theater with double curved atomized glass as façade to draw the surrounding waterscape, not only creates a pleasant visual effect, but also greatly reduces the thermal radiation of the glass surface, thus reducing indoor air conditioning energy consumption;

The theatre roof is housed solar photovoltaic panels which can automatically adjust the angles, and is fully used of renewable energy; the landscape design is utilized the original earthing in the base to create the natural organic urban landscape, making the Theatre construction decrease of 60% carbon emissions in earthmoving transport; the natural rainfall infiltration systems form a virtuous circle of water resources.

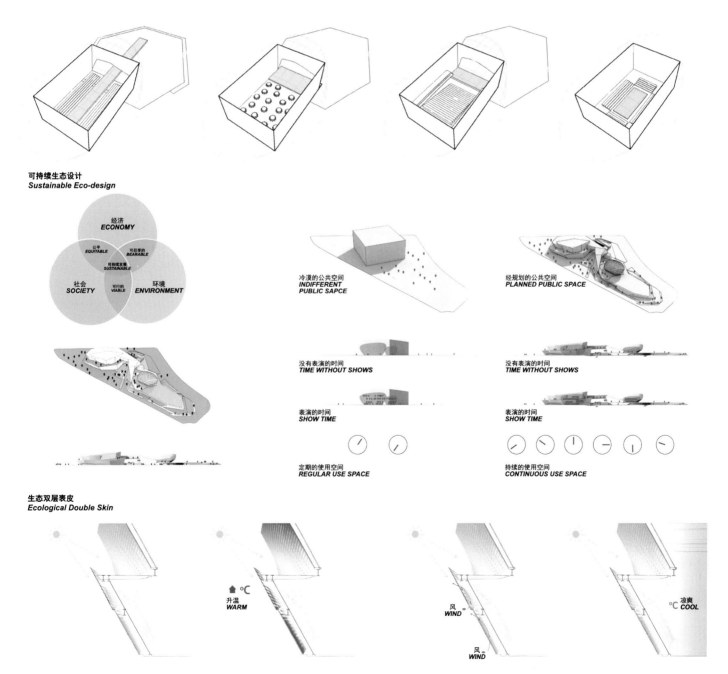

台湾台北艺术表演中心
Taipei Performing Arts Center

设计单位：Architects Collective ZT-GmbH	Designed by: Architects Collective ZT-GmbH
委托方：台湾台北市政府文化局	Client: Department of Cultural Affairs, Taipei City Government, Taiwan, R.O.C.
项目地址：台湾省台北市	Location: Taipei, Taiwan
建筑面积：39 125m²	Building Area: 39,125m²

绿色节能
太阳帽屋顶

项目概况

该建筑具有独特的外观，源于合并成声波的设计主题的城市文脉和功能标准。建筑物起伏的屋顶令人想起无尽的声波，就像一个健全的仪器从演艺中心辐射进入台北城。该剧院综合体满足各种形式的当代艺术的所有要求，也符合台湾的多元文化表现的需要。这是一个世界级的艺术场地，同时提供娱乐和最高专业品质的体验。

建筑设计

剧院综合体的设计就功能方面而言，是根据节能与绿色建筑的角度来进行设计的，并试图达到戏剧和建筑概念之间的合并。三个剧院是由公共大厅连接在一起的，就像一个容器的个别结构，它们共享维修车间及仓储。而其中的一个半室内空间是24小时开放的，不需要门票，与电影院的入口一起成为控制点。

与此同时，该建筑的设计不仅重视其城市和功能方面的发展，也旨在建造一个能适应和回应亚热带环境的建筑。各个剧院的屋顶像戴着太阳帽一样，并为建筑提供自然且高效的冷却效果。半封闭的普通大堂是遮蔽玻璃篷的一部分，开放到不需要加热或冷却的街道。

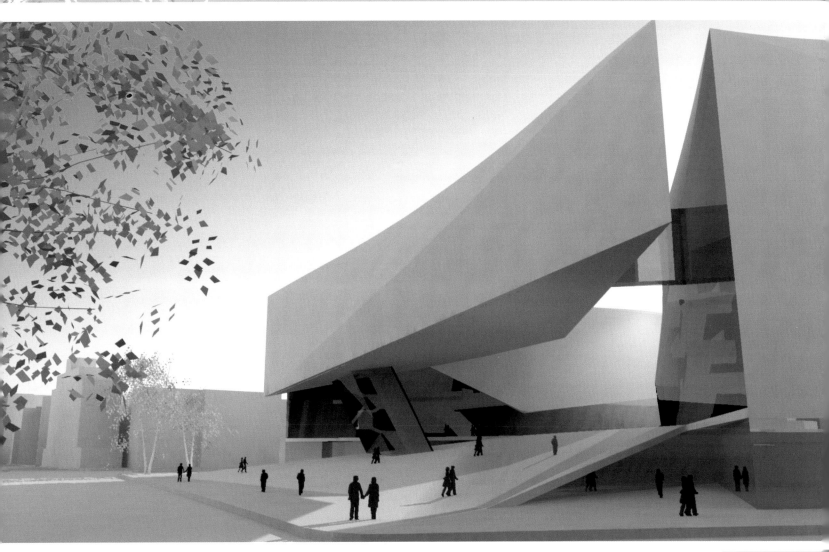

| 文体建筑方案集成 | CULTURAL AND SPORTS ARCHITECTURE PROGRAM INTEGRATION |

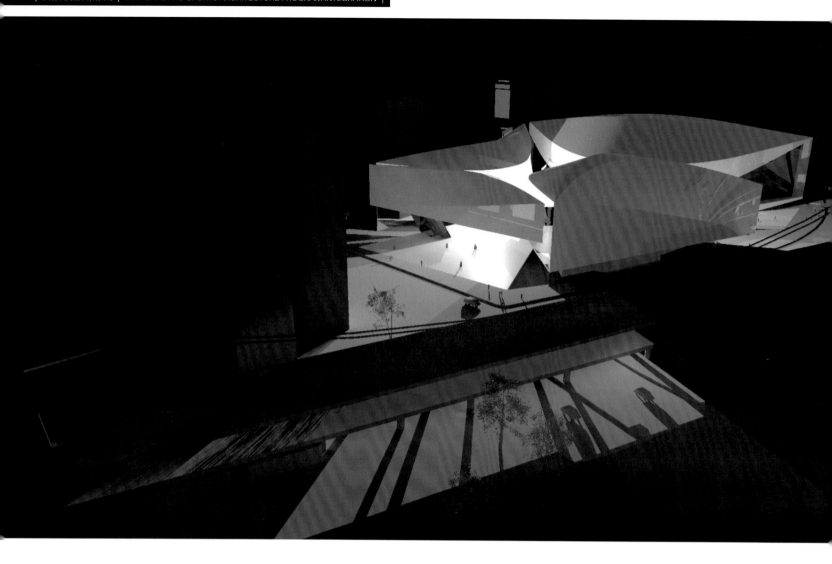

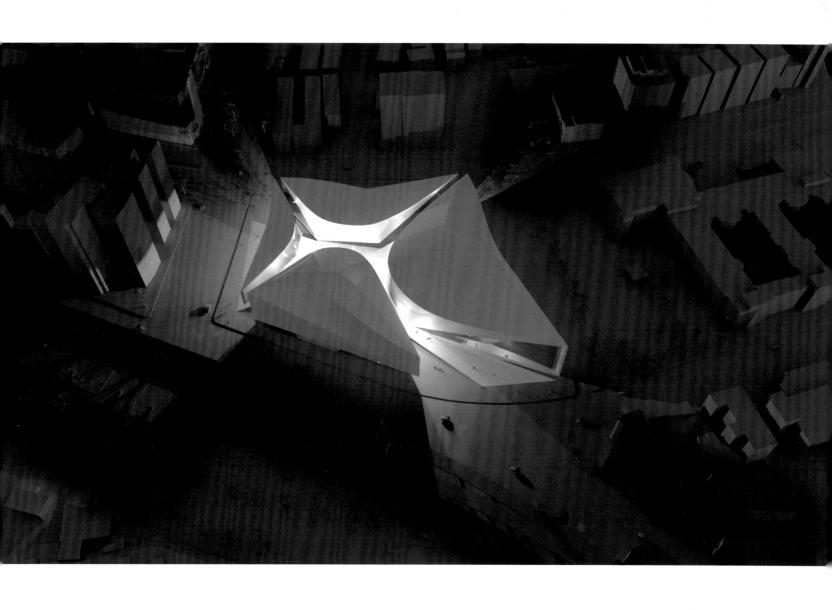
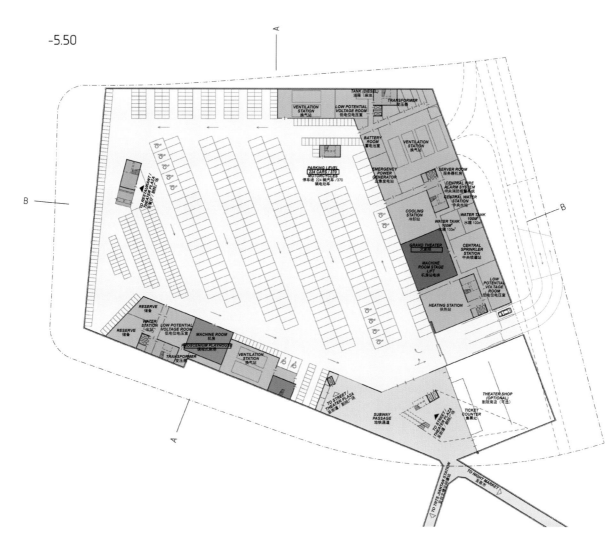

Green Saving Energy Sun-hat Roof

Project Overview

The building possesses a unique appearance that derives from the urban context and the functional criteria merged into the design motif of a sound wave. The buildings undulating roof is reminiscent of an endless sound wave that radiates from the Performing Arts Center into the city of Taipei like a sound instrument. The theatre complex fulfils all requirements of the various forms of the contemporary arts and complies with the needs of Taiwan's diverse performance culture as well. It's a world-class arts venue which provides both entertainment and the highest professional quality experience.

Building Design

The design of the theater complex on the terms of function is based on the perspective of energy efficiency and green building to be designed and tries to achieve a marriage between theatrical and architectural concepts. The three theatres are connected by the Common Lobby like a corresponding vessel, sharing repair shop and storage. A semi-indoor space that is 24-hour open and requires no ticket, with control points at the theatres entrance.

Moreover, the building design not only focuses on the development of its urban and functional context, but also aims to build a building adapting and reacting to the sub-tropical environment. The roof of the individual theaters is covered like sun hats providing natural and efficient cooling for the building. The semi-enclosed common lobby is a partially shaded glass canopy opening to the street that does not need to be heated or cooled.

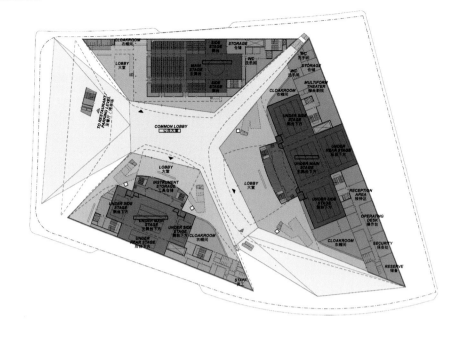

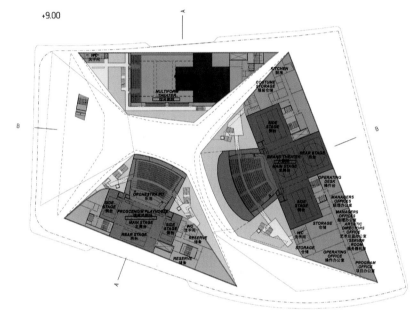

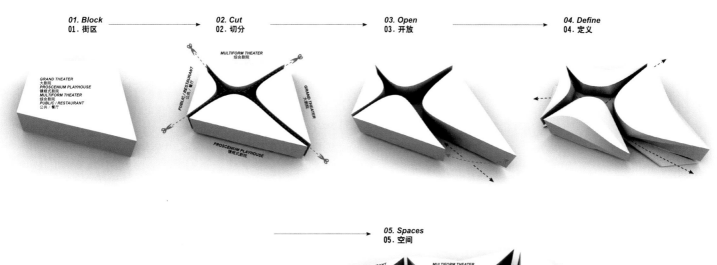

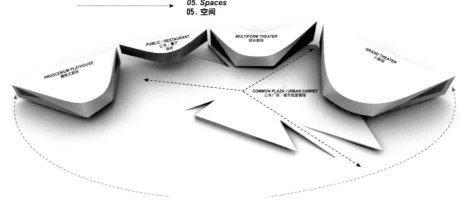

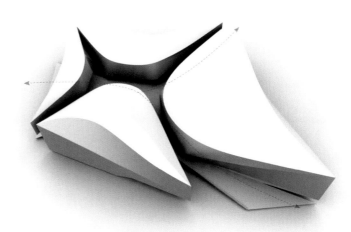

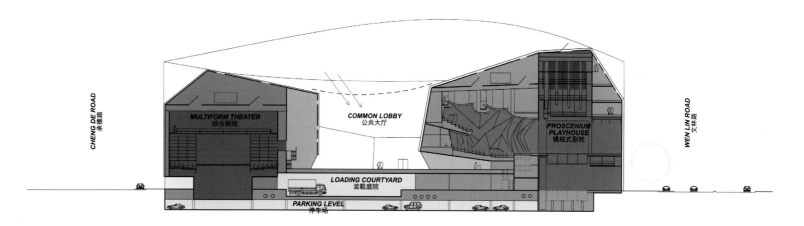
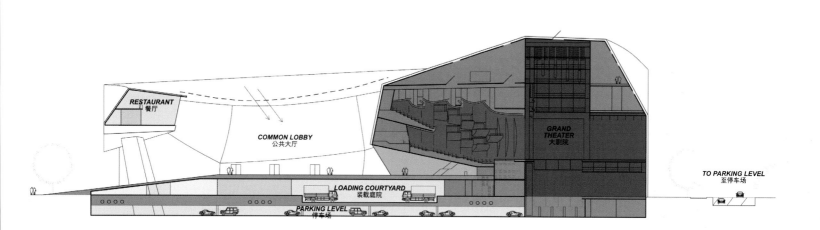

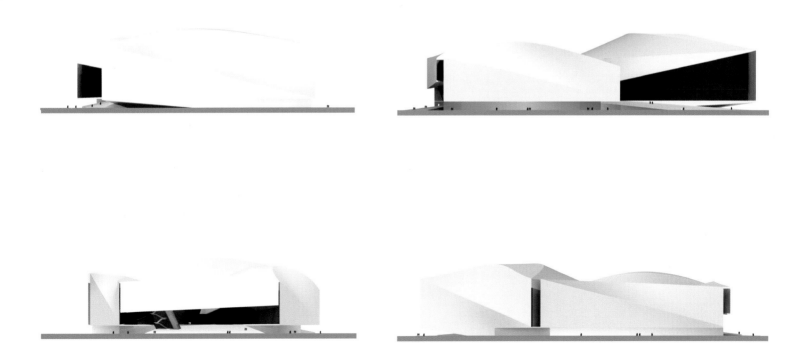

| 136–137 |

山东青岛文化艺术中心
Qingdao Culture and Art Center

设计单位：史蒂芬·霍尔建筑师事务所	Designed by: Steven Holl Architects
委托方：红岛开发建设厅 青岛市城市规划局	Client: Hongdao Development and Construction Department Qingdao Urban Planning Bureau
项目地址：山东省青岛市	Location: Qingdao, Shandong
项目面积：2 000 000m²	Area: 2,000,000m²
项目团队：Janine Biunno　　Bell Ying Yi Cai 　　　　　Nathalie Frankowski　Yu-Ju Lin 　　　　　Magdalena Naydekova　Yun Shi 　　　　　Wenying Sun　Manta Weihermann 　　　　　Yiqing Zhao	Project Team: Janine Biunno, Bell Ying Yi Cai, Nathalie Frankowski, Yu-Ju Lin, Magdalena Naydekova, Yun Shi, Wenying Sun, Manta Weihermann, Yiqing Zhao

艺术岛屿
光循环
绿色技术

项目概况

位于青岛新开发区的新艺术文化中心，将要辐射和周边未来规划的70万人口，规划建设四个博物馆。史蒂芬·霍尔建筑师事务所的中标方案中用"光循环"的新颖设计将四个博物馆用连续的水景、花园以及长廊等景观联系起来，方便游客在多个博物馆中的转换。

建筑设计

该项目旨在打造一个艺术岛屿，设计采用了三个雕刻立方体和四个小园林的形式组成的室外雕塑花园，五个阶梯状的反映池推动了景观，并通过天窗将光带到水平面之下。光循环和艺术岛概念促进公共空间的塑造，可供大型集会的中央广场是也一个大型水上花园，现代艺术博物馆形成了中央广场。北艺术岛包含了经典的艺术博物馆，和酒店一起在最高水平位置；南方艺术岛漂浮在大的南边反映池，可以表演艺术节目。

在光循环之下，所有水平画廊都接收了来自屋顶的自然光，可以用20%的屏幕以及停电选项控制。光循环的20m宽的部分允许侧面照到较低水平的画廊，并为并排的两个画廊提供空间，避免死胡同循环。

整个项目采用最可持续的绿色技术。放置在天窗之间的光伏电池将提供博物馆80%的电力需求。整个建筑体是在打磨海洋铝和彩色混凝土的基础上架构的。

Art Islands
Light Loop
Green Technology

Project Overview

The project is located in the new arts and cultural center of Qingdao New Development Zone, which will benefit the around future planning of 70 million people and be planned to build four museums. Steven Hall Architects with the novel design of the "light cycle" used in the bid-winning program makes the four museums link with a continuous waterscape, garden and corridor landscape, which is convenient for visitors to visit in the multiple museums.

Building Design

The project aims to build an art islands, and it takes the form of three sculpted cubes, and four small landscape art islands that form outdoor sculpture gardens. Five terraced reflecting pools animate the landscape and bring light to levels below via skylights. The Light Loop and Art Island concepts facilitate the shaping of public space. A great central square for large gatherings is at the center of the site overlooking a large water garden. The Modern Art Museum shapes the central square. The North Art Island contains the Classic Art Museum, with a hotel at its top levels, and the South Art Island, which floats over the large south reflecting pool, holds the Performing Arts Program. In the Light Loop, all horizontal galleries receive natural light from the roof that can be controlled with 20% screens as well as blackout options. The 20 meter wide section of the Light Loop allows side lighting to the lower level galleries, and provides space for two galleries side by side, avoiding dead-end circulation.

The entire project uses the most sustainable green technologies. Placed between the skylights on the Light Loop, photovoltaic cells will provide 80% of the museum's electrical needs. The whole building is an architecture that based on the simple monochrome of sanded marine aluminum and stained concrete.

| 文体建筑方案集成 | CULTURAL AND SPORTS ARCHITECTURE PROGRAM INTEGRATION |

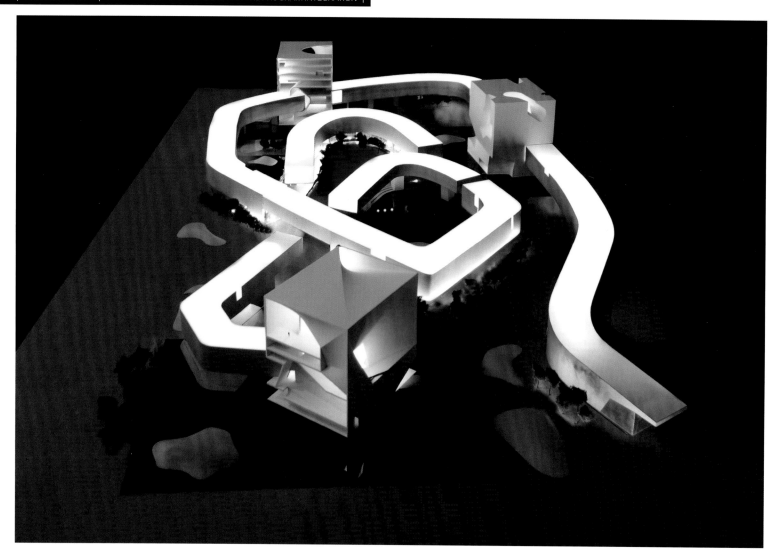

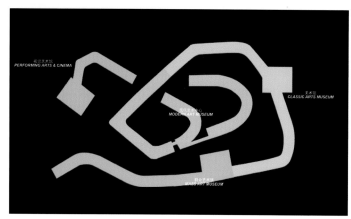

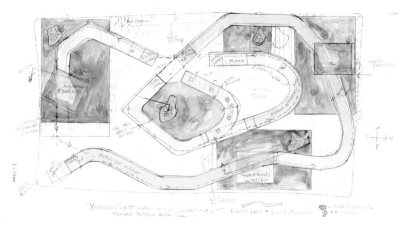

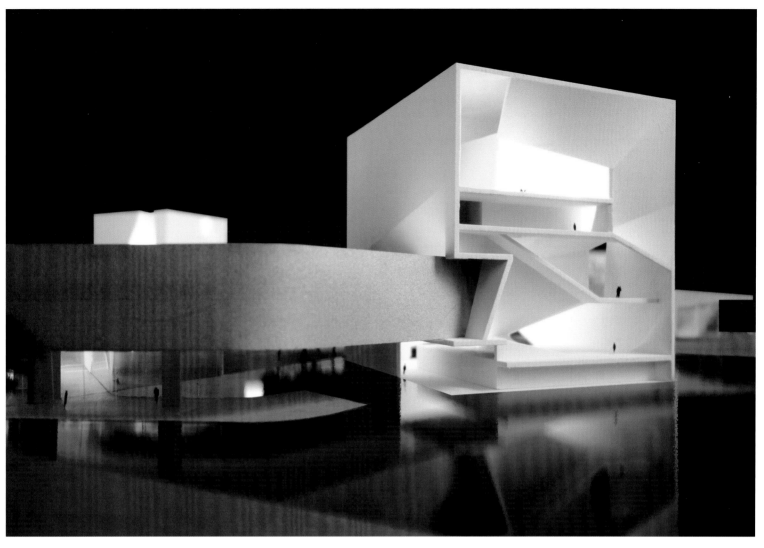

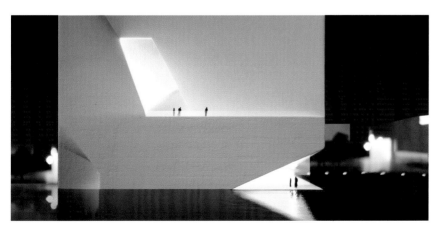
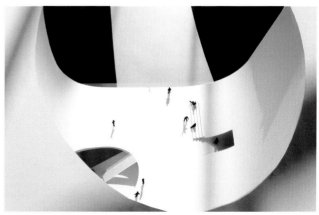

湖南长沙梅溪湖国际文化艺术中心
Changsha Meixi Lake International Culture & Arts Center

设计单位：Zaha Hadid Architects
委托方：长沙市梅溪湖实业有限公司
项目地址：湖南省长沙市
项目面积：115 000m²
渲染图纸：Zaha Hadid Architects

Designed by: Zaha Hadid Architects
Client: Changsha Meixihu Industry Co., Ltd.
Location: Changsha, Hunan
Area: 115,000m²
Renders and Drawings: Zaha Hadid Architects

高端艺术平台
文化荟萃

项目概况

长沙梅溪湖国际文化艺术中心，预计 2015 年建成，年接待能力将达到 250 场次约 31 万人次。项目总投资超过 20 亿元，总用地面积 10 万 m²，包括 3.5 万 m² 的大剧院和 2.5 万平米的艺术馆两大主体功能，大剧院由 1 800 个座位的主演出厅和 500 个座位的多功能小剧场组成；艺术馆由 12 个展厅组成，展厅面积达 1 万平米，能承接世界一流的大型歌剧、舞剧、交响乐等高雅艺术表演。

项目建成后，将是湖南省规模最大、功能最全、全国领先、国际一流的国际文化艺术中心，填补全市和全省高端文化艺术平台的空白。

建筑设计

国际文化艺术中心体现了一种独特的公共节点和空间：一个大剧院、一个当代艺术博物馆、一个多功能大厅及配套设施。中环广场是由这些独立建筑物的相对位置产生的，并提供了强大的城市经验，据此，来自场地各方的行人游客相交流动。

大剧院是长沙国际文化艺术中心的焦点。这有1 800个座位，是全市最大的表演场地。该设计为了承载国际标准性的表演，建筑将包含内部功能所必需的，如大堂、衣帽间、酒吧、餐厅和贵宾接待，以及所需要的辅助功能，如行政管理、排练室、后台后勤、更衣间、化妆室和衣柜。

小剧场（多功能厅）以其灵活性为特点。由于具有500个座位的最大容量，它可以适应并转化到不同的配置。因此，它可以适应各种功能和表演，从宴会和商业活动跨越到小剧、时装表演和音乐。因此，这里也是一个商业的吸引力磁场，它从公共场地无缝对接到零售区和餐饮设施，是设定在一个开放并连接至地下一层的渐渐下沉的庭院。

博物馆（艺术馆）的三个流动花瓣的组成围绕其内部的中心中庭，与各种画廊空间的拼接物以完全无缝的方式进行并列。景观和阳台朝向外部，其目的是保证该场地独特的地理位置和周围的景色映入画廊空间。面向梅溪湖路的外部广场顾及到户外雕塑、展览和活动，将延伸到一个广阔的户外空间。

为体现功能性、优雅和创新的价值观，长沙梅溪湖国际文化艺术中心的目标是成为新文化和公民节点的长沙城，并成为全球文化旅游胜地。

High-end Art Platform Cultural Extravaganza

Project Overview

The International Culture & Arts Center is expected to be built in 2015, and the annual reception capacity will reach 250 sessions with approximately 31 million. The total investment is more than 20 billion with a total land area of 100,000m², including 35,000m² of the Grand Theatre and 25,000m² two main functions of the Art Gallery. The Grand Theatre consists of the main performance hall with a total capacity of 1800 seats and the multifunctional little theater with 500 seats; the Art Museum consists of 12 exhibition halls with the area of 10,000m², which can be undertake world-class large-scale opera, ballet, symphony and other high art performances.

After the completion of the project, it will be the world-class international art and cultural center with the largest scale, the most complete functions and the nation's leading in Hunan Province to fill the blank of the high-end arts and culture platforms between the city and the province.

Planning Design

The International Culture & Arts Center embodies a unique variety of civic nodes and spaces: a Grand Theatre, a Contemporary Art Museum, a Multipurpose Hall and supporting facilities. The central plaza is generated by the relative position of these separate buildings and offers a strong urban experience whereby the flow of pedestrian visitors that come from all sides of the site intersect and meet.

The Grand Theatre is the focal point of the Changsha International Culture & Arts Center. It is the largest performance venue in the city with a total capacity of 1800 seats. Designed to host world-standard performances, the building will contain all the necessary front of house functions, such as lobbies, cloakrooms, bars, restaurants, and VIP hospitality, as well as the required ancillary functions, such as administration, rehearsal rooms, backstage logistics, dressing and make-up rooms, and wardrobe.

The Small Theatre (Multipurpose Hall) is characterized by its flexibility. With a maximum capacity of 500 seats, it can be adapted and transformed to different configurations. It can therefore accommodate a broad range of functions and shows that span from banquets and commercial events to small plays, fashion shows and music. So, this venue is also a commercial attraction, and it shares seamless public access to retail areas and restaurant facilities, which are seated in an open and gently sunken courtyard linking visitors to and from the basement level.

The Museum's composition of three fluid petals around its internal central atrium, juxtaposes of the various patchworks of gallery spaces in a truly seamless fashion. With outward views and balconies to its exteriors, it aims to engage the site's unique location and surrounding views into some of its gallery spaces. An external plaza which faces Meixi Lake Road allows for outdoor sculptures, exhibitions and events to be extended to an expansive outdoor space.

Embodying values of functionality, elegance and innovation, the Changsha Meixi Lake International Culture & Arts Center aims to become the new cultural and civic node for the city of Changsha, and well as global cultural destination.

山东济南文化综合体
Ji'nan Cultural Complex

设计单位：法国 AS. 建筑工作室	Designed by: AS. Architecture-Studio
委托方：济南市西区建设投资有限公司	Client: Jinan West Zone Construction Investment Co., Ltd.
项目地址：山东省济南市	Location: Ji'nan, Shandong
建筑面积：380 000m²	Building Area: 380,000m²
图片版权：ARCHITECT: AS. Architecture-Studio PHOTOGRAPHER: Olivier Marceny	Photography Copyright: ARCHITECT: AS. Architecture-Studio PHOTOGRAPHER: Olivier Marceny

文化泉源
城市新形象

项目概况

山东济南文化综合体坐落在山东的省会,即山东的政治、经济、商业、金融、技术和文化中心——济南市,交通区位优势非常明显。项目由图书馆、美术馆、群众艺术馆、书城、影城以及其他公共配套共同构成以文化办公为主体的城市综合体。

项目所在的西站片区由于奥运会而成为热门地区,它有着巨大的发展潜能,主要是因为未来的京沪高铁项目和西站的总体规划。

规划设计

济南西客站片区核心区处于站前城市功能主轴与蜡山河交点处,作为济南市重要的文化建筑群,三馆设计既要延续济南的文化脉络,又要体现济南新时代大发展的现代化风貌。

本案建筑造型概念源于泉城济南涌动的泉水,泉水象征了城市文化的源泉,建筑形体高低起伏,充满张力,表现了泉水的涌动和文化的灵动之势。蜡山河是这一片地区一个重要的景观元素。该地区由三大组成元素构成,由两条大路分隔开来。

此外,该项目在西南部打算建一个宽敞的广场,突出场址的入口,一方面为这个"统一"的建筑强调场址的价值,另一方面为济南创造新的城市形象。

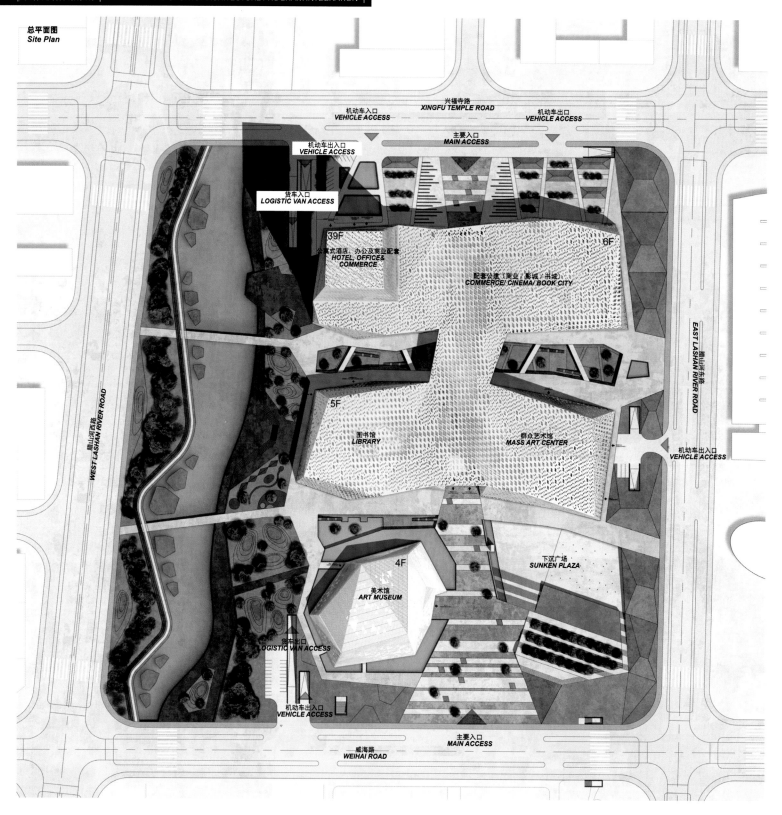

Culture Springhead New City Image

Project Overview

Jinan Cultural Complex is located in the provincial capital of Shandong, its political, economic, commercial, financial, technical and cultural center—Jinan City, so the traffic advantage is obvious. The project is a urban complex with the culture as the main office composed by the libraries, art galleries, public art, bookstores, cinemas and other public facilities.

The project is located in the West Station District, which has become very popular due to the Olympic Games. This area has a huge development potential, mainly due to the future Beijing-Shanghai high speed railway project and to the Western Railway Station master-plan.

Planning Design

The core area of Jinan West Station District is located at the intersection of the city function spindle and Lashan River. As Jinan important cultural buildings, the design of the three museums both continues Jinan cultural context, and reflects the modern style of Jinan new era of development. The architectural modeling concept of the Complex is derived from the swarming spring water of springhead Jinnan. Because the spring water symbolizes the source of urban culture, and architectural forms go up and down and are full of tension, so the project shows the power of surging spring and Smart culture. Lashan River constitutes an essential landscape element within this area. The area is made up of three big elements, which are separated by two main roads.

In addition, the project intends to build a spacious plaza in the southwest to highlight the entrance of the site. On one hand, it emphasizes the value of the site for the "unification" of the building; on the other hand it creates a new image for the city of Jinan.

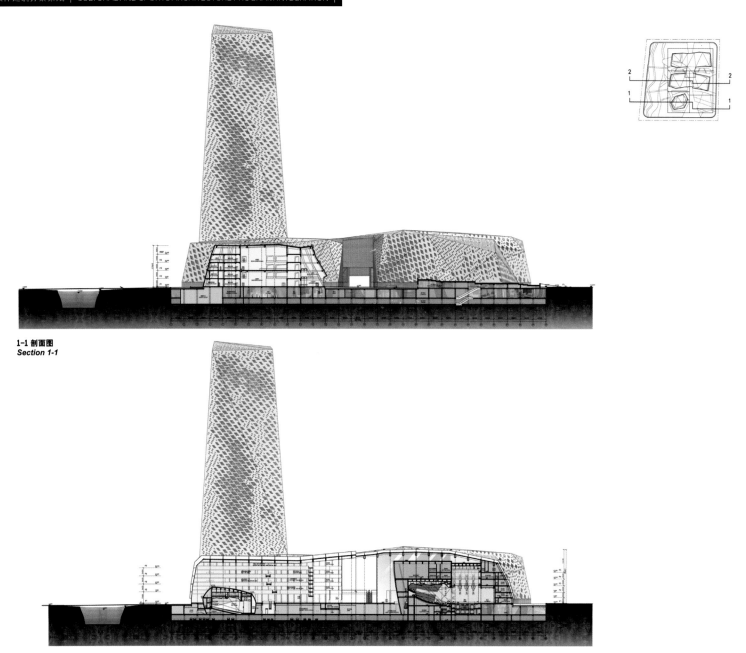

1-1 剖面图
Section 1-1

2-2 剖面图
Section 2-2

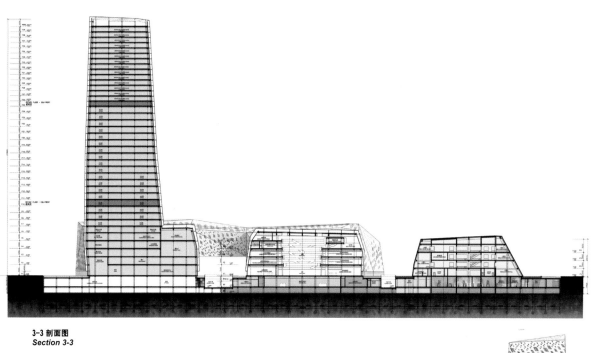

3-3 剖面图
Section 3-3

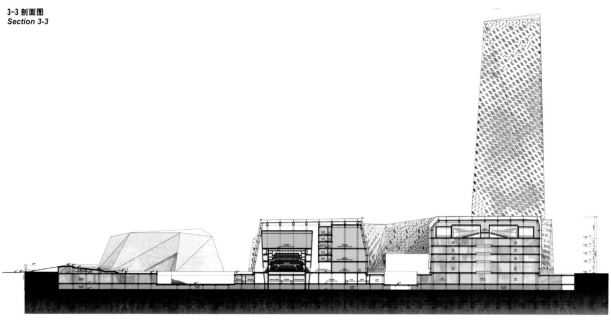

4-4 剖面图
Section 4-4

美国纽约州立大学石溪分校西门子几何物理中心
SUNY Stony Brook Simons Center for Geometry and Physics

设计单位：Perkins Eastman
委托方：纽约州立大学石溪分校
项目地址：美国纽约
建筑面积：3 661.68m²
项目摄影：Paúl Rivera/ArchPhoto 版权

Designed by: Perkins Eastman
Client: State University of New York at Stony Brook
Location: New York, USA
Building Area: 3,661.68m²
Photography: Copyright Paúl Rivera/ArchPhoto

学术建筑
二元性反射
百叶窗玻璃

项目概况

该中心是一个独特的、可持续发展的五层学术建筑，位于石溪大学校园。其建设的主要目的是在几何和物理理论领域，为最高级别的思想家思考、工作和合作提供一个地方。新中心的建筑将主要由教师、博士后和研究生使用，为此，该方案提供了教师办公室和讨论的空间，但没有太多的教学空间。

新设施代表了几何和物理之间尖端想法的融合，它的主要目标是创造一个充满活力的环境互动、学术研究，以及这两个学派之间的合作，提供一个推动科学界限的十字路口。

建筑设计

项目两个截然不同体块的合并表达了该中心性质的二元性反射。它提供了编程功能的明确指定。同时，在直杆砖结构上也涉及到了相邻数学和物理建筑物的研究范围，而弯曲的百叶窗玻璃带为校园创建了一个灯塔。室内中庭在3个层次的办事处之间创造了连接。在这个三层楼的空间与一个南向天窗和顶层反光弧形天花板都有自然光线的渗透。

五楼之上的模块化绿色屋顶和二楼的刨床致力于项目的绿地，指定设计土壤的 Live Roof 和矮浅根景天属植物和长生草属植物的融合营造了一个自然的地毯。南侧的玻璃幕墙的设计最大限度地控制阳光，同时与南部的景观而存在。一个水平遮阳百叶系统包括维护炉排让更多的阳光在冬季渗透到内部而在夏季减少曝露。一个大4 000加仑的雨水收集箱被设计安装在该场址的西侧。虽然该场址是相当有限的，但是能够最大限度地利用屋顶排水进行雨水收集，以便提供厕所和灌溉用水。

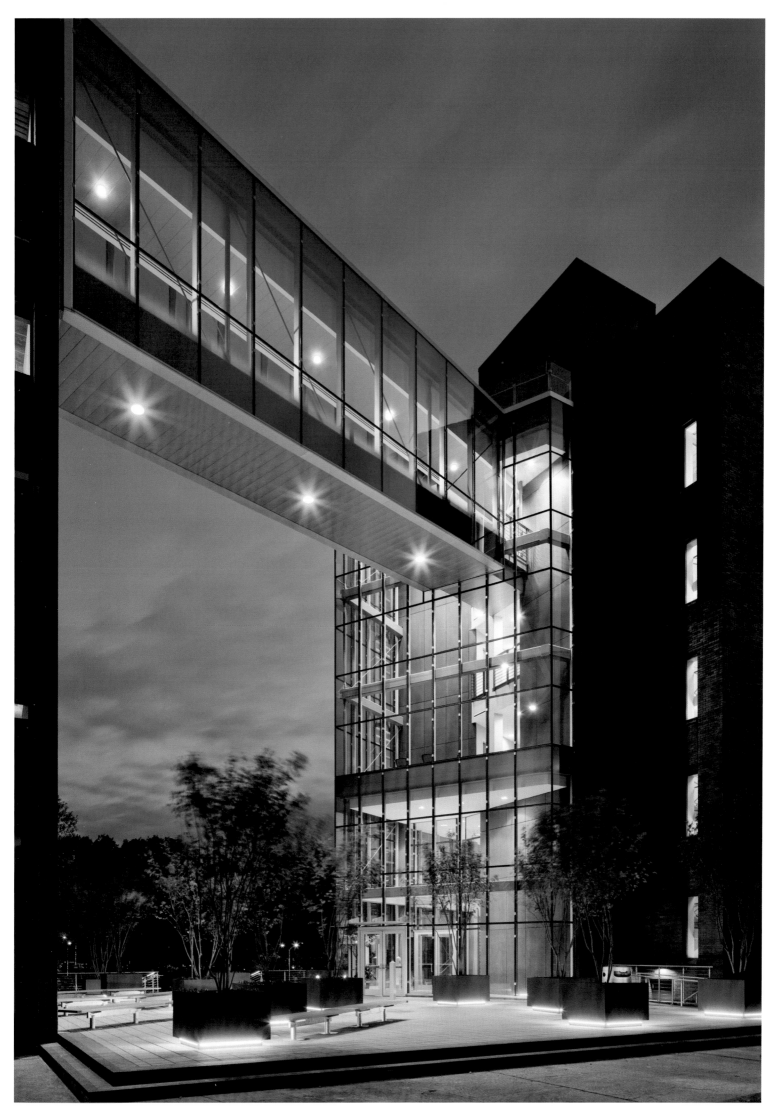

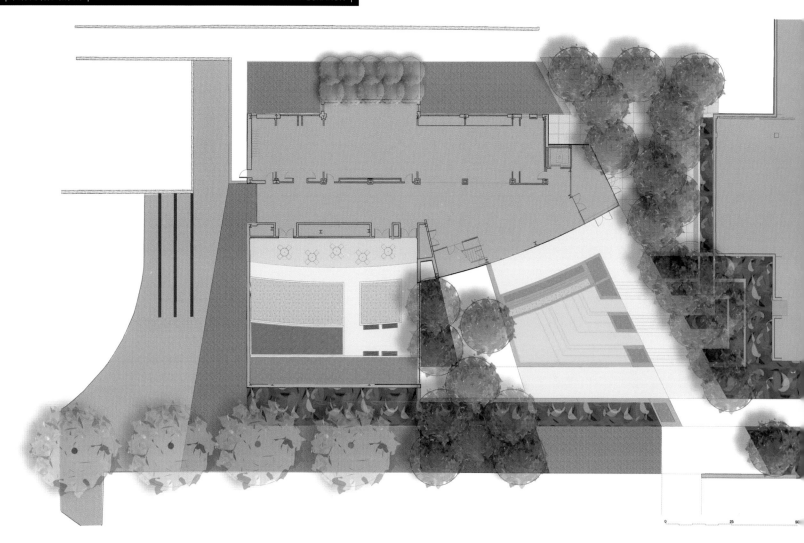

Academic Building Duality Reflection Shutters Glass

Project Overview

The center is a unique, environmentally sustainable five-storey academic building located on the Stony Brook University campus. The Center's primary purpose is to provide a place for the highest-level thinkers in the fields of geometry and theoretical physics to think, work and collaborate. The new Center building will be used primarily by faculty members, post-doctoral and graduate students. To that end, the program provides faculty offices and discussion spaces, but not much instructional space.

The new facility represents the convergence of cutting-edge ideas between geometry and physics. Its main goals are to create a dynamic environment for interaction, academic study and collaboration among the two schools of thought, providing a crossroads that pushes the boundaries of science.

Building Design

The merging of two strongly distinct masses of the project expresses a duality reflection of the Center's nature. It provides clear designation of programmatic functions. At the same time, the straight brick bar architecturally relates to the context of the adjacent Math and Physics Buildings while the curved louvered glass band opens up and creates a beacon to the campus. The interior atrium creates a connection among the three levels of offices. Natural day-lighting is provided in this three-story space with a south-facing clerestory and reflective curved ceilings at the top floor.

A modular green roof above floor 5 and planers on floor 2 contribute toward the project's green space; Live-Roof-specified engineered soil and a blend of low-growing shallow root sedums and sempervivums create a natural carpet. The glass façade on the south side was designed to maximize sun control while maintaining the views to the south. A horizontal sun-shade louver system that includes maintenance grates allows more sun to permeate the interior during the winter months and less exposure in the summer months. A large, 4,000-gallon rain water harvesting tank was designed to be installed on the west side of the site. While the site is quite constrained, the team was able to maximize the collection of rainwater from roof drains in order to provide water for toilets and irrigation.

| 文体建筑方案集成 | CULTURAL AND SPORTS ARCHITECTURE PROGRAM INTEGRATION |

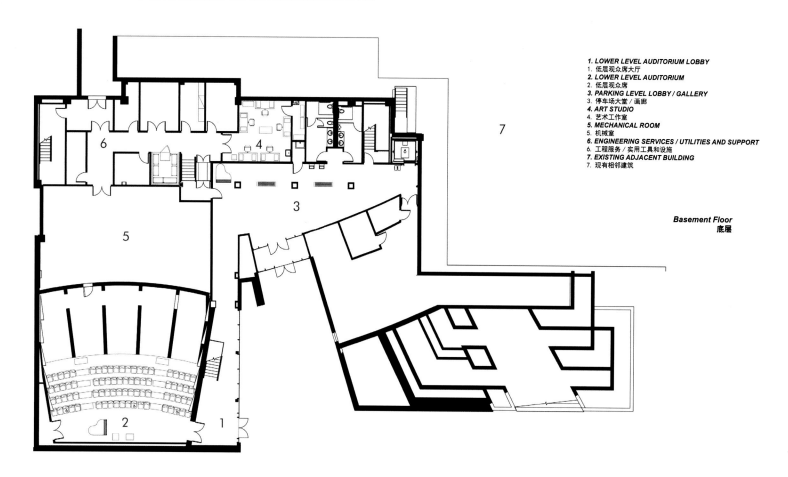

1. **LOWER LEVEL AUDITORIUM LOBBY**
   1. 低层观众席大厅
2. **LOWER LEVEL AUDITORIUM**
   2. 低层观众席
3. **PARKING LEVEL LOBBY / GALLERY**
   3. 停车场大堂 / 画廊
4. **ART STUDIO**
   4. 艺术工作室
5. **MECHANICAL ROOM**
   5. 机械室
6. **ENGINEERING SERVICES / UTILITIES AND SUPPORT**
   6. 工程服务 / 实用工具和设施
7. **EXISTING ADJACENT BUILDING**
   7. 现有相邻建筑

Basement Floor
底层

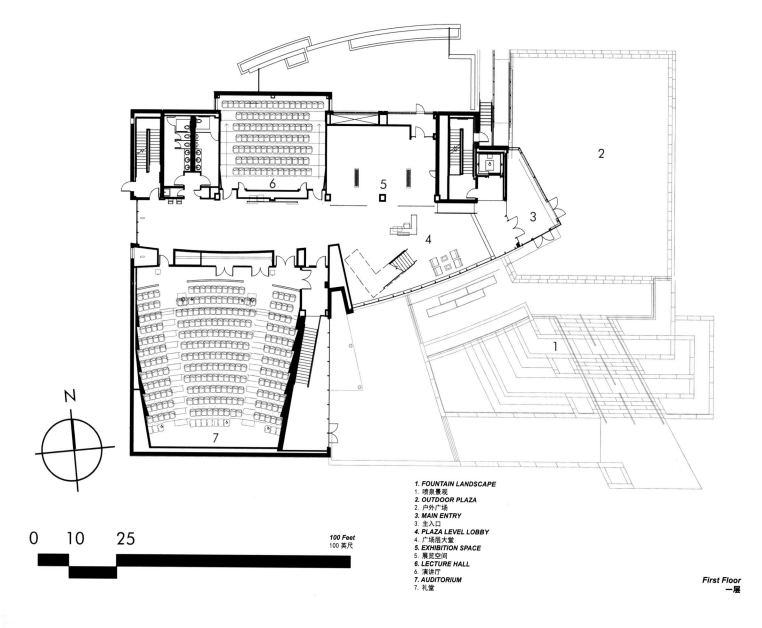

1. **FOUNTAIN LANDSCAPE**
   1. 喷泉景观
2. **OUTDOOR PLAZA**
   2. 户外广场
3. **MAIN ENTRY**
   3. 主入口
4. **PLAZA LEVEL LOBBY**
   4. 广场层大堂
5. **EXHIBITION SPACE**
   5. 展览空间
6. **LECTURE HALL**
   6. 演讲厅
7. **AUDITORIUM**
   7. 礼堂

First Floor
一层

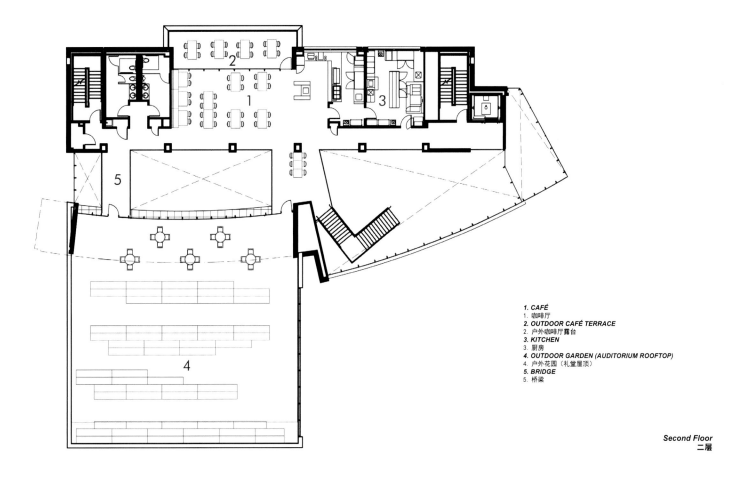

1. *CAFÉ*
1. 咖啡厅
2. *OUTDOOR CAFÉ TERRACE*
2. 户外咖啡厅露台
3. *KITCHEN*
3. 厨房
4. *OUTDOOR GARDEN (AUDITORIUM ROOFTOP)*
4. 户外花园（礼堂屋顶）
5. *BRIDGE*
5. 桥梁

Second Floor
二层

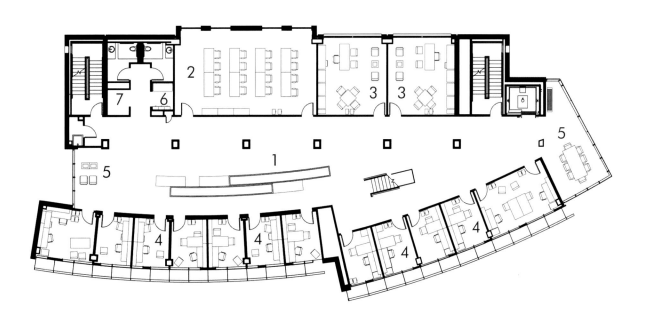

1. *ATRIUM*
1. 中庭
2. *SEMINAR ROOM*
2. 会议室
3. *FACULTY OFFICE*
3. 教研办公室
4. *VISITOR'S OFFICE*
4. 游客办公室
5. *DISCUSSION NOOK*
5. 讨论区
6. *WORK / COPY AREA*
6. 工作室／打印室
7. *PANTRY*
7. 餐具室

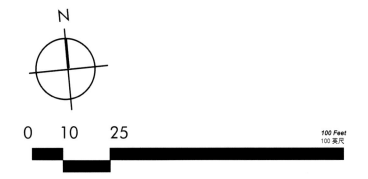

Third Floor
三层

| 文体建筑方案集成 | CULTURAL AND SPORTS ARCHITECTURE PROGRAM INTEGRATION |

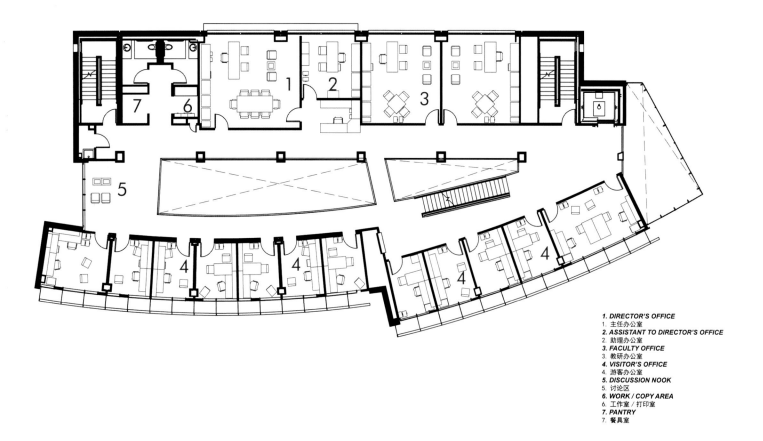

1. **DIRECTOR'S OFFICE**
1. 主任办公室
2. **ASSISTANT TO DIRECTOR'S OFFICE**
2. 助理办公室
3. **FACULTY OFFICE**
3. 教研办公室
4. **VISITOR'S OFFICE**
4. 游客办公室
5. **DISCUSSION NOOK**
5. 讨论区
6. **WORK / COPY AREA**
6. 工作室 / 打印室
7. **PANTRY**
7. 餐具室

Fourth Floor
四层

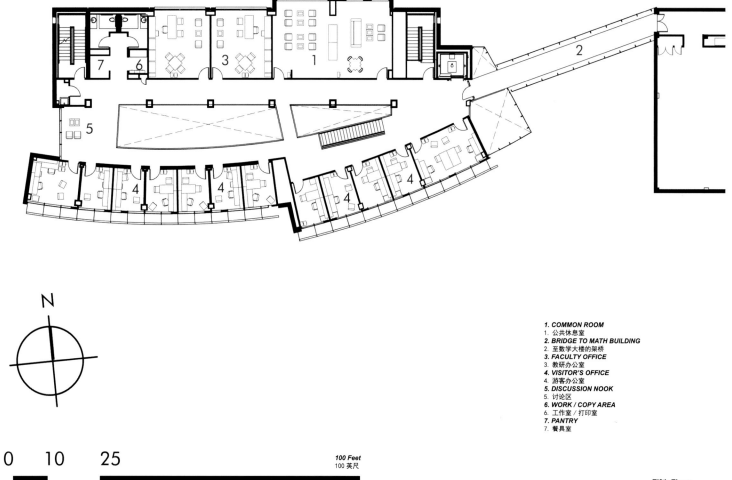

1. **COMMON ROOM**
1. 公共休息室
2. **BRIDGE TO MATH BUILDING**
2. 至数学大楼的架桥
3. **FACULTY OFFICE**
3. 教研办公室
4. **VISITOR'S OFFICE**
4. 游客办公室
5. **DISCUSSION NOOK**
5. 讨论区
6. **WORK / COPY AREA**
6. 工作室 / 打印室
7. **PANTRY**
7. 餐具室

Fifth Floor
五层

美国加州洛杉矶西方学院全球事务中心
Los Angeles Occidental College Center for Global Affairs

设计单位：Belzberg Architects Group
项目地址：美国加利福尼亚州洛杉矶
建筑面积：2 787.10m²
项目团队：Chris Sanford　Susan Nwankpa　Ashley Coon
　　　　　Chris Arntzen　Cory Taylor　David Cheung

Designed by: Belzberg Architects Group
Location: Los Angeles, California, USA
Building Area: 2,787.10m²
Project Team: Chris Sanford, Susan Nwankpa, Ashley Coon,
Chris Arntzen, Cory Taylor, David Cheung

空间重叠
磁化白玻璃墙面
社区座位

项目概况

建筑师在保留原有建筑外观的基础上，同时从根本上重新思考建筑内部的功能——现代教学的真正内涵、教学方法的可视化和网络化以及参与者的技术性学习等。该建筑的改造不是简单的外壳技术更换，它展现出来的效果是翻天覆地的变化，是在原有基础上的一种变革，或许可以说，该建筑已经成为技术本身的"代名词"。

建筑设计

鉴于技术的普遍性和教育在当代的应用，建筑师约翰逊·霍尔认为，未来的学习环境不再仅仅局限于单独的物理空间，未来将更加注重参与者的心灵契合，因此，需要缔造一个能让参与者身临其境的学习环境。

为了实现"身临其境"的学习环境，建筑的原始布局被改造成多层重叠的空间，这种转变设想是把建筑作为一个整体的学习环境，而不是多个教室的集合。运动和过渡区域得到增强和激活，摆脱了过去僵化的功能定位，教室被重新设计为全方位的、灵活的、透明的，并通过空间连接关系以及材料的透明和半透明性，使教育本身的行为变为可见。走廊，在原有的建筑结构中是狭窄的，局限于纯粹的流通，现在已加宽并重新设计成灵活的、光线充足的教育空间和民主化的公共空间，在这里可以自由探讨、彼此交流。为了与总体的、身临其境的学习环境的指导理念保持一致，建筑内部设计均采用磁化白玻璃墙面和灵活的社区座位形式，进一步展现自由、灵活的学习氛围。

| 文体建筑方案集成 | CULTURAL AND SPORTS ARCHITECTURE PROGRAM INTEGRATION |

Spatial Overlap
Magnetized White Glass Wall
Community Seats

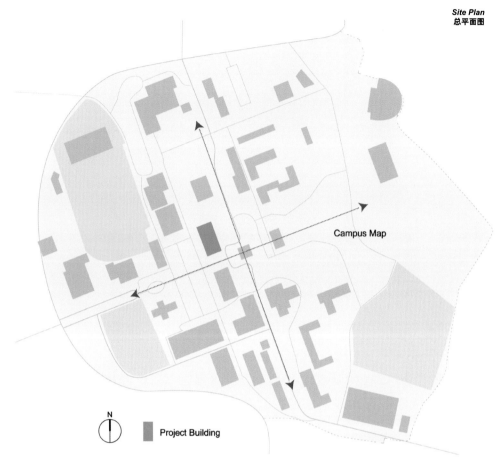

Site Plan
总平面图

Project Overview

The architects based on retaining the original building appearance, while fundamentally rethinking the interior functions—the true connotation of modern-day teaching, the visualization of teaching method and networked and participatory technology study and so on. The reconstruction of the building is rather than simply housing technology, the effect from it is earthshaking change, so it is a transformation based on the original basis, maybe the building has become the "synonym" for technology itself.

Building Design

Given the pervasive nature of technology and its contemporary applications in education, architect Johnson Hall thought that the future context of learning is no longer confined to physical space alone, it will be pay more attention to the metal simpatico from the participants. Therefore, there need to build an immersive learning environment allowing participants.

To achieve an "immersive" learning environment, the building's original layout was transformed into multilayered and overlapping spaces. This paradigm shift conceives of the entire building as a total learning environment, rather than a collection of multi classrooms. Areas of movement and transition are enhanced and activated, being rid of the past rigid functionality. Classrooms were redesigned to be all-around, flexible, and transparent, making visible the act of education itself, through spatial adjacency as well as material transparency and translucency. Hallways, which in the original building configuration were narrow and limited purely to circulation, have been widened and redesigned to serve as flexible, light-filled educational spaces and democratization public space where you can freely discuss and communicate with each other. In keeping with the guiding concept of the total, immersive learning environment, the exterior designs of the building adopt magnetized white glass and flexible community seating to further show free and flexible study atmosphere.

Floor Plans
楼层平面图

Ground Floor
底层

Second Floor
二层

Third Floor
三层

Fourth Floor
四层

1. **MEDIA WALL & LOBBY**
1. 媒体墙及大堂
2. **INNOVATION LAB**
2. 创新实验室
3. **OFFICES**
3. 办事处
4. **CLASSROOM**
4. 教室
5. **AUDITORIUM**
5. 礼堂
6. **BALCONY**
6. 阳台

EXISTING CONDITION:
现有条件
INSULAR LEARNING SPACES
独立的学习空间

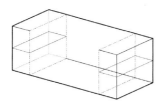

NEW CONDITION:
新条件
INTRODUCTION OF COMMUNAL INFORMAL LEARNING SPACE
引进公共的非正式学习空间

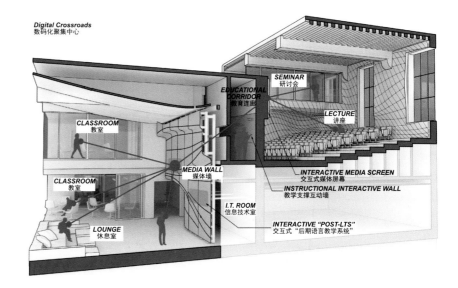

Digital Crossroads
数码化聚集中心

- SEMINAR 研讨会
- EDUCATIONAL CORRIDOR 教育连廊
- LECTURE 讲座
- CLASSROOM 教室
- INTERACTIVE MEDIA SCREEN 交互式媒体屏幕
- INSTRUCTIONAL INTERACTIVE WALL 教学支撑互动墙
- CLASSROOM 教室
- MEDIA WALL 媒体墙
- I.T. ROOM 信息技术室
- INTERACTIVE "POST-LTS" 交互式"后期语言教学系统"
- LOUNGE 休息室

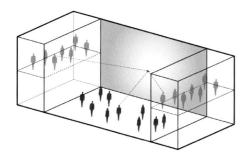

FORMAL GESTURE 1:
正规姿态 1:
ACTIVATE THE PRIMARY SURFACE OF THE COMMUNAL SPACE WITH INTERACTIVE MEDIA
激活公共空间与互动媒体的主表面

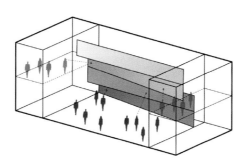

FORMAL GESTURE 2:
正规姿态 2:
SUBDIVIDE SURFACE TO CREATE LOCALIZED LEARNING EXPERIENCES
细分表面来创建本地化的学习经验

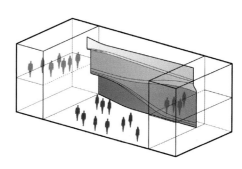

FORMAL GESTURE 3:
正规姿态 3:
SUBDIVIDE SURFACE TO CREATE LOCALIZED LEARNING EXPERIENCES
细分表面来创建本地化的学习经验

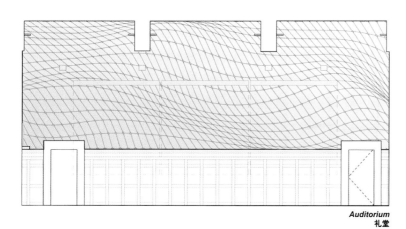

Auditorium
礼堂

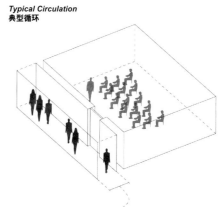

Typical Circulation
典型循环

MINIMUM REQUIRED CIRCULATION
最小所需循环

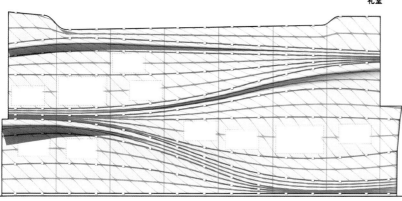

Media Wall
媒体墙

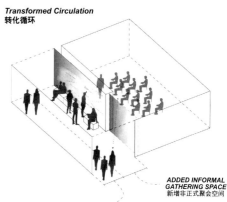

Transformed Circulation
转化循环

ADDED INFORMAL GATHERING SPACE
新增非正式聚会空间

MINIMUM REQUIRED CIRCULATION
最小所需循环

伊朗扎布尔医科大学总部
Headquarter of Zabol University of Medical Sciences

设计单位：New wave Architecture
委托方：扎布尔大学
项目地址：伊朗扎布尔
总建筑面积：15 000m²

Designed by: New wave Architecture
Client: University of Zabol
Location: Zabol, Iran
Gloss Floor Area: 15,000m²

倾斜的柱子
半透明枢纽
三角形图案

项目概况

伊朗扎布尔医科大学总部试图作为大学校园的焦点出现，它与周围环境恰当融合，并符合伊朗干旱和炎热的地区气候环境。

建筑设计

在建筑的中央部分，配置了一个玻璃状的半透明枢纽，旨在借助其实体性和透明度的组合概念来表达该建筑一个既现代又显著的特征。

建筑体底下倾斜的柱子，沿着项目动态过渡进行协调。它冷静地从周围上升，形成了建设的核心，并最终引导进入一个绿色的平整表面。建筑立面的角面通过减少阳光辐射，从而刻意去控制能量和通风。抽象演绎的建筑外观上的三角形图案来源于帕提亚时期"Koohe-Khaje 锡斯坦"的建筑遗产图案，体现了门面的轮回设计。

设计为了满足项目要求的15 000m²的总面积，每个楼层的主要部分被分配到不同的部门，围绕核心开设办事处和单独的客房，而垂直辅助功能是通过电梯和楼梯铺设的该项目北部部分而保持的。

| 文体建筑方案集成 | CULTURAL AND SPORTS ARCHITECTURE PROGRAM INTEGRATION |

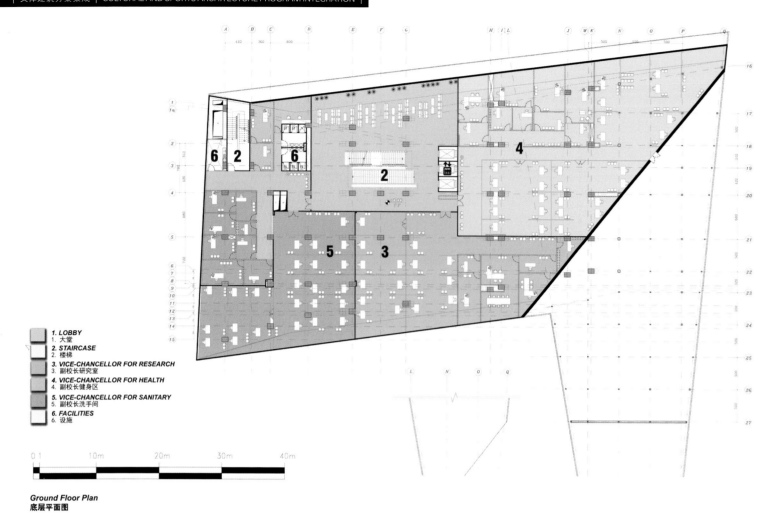

1. **LOBBY**
   大堂
2. **STAIRCASE**
   楼梯
3. **VICE-CHANCELLOR FOR RESEARCH**
   副校长研究室
4. **VICE-CHANCELLOR FOR HEALTH**
   副校长健身区
5. **VICE-CHANCELLOR FOR SANITARY**
   副校长洗手间
6. **FACILITIES**
   设施

Ground Floor Plan
底层平面图

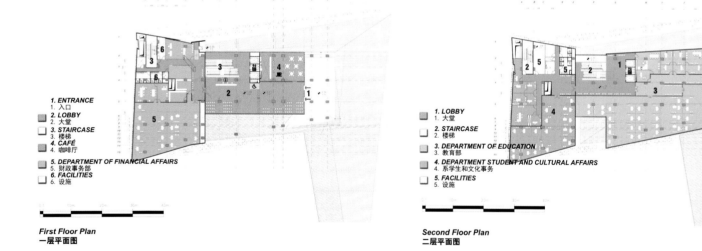

1. **ENTRANCE**
   入口
2. **LOBBY**
   大堂
3. **STAIRCASE**
   楼梯
4. **CAFÉ**
   咖啡厅
5. **DEPARTMENT OF FINANCIAL AFFAIRS**
   财政事务部
6. **FACILITIES**
   设施

First Floor Plan
一层平面图

1. **LOBBY**
   大堂
2. **STAIRCASE**
   楼梯
3. **DEPARTMENT OF EDUCATION**
   教育部
4. **DEPARTMENT STUDENT AND CULTURAL AFFAIRS**
   系学生和文化事务
5. **FACILITIES**
   设施

Second Floor Plan
二层平面图

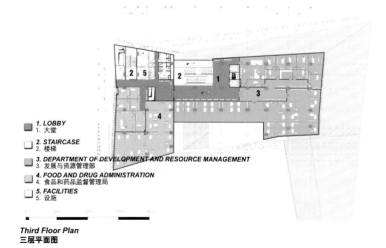

1. **LOBBY**
   大堂
2. **STAIRCASE**
   楼梯
3. **DEPARTMENT OF DEVELOPMENT AND RESOURCE MANAGEMENT**
   发展与资源管理部
4. **FOOD AND DRUG ADMINISTRATION**
   食品和药品监督管理局
5. **FACILITIES**
   设施

Third Floor Plan
三层平面图

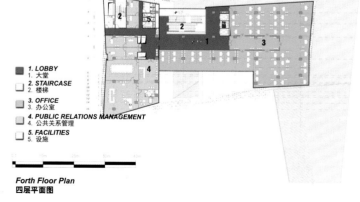

1. **LOBBY**
   大堂
2. **STAIRCASE**
   楼梯
3. **OFFICE**
   办公室
4. **PUBLIC RELATIONS MANAGEMENT**
   公共关系管理
5. **FACILITIES**
   设施

Forth Floor Plan
四层平面图

Tilt Column
Translucent Hub
Triangular Pattern

Project Overview

The building is attempting to be appeared as a focal point of the university campus, integrating with surroundings and being in lined with arid and hot climatic circumstances in the region of Iran.

Building Design

In central part of the project, it configures a translucent hub of glass and dedicates an authentic combination of solidness and transparency to express a modern and dominant character.

The inclined pillars under the building coordinate with a dynamic transition along the project. It rises soberly from surrounding, forms the core of the building and finally leads them to a green flat surface. Angular faces of the façade diminish sunlight radiation deliberately to control the energy and ventilation. The triangular motifs in the abstract deduction of the shape are from "Koohe-Khaje Sistan" in Parthian period, reflecting the rebirth design of the façade.

In order to meet the requirement of the project of total area of 15,000m^2, the main part of each floor is designed to assign to different departments, and around the core to set up offices and separate rooms, while the vertical auxiliary functions are maintained by the northern part of the project where has been paved elevators and stairs.

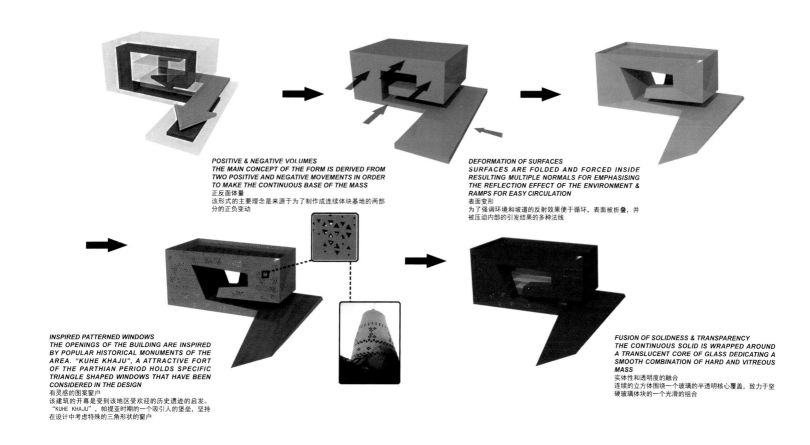

POSITIVE & NEGATIVE VOLUMES
THE MAIN CONCEPT OF THE FORM IS DERIVED FROM TWO POSITIVE AND NEGATIVE MOVEMENTS IN ORDER TO MAKE THE CONTINUOUS BASE OF THE MASS
正反面体量
该形式的主要理念是来源于为了制作成连续体块基地的两部分的正负变动

DEFORMATION OF SURFACES
SURFACES ARE FOLDED AND FORCED INSIDE RESULTING MULTIPLE NORMALS FOR EMPHASISING THE REFLECTION EFFECT OF THE ENVIRONMENT & RAMPS FOR EASY CIRCULATION
表面变形
为了强调环境和坡道的反射效果便于循环，表面被折叠，并被压迫内部的引发结果的多种法线

INSPIRED PATTERNED WINDOWS
THE OPENINGS OF THE BUILDING ARE INSPIRED BY POPULAR HISTORICAL MONUMENTS OF THE AREA. "KUHE KHAJU", A ATTRACTIVE FORT OF THE PARTHIAN PERIOD HOLDS SPECIFIC TRIANGLE SHAPED WINDOWS THAT HAVE BEEN CONSIDERED IN THE DESIGN
有灵感的图案窗户
该建筑的开幕是受到该地区受欢迎的历史遗迹的启发。"KUHE KHAJU"，帕提亚时期的一个吸引人的堡垒，坚持在设计中考虑特殊的三角形状的窗户

FUSION OF SOLIDNESS & TRANSPARENCY
THE CONTINUOUS SOLID IS WRAPPED AROUND A TRANSLUCENT CORE OF GLASS DEDICATING A SMOOTH COMBINATION OF HARD AND VITREOUS MASS
实体性和透明度的融合
连续的立方体围绕一个玻璃的半透明核心覆盖，致力于坚硬玻璃体块的一个光滑的组合

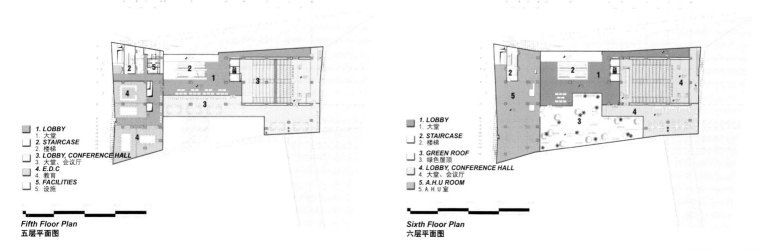

1. LOBBY
   1. 大堂
2. STAIRCASE
   2. 楼梯
3. LOBBY, CONFERENCE HALL
   3. 大堂、会议厅
4. E.D.C
   4. 教育
5. FACILITIES
   5. 设施

Fifth Floor Plan
五层平面图

1. LOBBY
   1. 大堂
2. STAIRCASE
   2. 楼梯
3. GREEN ROOF
   3. 绿色屋顶
4. LOBBY, CONFERENCE HALL
   4. 大堂、会议厅
5. A.H.U ROOM
   5. A.H.U 室

Sixth Floor Plan
六层平面图

美国纽黑文福特学院科技大厦
The Foote School Science and Technology Building

设计单位：玛丽安·汤普森建筑事务所
项目地址：美国纽黑文市
建筑面积：1 126m²
项目获奖：2013年波士顿社会建筑师可持续设计奖
　　　　　2013年波士顿社会建筑师教育设施奖

Designed by: Maryann Thompson Architects
Location: New Haven, USA
Building Area: 1,126m²
Awards: 2013 Boston Society of Architects Sustainable Design Award
　　　　2013 Boston Society of Architects Education Facilities Award

可再生理念
学校新中心

项目概况

福特学院科技大厦位于美国纽黑文市，由玛丽安·汤普森建筑事务所设计的一个具有最新水平的可持续设施。该建筑既使用常识被动策略，也使用活跃的可再生能源系统来创建一个用于学生的动态教学工具的建筑。

建筑设计

建筑深的悬臂结构和木制百叶包裹在南部和西部的曝露处，在炎热的月份防止热量的增加，同时在冬天吸收阳光。教室和实验室都镶有大的玻璃墙，使得在日常阳光下孩子的认识能够提高，并通过采光减少电力照明的使用。

面向操场的百叶窗式遮阳篷有三重作用，将光线漫射进入西边的教室，在操场创建一个入口，同时还作为太阳能光热热水系统操作，为建筑提供90%的热水。可操作的窗户设置在墙的高处，在吊扇的协助下，通过"烟囱效应"鼓动交叉通风，减少了需要空调的日子。

场地北部是九年制义务校园中心，毗邻现有建筑，与较落后学校举办的外部空间细粒度网络形成了鲜明的对比，这些网络以前作为浮动对象紧靠现有球场进行扩张。为合并这些现有建筑物，新的科技大厦将其本身进行编织，并扩展九年制义务校园的外部空间的循环脊椎和网络。通过细微的语调显现并回应现有的结构，该大厦形成了一系列新的联合的外部空间，并为在校园北侧的上层学校创建了一个新的中心。

所有的科学实验室都可直接进入到外部，与二楼的两个实验室用作外部教室在屋顶甲板上对外开放。在它的南部边缘，甲板通过阶梯式花园露台得到满足，可使地平面上升到二楼，也可以用作一个非正式的露天剧场。

| 文体建筑方案集成 | CULTURAL AND SPORTS ARCHITECTURE PROGRAM INTEGRATION |

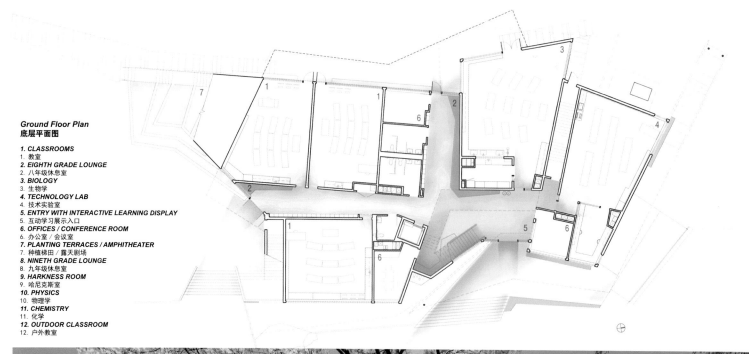

Ground Floor Plan
底层平面图

1. CLASSROOMS
1. 教室
2. EIGHTH GRADE LOUNGE
2. 八年级休息室
3. BIOLOGY
3. 生物学
4. TECHNOLOGY LAB
4. 技术实验室
5. ENTRY WITH INTERACTIVE LEARNING DISPLAY
5. 互动学习展示入口
6. OFFICES / CONFERENCE ROOM
6. 办公室 / 会议室
7. PLANTING TERRACES / AMPHITHEATER
7. 种植梯田 / 露天剧场
8. NINETH GRADE LOUNGE
8. 九年级休息室
9. HARKNESS ROOM
9. 哈尼克斯室
10. PHYSICS
10. 物理学
11. CHEMISTRY
11. 化学
12. OUTDOOR CLASSROOM
12. 户外教室

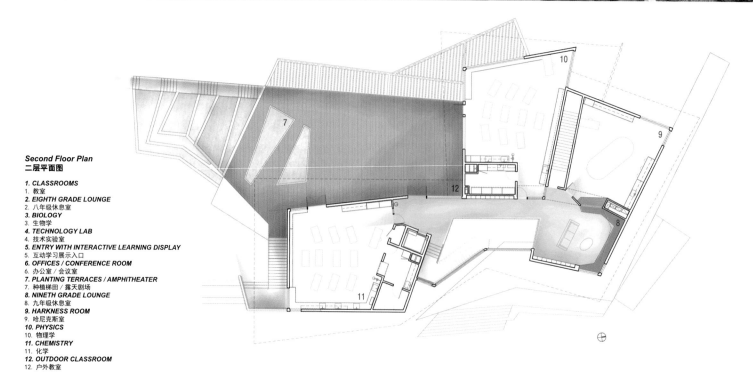

Second Floor Plan
二层平面图

1. CLASSROOMS
1. 教室
2. EIGHTH GRADE LOUNGE
2. 八年级休息室
3. BIOLOGY
3. 生物学
4. TECHNOLOGY LAB
4. 技术实验室
5. ENTRY WITH INTERACTIVE LEARNING DISPLAY
5. 互动学习展示入口
6. OFFICES / CONFERENCE ROOM
6. 办公室 / 会议室
7. PLANTING TERRACES / AMPHITHEATER
7. 种植梯田 / 露天剧场
8. NINETH GRADE LOUNGE
8. 九年级休息室
9. HARKNESS ROOM
9. 哈尼克斯室
10. PHYSICS
10. 物理学
11. CHEMISTRY
11. 化学
12. OUTDOOR CLASSROOM
12. 户外教室

Renewable Idea
New Center School

Project Overview

The Foote School Science and Technology Building located in New Haven was designed by Maryann Thompson Architects to be a state of the art sustainable facility. The building uses both common sense passive strategies as well as active renewable energy systems to create a building that is a dynamic teaching tool for the students.

Building Design

Deep overhangs and wood louvers wrap the southern and western exposures, preventing heat gain in the warm months while allowing sunlight in during the winter. The classrooms and labs are lined with large walls of glass, allowing a heightened awareness in the child of the daily traverse of the sun and reducing the use of electric lighting through day-lighting.

A louvered awning facing the playfield does triple duty, diffusing the daylight entering the western classrooms and creating a threshold to the field while also operating as a solar thermal hot water system that provides 90% of the hot water usage for the building. Operable windows set high in the walls; assisted by ceiling fans, encourage cross ventilation through the "stack effect" reducing the days that conditioned air is needed.

The site, to the north of the K-9 campus center, is adjacent to existing buildings which, in sharp contrast to the fine grained network of the held exterior spaces of the lower school, previously floated as objects next to the expanse of the play field. Incorporating these existing buildings, the new Science and Technology Building knits itself into and extends the K-9 campus' circulation spine and network of exterior spaces. Hugging and responding to the existing structures through subtle inflections, it forms a new series of linked exterior rooms and creates a new center for the Upper School at the north side of the campus.

All of the Science labs have direct access to the exterior, with the two labs on the second floor opening out to a roof deck used as an exterior classroom. At its southern edge the deck is met by stepped garden terraces that bring the ground plane up to the second floor and also serve as an informal amphitheater.

美国德克萨斯大学教育和商业大楼
The University of Texas Education and Business Complex

设计单位：阿特金斯
委托方：德克萨斯大学
项目地址：美国德克萨斯州
建筑面积：9 522.25m²

Designed by: Atkins
Client: The University of Texas at Brownsville
Location: Texas, USA
Building Area: 9,522.25m²

复古元素
浅色系外墙
拱形门窗

项目概况

该建筑毗邻洛萨诺瑞斯卡，通过一座步行桥连接主校区，建筑面积接近 10 000m²，是一栋综合性的大楼。建筑内部配备了实验室和其他一些服务性的行政办公室，旨在为学术研究和日常的行政办公提供一个舒适、安静的环境。

建筑设计

德克萨斯大学教育和商业大楼是使用高危传输的施工管理的方法完成的，建筑体内配备了一个新的中央热能制冷机房。

整个建筑多采用复古元素来展现浓厚的学术氛围：浅色系的建筑外观给人一种宁静、祥和的感觉，有利于为莘莘学子和学者提供一个思考的环境；建筑的门窗均采用拱形设计，线条构造柔和，有利于自然光线的渗透，进而增加建筑内部空间的光亮度，形成一种敞亮、明快的感觉，衬托出青年人求学的积极向上、朝气蓬勃的精神。

此外，校区内绿色植被丰富，大面积的草坪横卧其中，为学生们开展户外活动提供了一个便捷舒适的场所。

Retro Elements
Light-colored Façades
Arched Windows and Doors

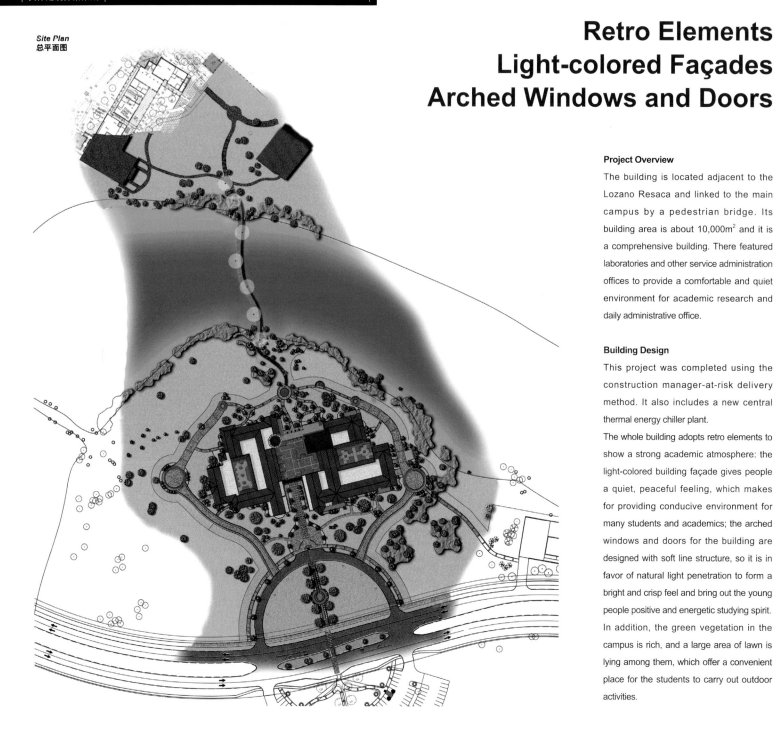

Site Plan
总平面图

Project Overview
The building is located adjacent to the Lozano Resaca and linked to the main campus by a pedestrian bridge. Its building area is about 10,000m² and it is a comprehensive building. There featured laboratories and other service administration offices to provide a comfortable and quiet environment for academic research and daily administrative office.

Building Design
This project was completed using the construction manager-at-risk delivery method. It also includes a new central thermal energy chiller plant.

The whole building adopts retro elements to show a strong academic atmosphere: the light-colored building façade gives people a quiet, peaceful feeling, which makes for providing conducive environment for many students and academics; the arched windows and doors for the building are designed with soft line structure, so it is in favor of natural light penetration to form a bright and crisp feel and bring out the young people positive and energetic studying spirit. In addition, the green vegetation in the campus is rich, and a large area of lawn is lying among them, which offer a convenient place for the students to carry out outdoor activities.

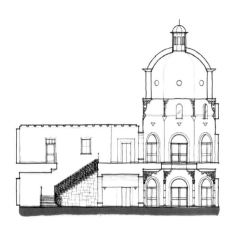

Inside The Tower Lobby
塔楼大堂内部

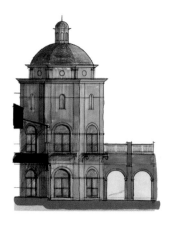

Tower Lobby – View Looking North
塔楼大堂 —— 北部视图

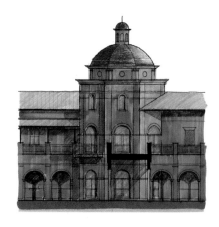

Tower Lobby – View at Central Courtyard
塔楼大堂 —— 中央庭院视图

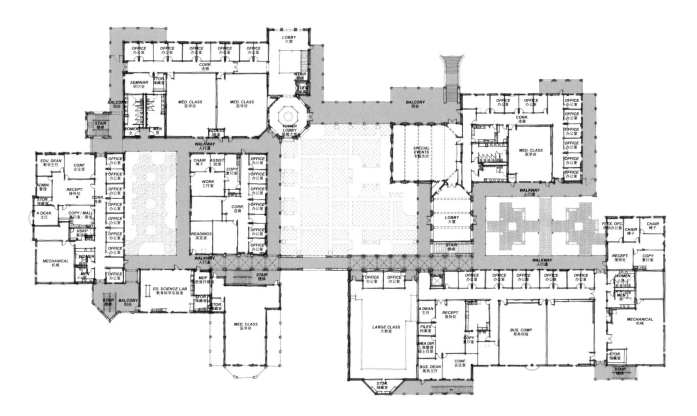

First Floor Plan
一层平面图

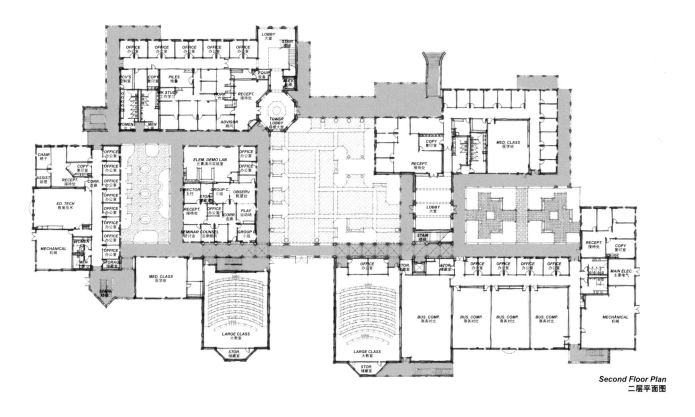

Second Floor Plan
二层平面图

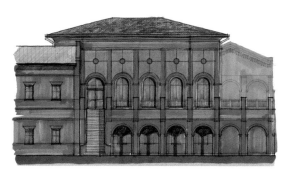

Special Events – View at Resaca
专题活动 —Resaca 视图

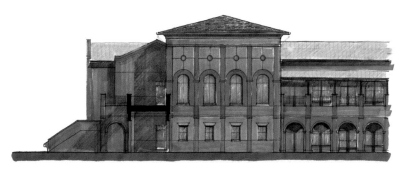

Special Events –View at Central Courtyard
专题活动 — 中央庭院视图

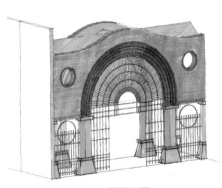

Entry Arch
入口拱门

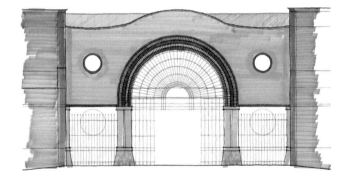

ENTRY GATE
入口门

ENTRY GATE
入口门

ENTRANCE FROM BOULEVARD
林荫大道入口

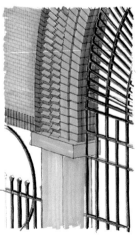

BRICK DETAIL
砖块细节

斯洛文尼亚卢布尔雅那国家和大学图书馆
Ljubljana National and University Library

设计单位：Waltritsch a+u Architetti Urbanisti Trieste (Itay)
　　　　　Guarnieri Architects London (UK)
　　　　　Piero Ongaro Trieste (Italy)
　　　　　Ove Arup Milano (Italy)
委托方：斯洛文尼亚国家和大学图书馆
项目地址：斯洛文尼亚卢布尔雅那
总建筑面积：20 000m²

Designed by: Waltritsch a+u Architetti Urbanisti Trieste (Itay)
　　　　　　Guarnieri Architects London (UK)
　　　　　　Piero Ongaro Trieste (Italy)
　　　　　　Ove Arup Milano (Italy)
Client: Slovenia National and University Library
Location: Ljubljana, Slovenia
Gross Floor Area: 20,000m²

粘合剂
V 形柱

项目概况

这个新的国家和大学图书馆建立在一个重要的考古遗址之上，作为几所大学建筑之间以及城市不同部分之间新的粘合剂，这种情况为获得该位置的资格开辟了新的机会：城市文脉的枢轴。该项目的建造将作为一个强调远古城市重要性的设置和一个连接活跃城市的平台，由此而产生21世纪的文化偶像。

建筑设计

该建筑渗透在街道层面，在不同的部分之间作为一个城市连接器（市区房）。图书馆设施从更高层次开始，在计划中利用了三个领先结构的形状。地板正在慢慢扩大其表面，同时高度也在增加，也为中心的室内空间留下了空间。图书馆体块被视为固体体块，在更广泛的城市背景下作为公共建筑巩固了它的存在。

图书馆楼层堆叠的体块也足以从考古遗址的水平分离，这样不会与它竞争。该结构是为了维护该址的考古存在和现有的"轴节"和"Decumano"轨迹轴而设计的，通过V形柱手段尽可能少的干扰考古学遗迹。顶部的建筑结构，被悬臂部分所特定化，这一设计，包括一系列的传输结构。

| 文体建筑方案集成 | CULTURAL AND SPORTS ARCHITECTURE PROGRAM INTEGRATION |

Mezzanine Level
夹层

Level 2
二层

Level 3
三层

Level 1
一层

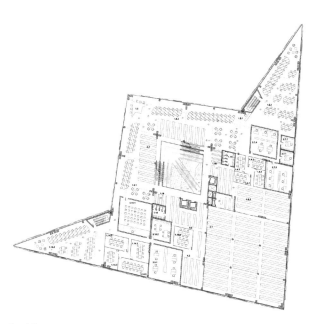

Level 4
四层

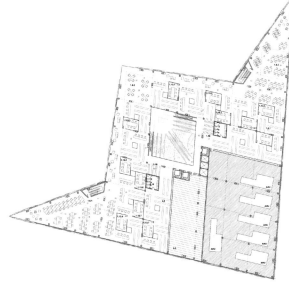

Level 5
五层

Urban and Cultural Context / The Urban Path
城市和文化环境 / 市区道路

THE URBAN PATH
市区道路

NUK II
卢布尔雅那图书馆 II

Adhesives
V-shaped Columns

Project Overview

The new National and University Library is built on the top of an important archaeological site at the edge of the capital's historical hart. This condition opens up new opportunities to qualify the area as the new binder between several university buildings, as well as between different parts of the city: a pivot of the urban context. The project act as a setting, which emphasizes the importance of the ancient past of the city, acts as an active urban connecting platform, and produces a cultural icon of the 21st century.

Building Design

The building is very permeable at the street level, acting as an urban connector (the urban room) between different parts. Library facilities start at higher levels, taking the shape of a three headed structure in plan. Floors are slowly enlarging their surface while growing in height, leaving space for a central indoor patio as well. The library volume is treated as a solid mass strengthening its presence as a public building in the wider urban context.

The mass of the stack of library floors is also sufficiently detached from the level of the archaeological site so not to compete with it. The structure is designed to maintain the archaeological presences of the site and the existing path axes which are the "Cardo" and the "Decumano", interfering as less as possible with the archaeology remains by means of V shaped columns. The building's structure on top, characterized by cantilever portions, has been designed including a series of transfer structures.

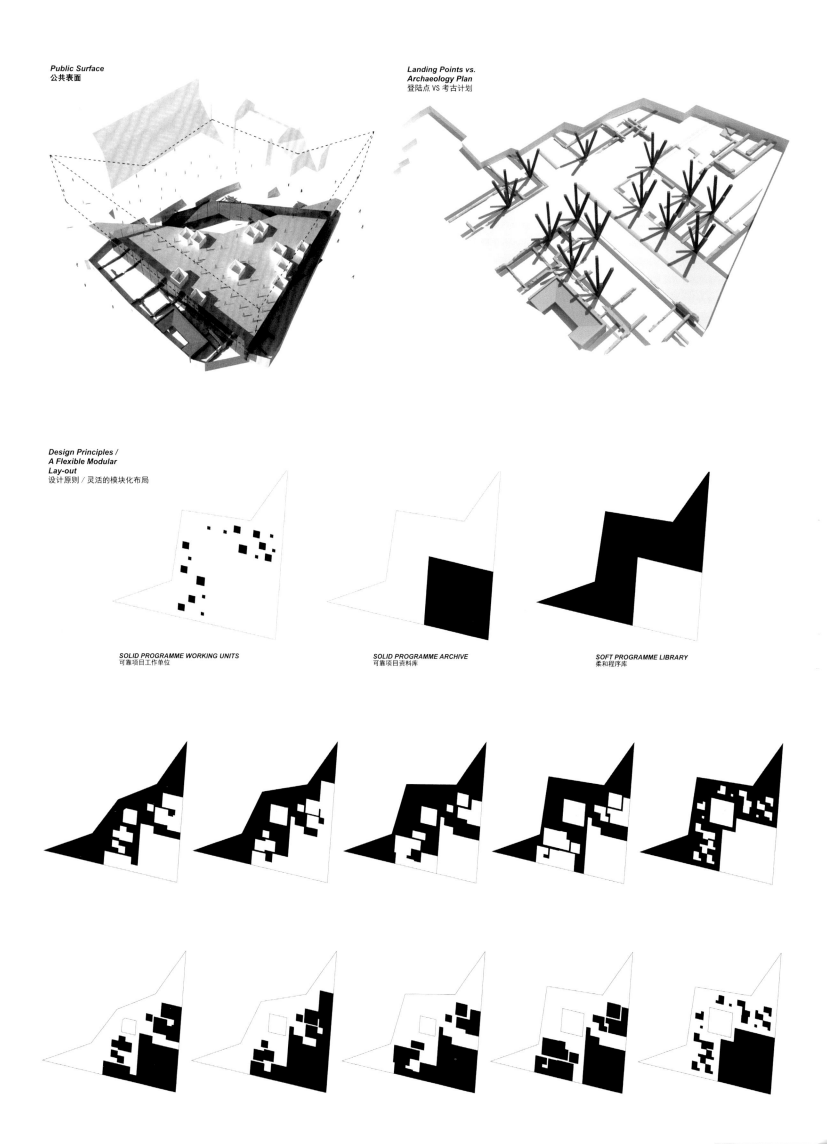

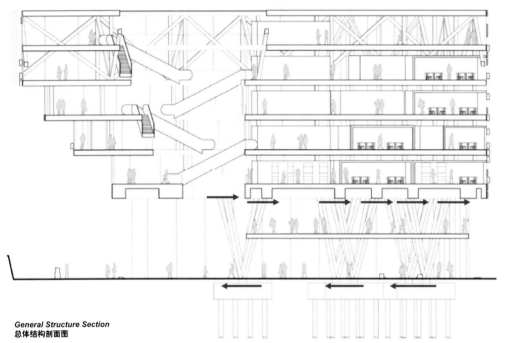

General Structure Section
总体结构剖面图

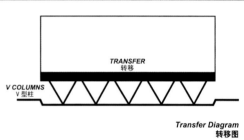

Transfer Diagram
转移图

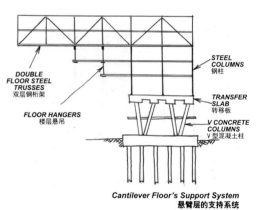

Cantilever Floor's Support System
悬臂层的支持系统

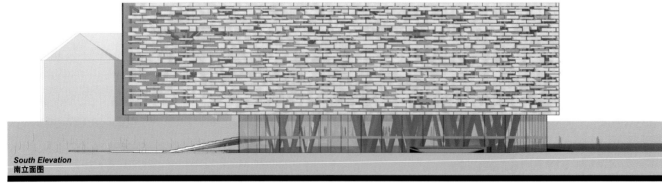

South Elevation
南立面图

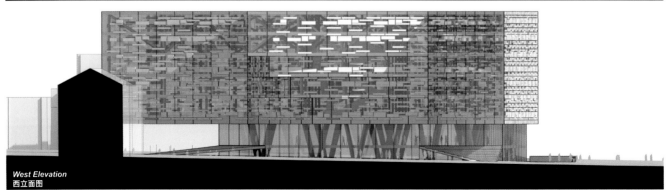

West Elevation
西立面图

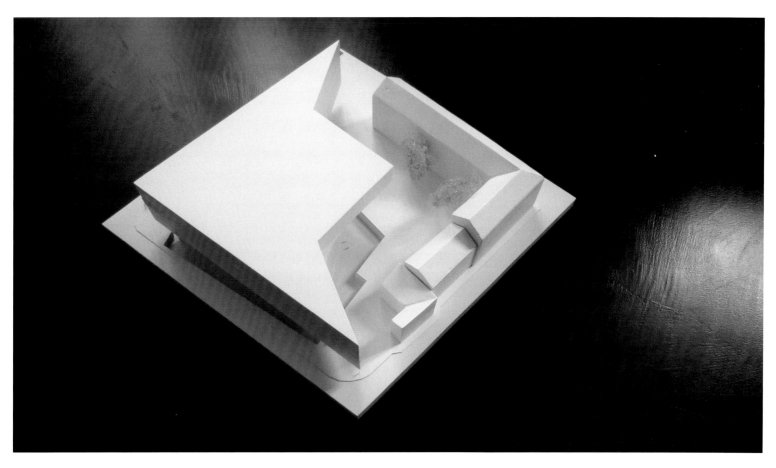

荷兰阿默斯福特文化中心"Eemhuis"
Amersfoort Cultural Center "Eemhuis"

设计单位: Neutelings Riedijk Architects	***Designed by:** Neutelings Riedijk Architects*
委托方: 阿默斯福特市	***Client:** City of Amersfoort*
项目地址: 荷兰阿默斯福特市	***Location:** Amersfoort, Netherlands*
项目面积: 16 000 m²	***Area:** 16,000m²*

**文化集聚
城市广场
皇冠**

项目概况

Eemhuis 是荷兰阿默斯福特的综合性文化建筑，它集市图书馆、会展中心、文物档案中心和舞蹈、音乐及视觉艺术学校于一身。建筑坐落在靠近市中心的城市重建区，城市公共领域从四面八方汇聚于此。

建筑设计

Eemhuis 地面层设有封闭式的公共广场，广场处有一个大型的咖啡厅和通向多个功能区的入口。地面层广场的旁边是会展中心，大的中心展厅一部分嵌入地下，两边是纵列的小型展览室。梯田式的图书馆从城市广场延伸而来，将人们引领至上面的主图书馆。图书馆在楼梯尽头散开成一个巨大的开放空间，这里书架林立、从阅读区和学习区都能够观赏城市景观。盘旋在图书馆上端的是档案室，也是 Eemhuis 的"天花板"。最顶端的阁楼是艺术学校。艺术学校三个相互独立的悬臂式区域（戏剧&舞蹈、视觉艺术和音乐）如同加冕于建筑综合体上的三顶皇冠。

根据总体规划，建筑的整个立面由三个部分组成。第一层的基座由 30cm 的加长型釉面砖筑成，强化了水平的线条感。二层由玻璃板构成，可以将光线引入建筑内部，提高透明度。三层的"皇冠"部分由半球体点缀的金属板制成，半球体的设计能有效减少悬臂体量受到北部风雨的侵蚀。

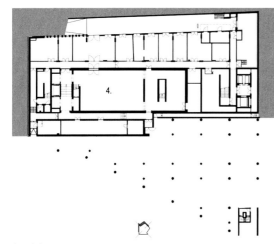

Level -1
负一层

1. *LIBRARY*
1. 图书馆
2. *HERITAGE ARCHIVES*
2. 文物档案
3. *ARTS SCHOOL*
3. 艺术学校
4. *EXHIBITION CENTER*
4. 展览中心

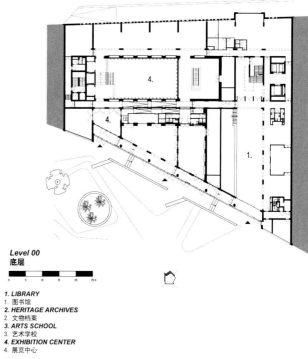

Level 00
底层

1. *LIBRARY*
1. 图书馆
2. *HERITAGE ARCHIVES*
2. 文物档案
3. *ARTS SCHOOL*
3. 艺术学校
4. *EXHIBITION CENTER*
4. 展览中心

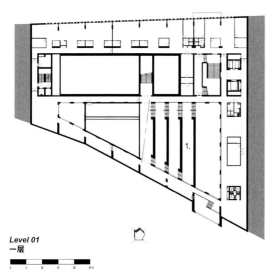

Level 01
一层

1. *LIBRARY*
1. 图书馆
2. *HERITAGE ARCHIVES*
2. 文物档案
3. *ARTS SCHOOL*
3. 艺术学校
4. *EXHIBITION CENTER*
4. 展览中心

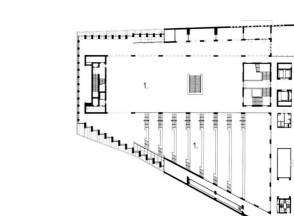

Level 02
二层

1. *LIBRARY*
1. 图书馆
2. *HERITAGE ARCHIVES*
2. 文物档案
3. *ARTS SCHOOL*
3. 艺术学校
4. *EXHIBITION CENTER*
4. 展览中心

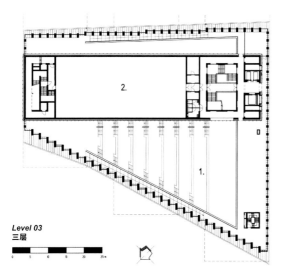

Level 03
三层

1. *LIBRARY*
1. 图书馆
2. *HERITAGE ARCHIVES*
2. 文物档案
3. *ARTS SCHOOL*
3. 艺术学校
4. *EXHIBITION CENTER*
4. 展览中心

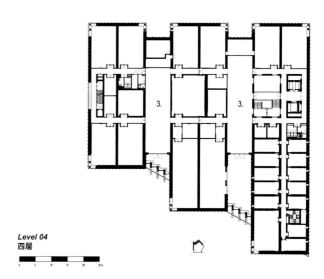

Level 04
四层

1. *LIBRARY*
1. 图书馆
2. *HERITAGE ARCHIVES*
2. 文物档案
3. *ARTS SCHOOL*
3. 艺术学校
4. *EXHIBITION CENTER*
4. 展览中心

Cultural gathering
City Square
Imperial Crown

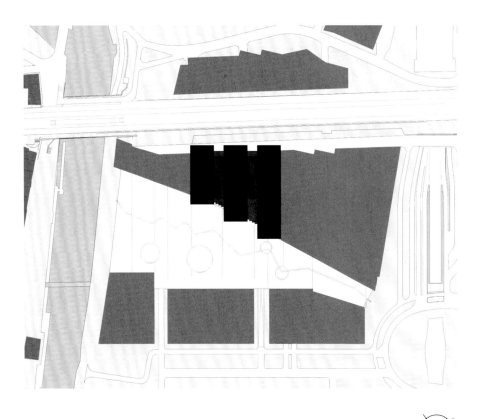

Project Overview

The Eemhuis combines a number of existing cultural institutes in the city of Amersfoort: the city library, the exposition center, the heritage archives and a school for dance, music and visual arts. It is located on an urban redevelopment area close to the city center. The public domain is continued into the interior of the building in all directions.

Building Design

At the ground floor, the public square becomes a covered plaza, with a grand café and entrances to the various functions. The exposition center is set directly off the square on the ground floor, with a large central exhibition hall that is half sunken in the ground and is surrounded by an enfilade of smaller exhibition rooms. The library is a plaza of stepped information terraces as a prolongation of the city square that brings the visitors up to the main library floor. On the top of the stairs the library spills into a vast open space with book stacks and reading and study areas overlooking the city. Above, it spirals the archive volume that forms the ceiling of this space. The attic of the building houses the arts school. The three arts departments (theatre & dance, visual arts and music) are each expressed separately as cantilevered beams that crown the complex.

The façades are composed of a classical tripartite as imposed by the master plan. The plinth is made of 30cm long elongated glazed bricks, reinforcing the horizontal lining. The crown of the building is made of metal panels with a dotted pattern of semi-spheres that enhance the alienating quality of the cantilevered volumes against the northern Dutch clouds.

| 文体建筑方案集成 | CULTURAL AND SPORTS ARCHITECTURE PROGRAM INTEGRATION |

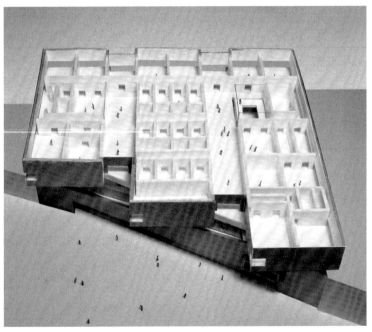

| 194–195 |

| 文体建筑方案集成 | CULTURAL AND SPORTS ARCHITECTURE PROGRAM INTEGRATION |

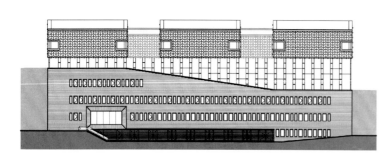
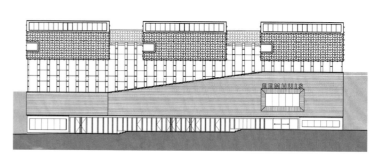

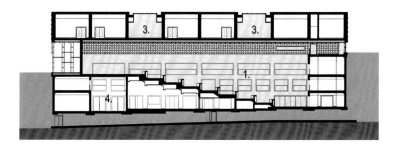

Section AA
剖面 AA

1. *LIBRARY*
   1. 图书馆
2. *HERITAGE ARCHIVES*
   2. 文物档案
3. *ARTS SCHOOL*
   3. 艺术学校
4. *EXHIBITION CENTER*
   4. 展览中心

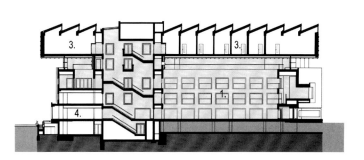

Section DD
剖面 DD

1. *LIBRARY*
   1. 图书馆
2. *HERITAGE ARCHIVES*
   2. 文物档案
3. *ARTS SCHOOL*
   3. 艺术学校
4. *EXHIBITION CENTER*
   4. 展览中心

韩国大邱 Gosan 公共图书馆（方案一）
Daegu Gosan Public Library

设计单位：Foundry of Space (FOS)
委托方：大邱广域市寿城区政府
项目地址：韩国大邱市
项目面积：2 080m²

Designed by: Foundry of Space (FOS)
Client: Daegu Metropolitan City Suseong-gu Office
Location: Daegu, South Korea
Area: 2,080m²

项目概况

本案将图书馆作为当地社区的文化和社会空间。项目的重点是获得构成对邻里日常生活重要组成部分的功能，形成具有吸引力和舒适度的配套设施。

大邱 Gosan 公共图书馆所提出的设计不仅服务于地方的基本实用性，而且知识和各种形式的新闻媒体都保存完好，并以一种更时尚的方式分享，提供社交平台，以重新定位项目本身是大邱最具生机和最有活力的现代场地。

建筑设计

场地位置被宜人的住宅和从大邱市中心分离出来的绿地所环绕。设计通过无缝融合图书馆的内外部空间与北部城市体块和公园的肌理使得这个优势充分发挥。通过提高街道水平之上的主要项目（主大厅、一般收集室、儿童房），以及降低地下文化与终身学习和服务空间，广泛地让周围的城市空间流经建筑物的开放场地，以及在下层举办的公众活动融入到环境之中。

扮演新角色
图书无限循环
结构、环保理念

建筑物结构系统的原理是受无穷形状的启发,上层较低的拱门大部分被悬挑出。该结构的复杂性还以戏剧性的方式明确了图书馆和环境之间"相互缠绕"的概念。它与周围环境的共存看起来好像它是由一系列巨大的线松散地缝合起来的,暗示大邱和纺织工业之间蓬勃发展的关系。

为符合城市的绿色建筑的活动,尤其是太阳能,所提出的设计部署光伏建筑一体化(BIPV)的先进技术,即在建筑外墙使用光伏材料。光伏玻璃与光扩散半透明膜被应用到图书无限循环的整个表面上,产生的能量作为电能的辅助源以便用于照明阅读空间。半透明的薄膜也有助于扩散直射的阳光,防止从外面看产生眩光,从而保持室内更均匀分布的照明。

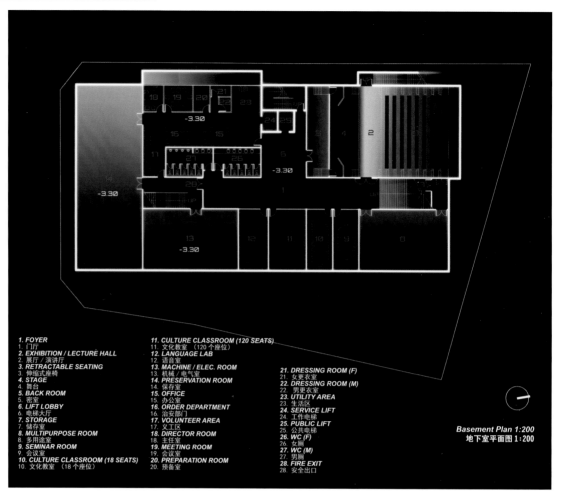

1. FOYER　　1. 门厅
2. EXHIBITION / LECTURE HALL　　2. 展厅 / 演讲厅
3. RETRACTABLE SEATING　　3. 伸缩式座椅
4. STAGE　　4. 舞台
5. BACK ROOM　　5. 密室
6. LIFT LOBBY　　6. 电梯大厅
7. STORAGE　　7. 储存室
8. MULTIPURPOSE ROOM　　8. 多用途室
9. SEMINAR ROOM　　9. 会议室
10. CULTURE CLASSROOM (18 SEATS)　　10. 文化教室(18 个座位)
11. CULTURE CLASSROOM (120 SEATS)　　11. 文化教室(120 个座位)
12. LANGUAGE LAB　　12. 语音室
13. MACHINE / ELEC. ROOM　　13. 机械 / 电气室
14. PRESERVATION ROOM　　14. 保存室
15. OFFICE　　15. 办公室
16. ORDER DEPARTMENT　　16. 治安部门
17. VOLUNTEER AREA　　17. 义工区
18. DIRECTOR ROOM　　18. 主任室
19. MEETING ROOM　　19. 会议室
20. PREPARATION ROOM　　20. 预备室
21. DRESSING ROOM (F)　　21. 女更衣室
22. DRESSING ROOM (M)　　22. 男更衣室
23. UTILITY AREA　　23. 生活区
24. SERVICE LIFT　　24. 工作电梯
25. PUBLIC LIFT　　25. 公共电梯
26. WC　　26. 女厕
27. WC (M)　　27. 男厕
28. FIRE EXIT　　28. 安全出口

Basement Plan 1:200
地下室平面图 1:200

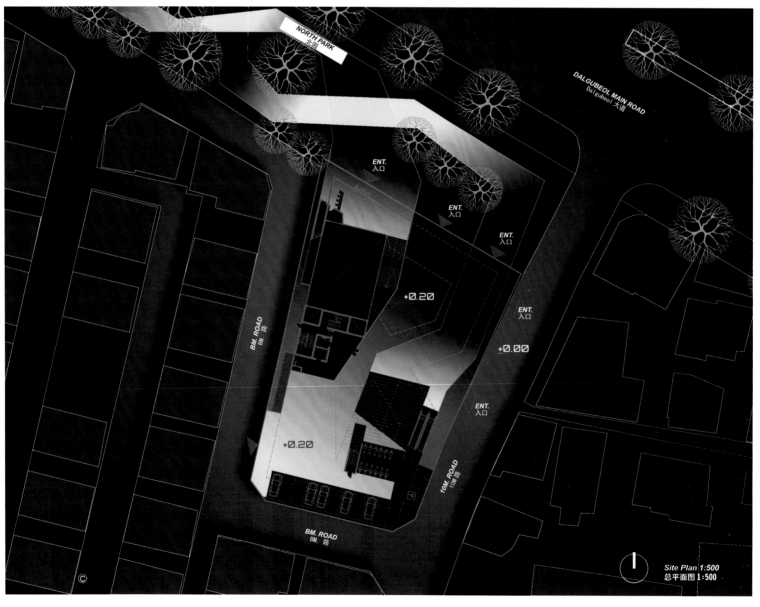

Site Plan 1:500
总平面图 1:500

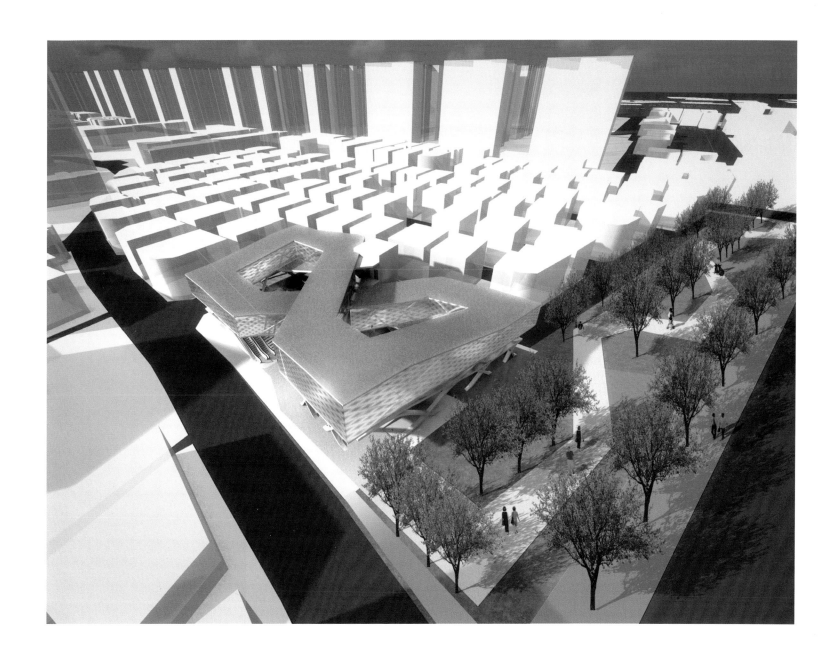

New Role
Books in Infinite Loop
Structure and Environmental Philosophy

Project Overview

The project makes the library as the cultural and social spaces in the local communities. The focus of the project is to gain a function which constitutes the important part of the neighbors' daily life, and to form attractive and comfortable facilities.

The proposed design of Daegu Gosan Public Library is not only to serve the fundamental practicality of a place where knowledge and various forms of information media are well preserved and openly shared in a more reachable fashion but also to provide a social platform where local and global events in the city can take place in order to reposition itself to be Daegu's most vibrant and dynamic contemporary venue.

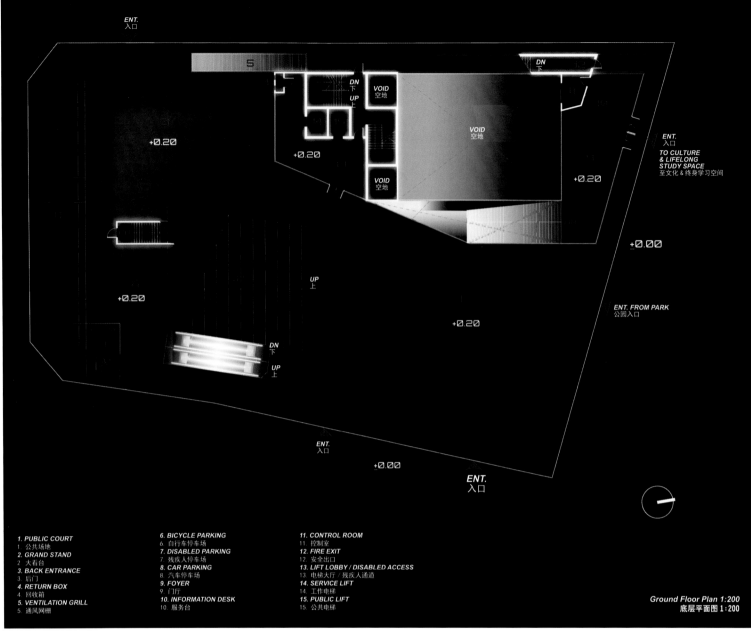

1. PUBLIC COURT	6. BICYCLE PARKING	11. CONTROL ROOM
1. 公共场地	6. 自行车停车场	11. 控制室
2. GRAND STAND	7. DISABLED PARKING	12. FIRE EXIT
2. 大看台	7. 残疾人停车场	12. 安全出口
3. BACK ENTRANCE	8. CAR PARKING	13. LIFT LOBBY / DISABLED ACCESS
3. 后门	8. 汽车停车场	13. 电梯大厅 / 残疾人通道
4. RETURN BOX	9. FOYER	14. SERVICE LIFT
4. 回收箱	9. 门厅	14. 工作电梯
5. VENTILATION GRILL	10. INFORMATION DESK	15. PUBLIC LIFT
5. 通风网栅	10. 服务台	15. 公共电梯

Ground Floor Plan 1:200
底层平面图 1:200

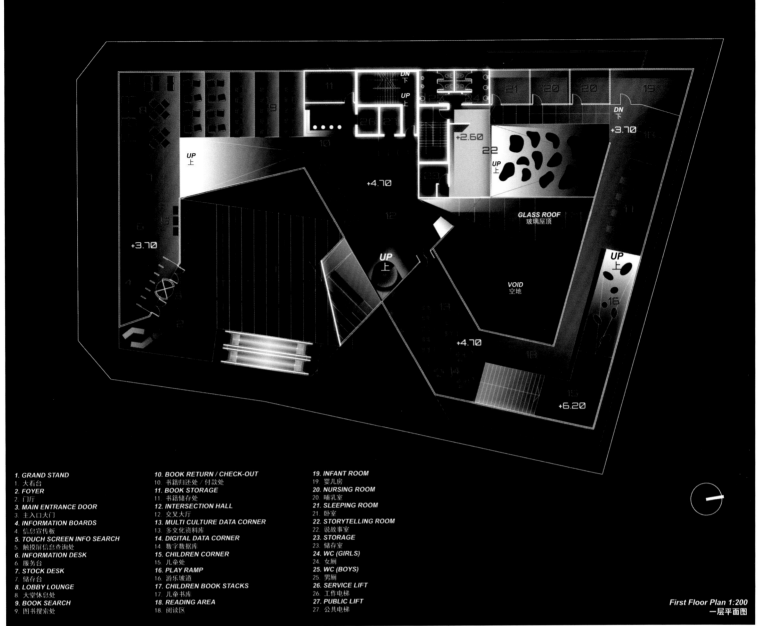

1. GRAND STAND 1. 大看台	**10. BOOK RETURN / CHECK-OUT** 10. 书籍归还处 / 付款处	**19. INFANT ROOM** 19. 婴儿房
2. FOYER 2. 门厅	**11. BOOK STORAGE** 11. 书籍储存处	**20. NURSING ROOM** 20. 哺乳室
3. MAIN ENTRANCE DOOR 3. 主入口大门	**12. INTERSECTION HALL** 12. 交叉大厅	**21. SLEEPING ROOM** 21. 卧室
4. INFORMATION BOARDS 4. 信息宣传板	**13. MULTI CULTURE DATA CORNER** 13. 多文化资料库	**22. STORYTELLING ROOM** 22. 说故事室
5. TOUCH SCREEN INFO SEARCH 5. 触摸屏信息查询处	**14. DIGITAL DATA CORNER** 14. 数字数据库	**23. STORAGE** 23. 储存室
6. INFORMATION DESK 6. 服务台	**15. CHILDREN CORNER** 15. 儿童处	**24. WC (GIRLS)** 24. 女厕
7. STOCK DESK 7. 储存台	**16. PLAY RAMP** 16. 游乐坡道	**25. WC (BOYS)** 25. 男厕
8. LOBBY LOUNGE 8. 大堂休息处	**17. CHILDREN BOOK STACKS** 17. 儿童书库	**26. SERVICE LIFT** 26. 工作电梯
9. BOOK SEARCH 9. 图书搜索处	**18. READING AREA** 18. 阅读区	**27. PUBLIC LIFT** 27. 公共电梯

First Floor Plan 1:200
一层平面图

1. **GENERAL COLLECTION ROOM**
   1. 一般收藏室
2. **READING AREA**
   2. 阅读区
3. **COPY CORNER**
   3. 复印处
4. **LOUNGE**
   4. 休息室
5. **DIGITAL DATA CORNER**
   5. 数字数据库
6. **SERVICE LIFT LOBBY**
   6. 工作电梯大堂
7. **WC (M)**
   7. 男厕
8. **WC (F)**
   8. 女厕
9. **SERVICE LIFT**
   9. 工作电梯
10. **PUBLIC LIFT**
    10. 公共电梯

Second Floor Plan 1:200
二层平面图 1:200

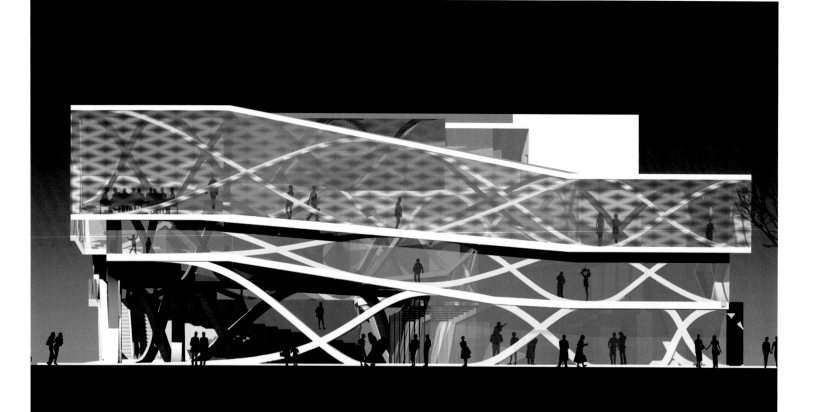

Building Design

The location of the site is surrounded by pleasant residential and green zones separated from the city center of Daegu. Our proposal makes the most of this advantage by seamlessly blend the internal and external spaces of the library with the urban fabric of the city block and the park in the North side. By lifting main programs (Main Hall, General Collection Room, Children Room) above the street level and also lowering Culture & Lifelong Study and Service Spaces to underground, it extensively allows surrounding urban space to flow through the building's open courts and also public events held underneath to flow back out to the context.

The principle of the building's structural system is inspired by the infinity shape. The upper floors are to be stacked upon the lower arches and cantilevered out extensively. The intricacy of the structure also articulates the concept of "intertwining" between the library and the context in a dramatic way. Its coexistence with the surrounding appears as if it was loosely stitched by a string of gigantic threads, hinting the flourishing relationship between Daegu and the textile industry.

To be in line with the city's campaign of the green buildings, particularly Solar Energy, the proposed design deploys the advanced technology of Building-integrated photovoltaic (BIPV), photovoltaic materials that are used on building façades. Photovoltaic glazing with light-diffuser translucent films are applied onto the entire surface of the Book Infinity Loop, producing energy as an ancillary source of electrical power to be used for lighting in the reading space. The translucent films also help diffuse the direct sun light and prevent glare from the outside, keeping the illumination of the interior more evenly distributed.

韩国大邱 Gosan 公共图书馆（方案二）
Daegu Gosan Public Library

设计单位：Synthesis Design + Architecture (SDA)
委托方：大邱广域市寿城区政府
项目地址：韩国大邱市
项目面积：3 100m²

Designed by: Synthesis Design + Architecture (SDA)
Client: Daegu Metropolitan City Suseong-gu Office
Location: Daegu, South Korea
Area: 3,100m²

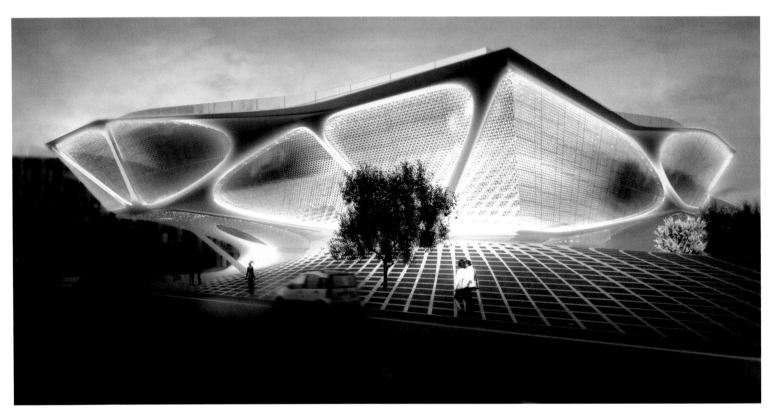

嵌入式交流
集体知识
动态网格

项目概况

项目位于韩国大邱市,针对竞赛提案,设计师提出了一个智能、开放、集成的图书馆体验,来取代以往的媒体存储方法,并通过无处不在的信息资源、集成家具和积极的公共社会空间,将图书馆的空间改变成一个混合的环境。该架构旨在通过自由形式的几何形状、开放计划和综合设施,体现了嵌入式的交流和集体知识的精神。

建筑设计

项目场地作为相邻的公园和城市肌理的延续进行膨胀、剥落和垂直延伸。这种不断的截面变化被阐述为光滑的垂直梯度,该梯度融合地面、坡道、楼梯、阳台和家具成为一个宜居的和人性化的景观,最终汇聚在一个露天的屋顶景观休息室和露台,可供俯瞰大邱城市。

设计将空间、地板和功能之间的门槛最小化,把图书馆当作是积极的、连续的和流动的社会、文化及知识话语场地。设计特意模糊楼层之间的界限,不断变动的外观结合了图书馆的多重空间,促进循环的、物理的和视觉的连接。

建筑物为一个现浇钢筋混凝土结构,在设计上完全与图书馆的几何形状结合起来。建筑物的中央核心提供它的主要参考结构点,通过整个结构垂直地连接。自由形式的几何结构定义了可移动的外观,统一了从这个中央核心悬挑出来的建筑,并由其内部互连(坡道)、周边互连(序列)和提升的地平面(基础)支持。

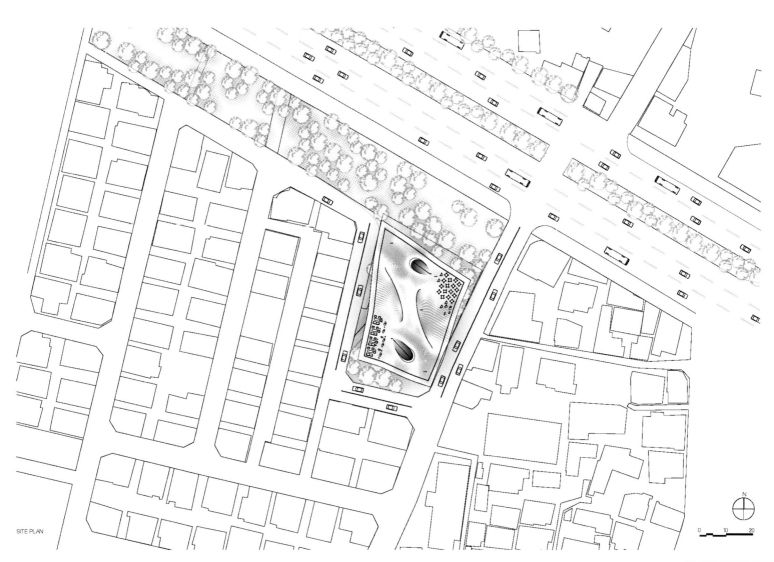

SITE PLAN

| 文体建筑方案集成 | CULTURAL AND SPORTS ARCHITECTURE PROGRAM INTEGRATION |

Open-source Exchange
Collective Knowledge
Dynamic Mesh

Project Overview

The proposal for the Daegu Gosan Public Library challenges the conventional understanding of the spatial and social experience of a public library as a series of discrete reading rooms with defined thresholds and cluttered stacks. The design has proposed an intelligent, open, and integrated library experience which supersedes the media storage methods of the past and changes the library space into a hybrid environment through ubiquitous information resources, integrated furnishings and active communal social spaces. The

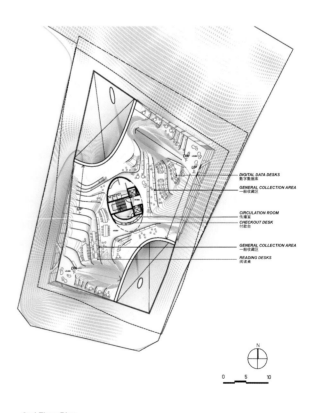

Ground Floor Plan
底层平面图

2nd Floor Plan
二层平面图

architecture is designed to enable and embody the spirit of open-source exchange and collective knowledge through free-form geometries, open plans and integrated amenities.

Building Design

Conceptually and literally, the ground field of the site swells, peels, and multiplies vertically as a continuation of the adjacent park and urban fabric. This constant sectional change is articulated as a smooth vertical gradient which merges floors, ramps, stairs, terraces and furnishings into an inhabitable and ergonomic landscape culminating in an open-air roof-scape lounge and terrace overlooking the city of Daegu.

The design has minimized the thresholds between spaces, floors, and functions to consider the library as an active, continuous and fluid field of social, cultural, and intellectual discourse. The boundaries between floors are blurred, as the continuously walk-able surface which unifies the many spaces of the library, facilitates circulatory, physical and visual connections both internally within the network of spaces and externally with the surrounding context.

The building is materialized as an in-situ reinforced concrete structure which, like all other aspects, has been designed to be fully integrated with the geometry of the library. The central core of the building provides its primary structural point of reference connecting vertically through the entire structure. The free-form geometry that defines the walk-able surfaces and unifies the building cantilevers out from this central core and is supported by its internal interconnections (ramps), perimeter interconnections (columns) and the lifted ground plane (foundation).

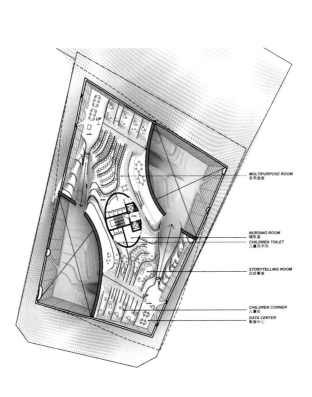

3rd Floor Plan
三层平面图

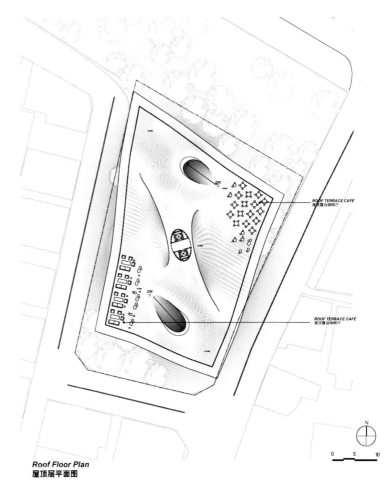

Roof Floor Plan
屋顶层平面图

| 文体建筑方案集成 | CULTURAL AND SPORTS ARCHITECTURE PROGRAM INTEGRATION |

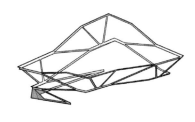
Initial Frame
原始框架

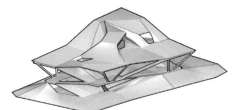
Simple Surfaces
简单曲面

First Iteration Mesh Relaxation
一层循环网格松弛

Second Iteration Mesh Relaxation
二层循环网格松弛

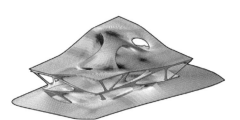
Third Iteration Mesh Relaxation
三层循环网格松弛

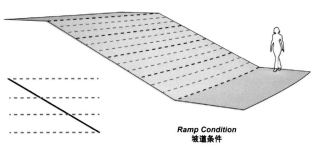
Ramp Condition
坡道条件

Stair Condition
楼梯条件

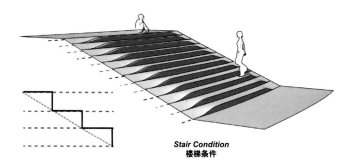
Seat / Bookshelf Condition
座椅 / 书架条件

Desk Condition
书桌条件

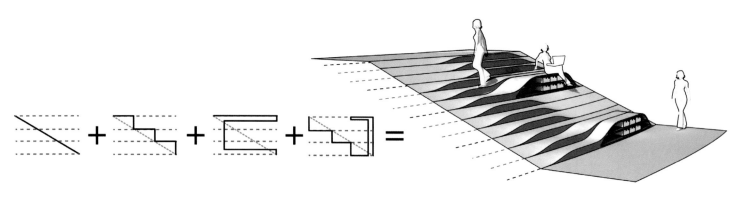
Integrated Condition
综合条件

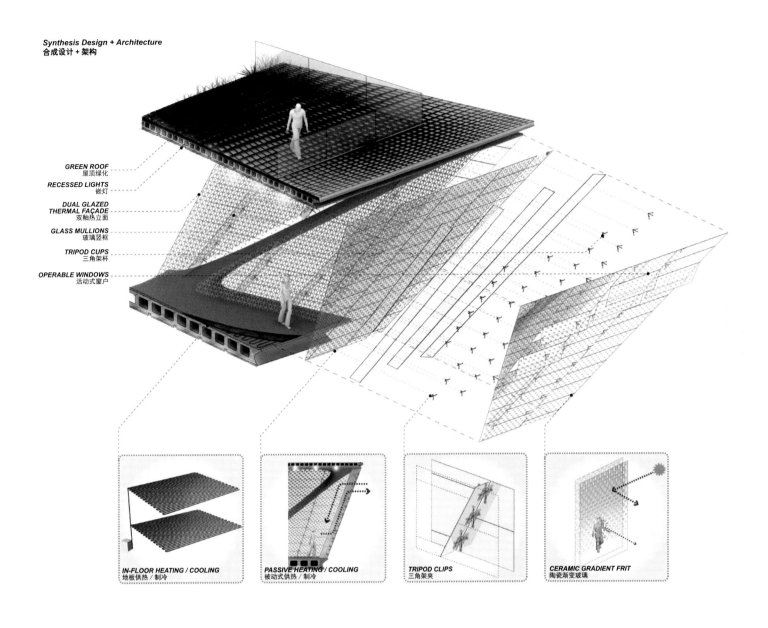
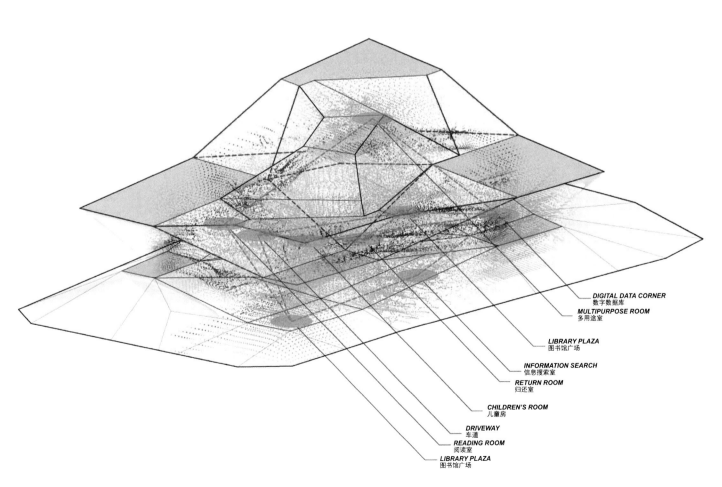

| 文体建筑方案集成 | CULTURAL AND SPORTS ARCHITECTURE PROGRAM INTEGRATION |

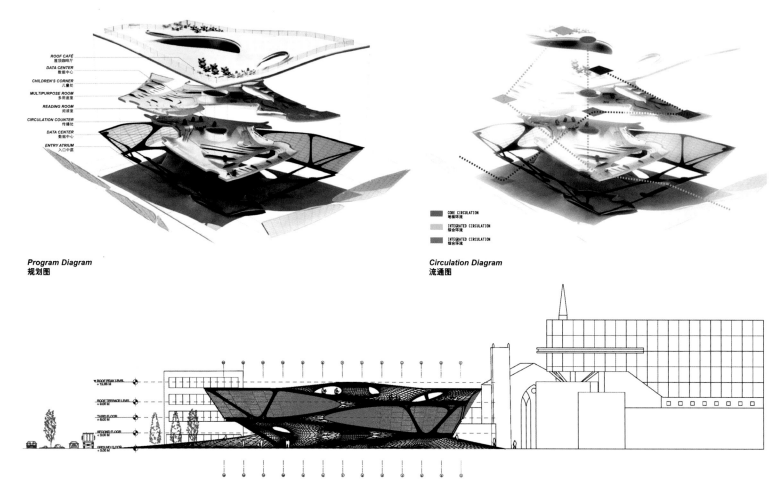

Program Diagram
规划图

Circulation Diagram
流通图

West Elevation
西立面

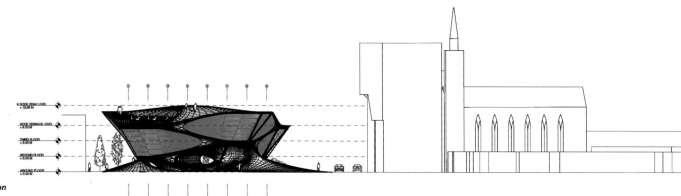

South Elevation
南立面

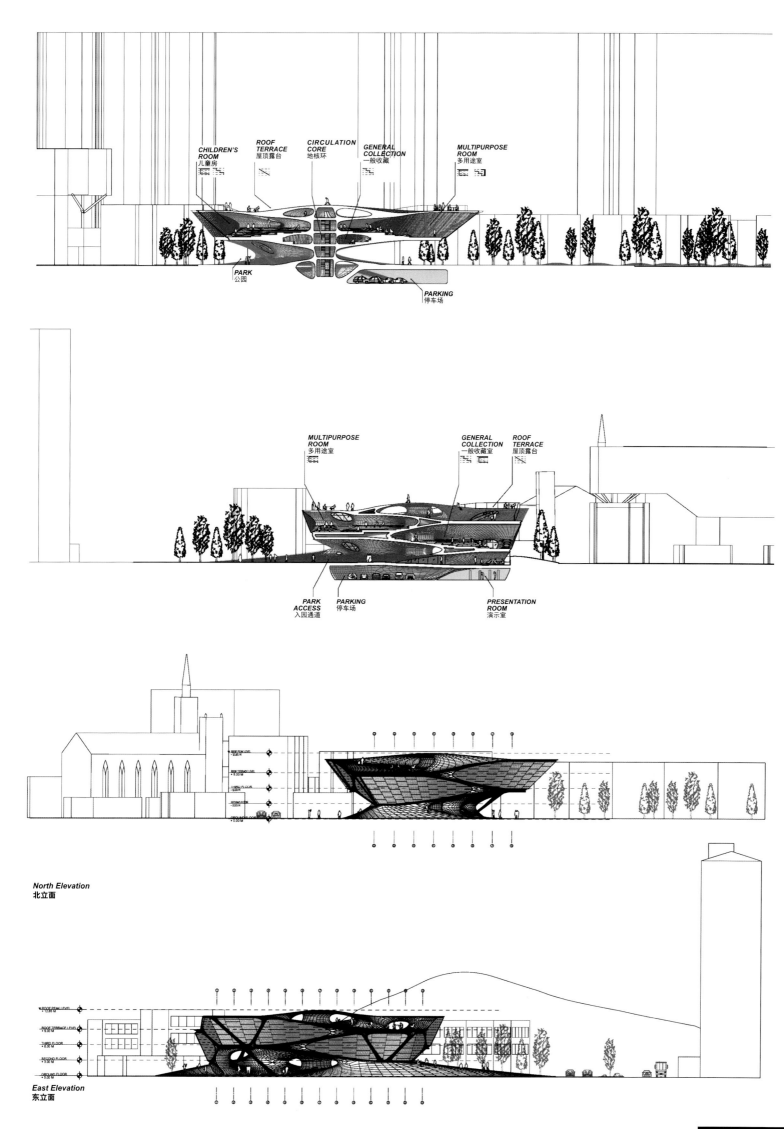

韩国大邱 Gosan 公共图书馆（方案三）
Daegu Gosan Public Library

设计单位：Disguincio&co srl OPENSYSTEMS	Designed by: Disguincio&co srl OPENSYSTEMS
委托方：大邱广域市寿城区政府	Client: Daegu Metropolitan City Suseong-gu Office
项目地址：韩国大邱市	Location: Daegu, South Korea
用地面积：3 500m²	Site Area: 3,500m²

突触
神经元
节能环保

项目概况

信息交换作为学习的一种方式，是人类发展和建立关系的基础。神经元每刻在我们的身体里这样做：它们通过突触进行信息交换。突触是一个过程，让整个神经系统的互连为刺激和处理信息作出回应。在这个思路的基础上，韩国大邱图书馆被设计为连接整个程序的空间。

建筑设计

该项目旨在生成一个组织，其中在公园和建筑之间形成连续性。因此，绿化带引入地块来绘制由三个不同体量组成的该建筑几何形状。因此，园区引导人们进入图书馆，振兴居委会，并建立新的运动坐标轴，使公民学习和求知。

整个建筑是由三个体量组成：侧面两个和由连接器连接的中心一个。侧面的两个体量在侧边更高，它们是从一楼而上。因此，它们侧面与在中间的体量相接，并创建一个立面朝向邻近的街道。中间的体量没有接触到地面，使得公园作为一个公共空间深入地块。与中间由玻璃制成的体量相比，两侧的体量有更紧凑的外形。它们有一个像素化的立面合成物，有一个从底部缩减到顶部的紧密坡度。周边百叶窗的设计强调垂直梯度和与公园水平的连接。

建筑物内的用途和空间的位置也影响更多的公众到更多的私人的梯度及更多的噪音到更安静的梯度。中间的体量充当中间连接器，也更加公开了。

此外，它的意图是该建筑为了用户互动和交换信息，其本身可能是一个学习的元素。在建筑中设有 LED 屏，其能量来自太阳能，而屋顶收集的雨水则用于绿化带植物的灌溉。此外，还有一种路面发电用于公园的照明。因此，用户不知不觉地变节能环保起来。

Site Plan
总平面图

Synapse
Neuron
Energy Saving

Project Overview

The exchange of information as a way of learning is the base of knowledge which makes humans develop and establish relationships. Neurons are doing it in our bodies at every moment. They exchange information through synapse. The synapse is a process that lets the entire nervous system be interconnected to respond to stimulus and process information. From the basis of this idea, the project has been designed to link the whole program spaces.

Building Design

The proposed project aims to generate a tissue in which there is continuity between the park and the building. Thus, the green belt enters into the plot to draw the building geometry consisting of three different volumes. Therefore, the park leads people into the library, revitalizing the neighborhood and establishing new axes of motion to bring citizens learning and knowledge.

The building is composed by three volumes: two in the laterals and one in the center connected by linkers. The volumes in the laterals are taller in the side and they are up from the ground floor. Thus, they flank the volume in the middle and create a façade facing to the adjacent streets. The volume in the center is not touching the ground level to let the park to continue into the plot as a public space. The volumes of the sides have a more compact appearance in comparison to the intermediate volume, which is made of glass. They have a pixelated façade composition and there is a gradient of compactness decreasing from bottom to top. The design of perimeter louvers emphasizes the gradient vertically and the connection to the park horizontally.

The location of uses and spaces within the building also responds to a gradient of more public to more private and of more noise to quieter. The volume in the middle acts as an intermediate connector and is more public.

Moreover, it is intended that the building can be itself an element of learning for the users to interact and exchange information. There are LED screens designed in the building whose energy comes from solar gains and the roofs collect rainwater for irrigation of the plants in the green belt. Furthermore, there is a kind of pavement power generator for lighting in the park. Thus, the library users become familiar with recycling and energy saving unconsciously.

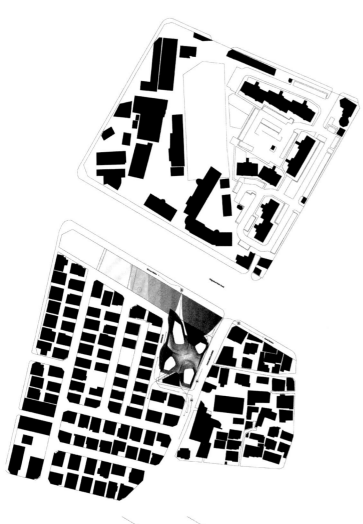

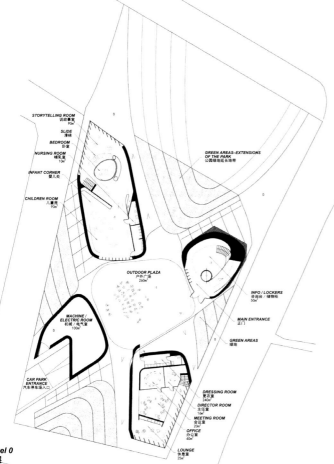

Level 0
底层

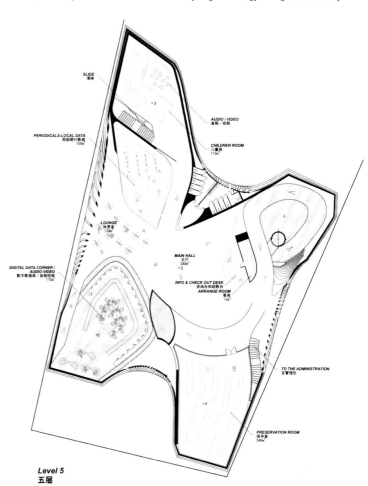

Level 5
五层

| 文体建筑方案集成 | CULTURAL AND SPORTS ARCHITECTURE PROGRAM INTEGRATION |

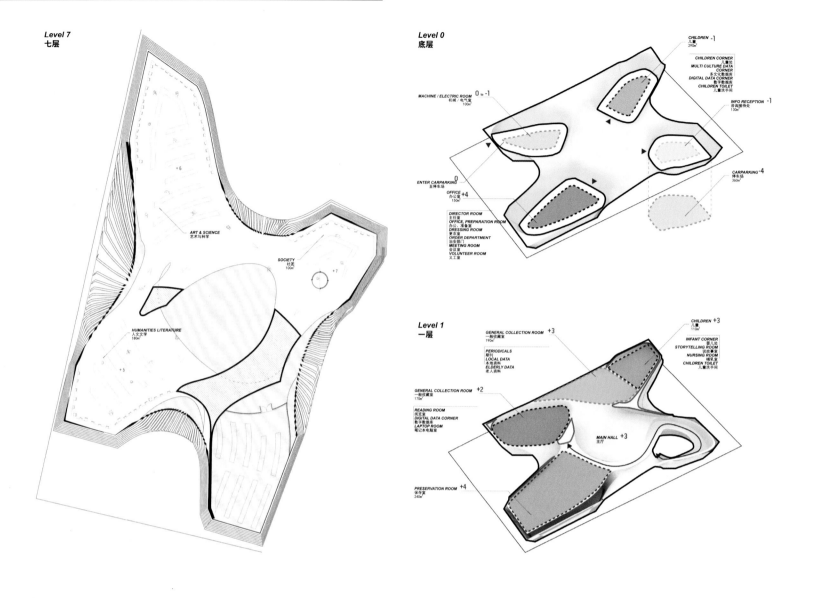

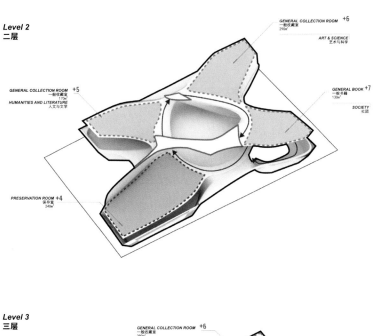
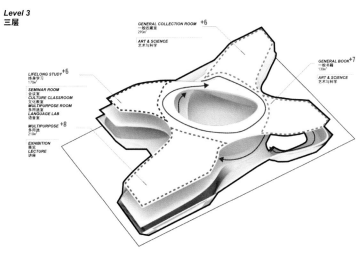
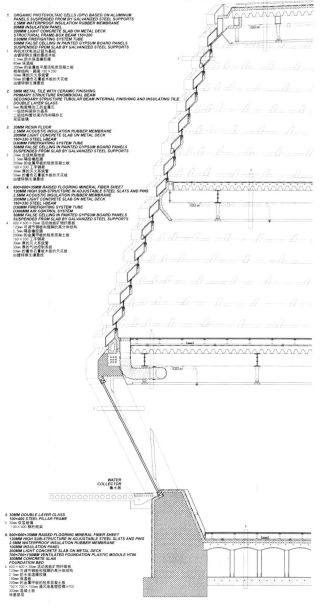
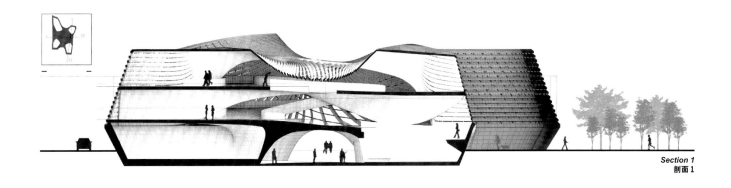

Section 1
剖面 1

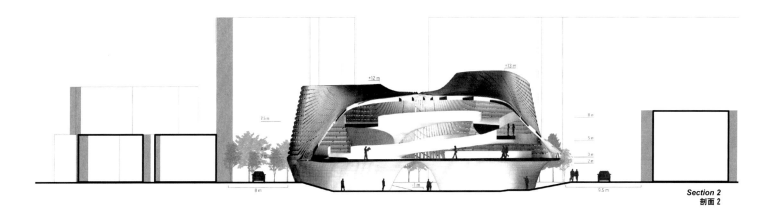

Section 2
剖面 2

内蒙古鄂尔多斯东胜基督教堂
Dong Sheng Protestant Church

设计单位：WVA architects LTD	Designed by: WVA architects LTD
委托方：基督教会	Client: Protestant Church
项目地址：内蒙古自治区鄂尔多斯市	Location: Ordos, Inner Mongolia
建筑面积：8 500m²	Building Area: 8,500m²

和平鸽
教堂建筑

项目概况

该方案位于中国内蒙古鄂尔多斯市东胜区，109国道与东纬四路的交界处。经过三年讨论，两年基地位置更改，当地政府与基督教协会达成共识，同意新教堂的建设。

此项目的主要概念是基于最具教堂特点的象征之一——和平鸽。早期的基督徒描绘的洗礼中，伴随着一只口中衔着橄榄枝的白鸽，这个图像成为一个和平的寓言。

建筑设计

建筑的形式让人联想到白色的鸽子在起飞前小憩。轮廓中体现出的和平从一开始就是含蓄的，不仅是因为平面图让人想起和平鸽，也因为它们之间的空间分布和联系：弯曲的墙壁贯穿始终，地形环抱着建筑以及在建筑中穿梭的人们。

教堂的正门藏在两个主要空间之间。对于空间的感受与理解由建筑外部景观一直延续到建筑物内部。教友们可以从一个大窗口欣赏主厅内部，而不需要走近来看。教会的特性之一是它能够最大程度地利用自然光。外立面上的窗户吸收大量的光，以改善室内光线条件，在不同的空间提供不同强度的光。主厅由一个大天窗提供光线，象征着连接空间和主建筑轴的和平鸽之心。室内空间是由两个有联系却相对独立的空间组成。以由主厅向外辐射状为轴依次布置功能空间，形成一系列的序列空间，由此而衍生出的景观环境也以同样的方式，向周围延伸出去。

"Dove of Peace" Church Building

Project Overview

The program is located in Dongsheng District of Ordos City, Inner Mongolia, China, and the junction of 109 East State Road and the Fourth East Weft. After three years of discussion and the base position changes for two years, the local government and the Christian Council reached a consensus to the new church building.

The main concept of the church is based on one of the most ecclesiastical symbols, the dove of peace. Early Christian's portrayed baptism accompanied by a dove holding an olive branch in its peak and used the image as an allegory of peace.

Building Design

The form of the building is reminiscent of a white dove resting before flying. The peace that passes the silhouette is implicit from the beginning not only because of the floor plan that remembers of a dove but also because the distribution of spaces and the relationship between them, always with curved walls and a topography that embraces both the building and the persons frequenting it.

The main entrance of the Church is hidden between the two main spaces. The space feelings and understanding has extended from the external landscape to the internal buildings. The faithful people can enjoy the main inside hall from a large window without the need to come closely to look. One of the characteristics of the church is its ability to maximize the use of natural light. Windows on the façade absorb a lot of light to improve indoor lighting conditions in order to provide different intensities of light in a different space.

The main hall by a large skylight provides light, symbolizing peace dove heart of the connecting spaces and the main shaft of the building. The interior space is composed of two linked but independent spaces, with radial outward from the main hall for the axis to layout function space, forming a series of sequence space, and thus the landscape environment brought by it in the same way is extending out to the surrounding.

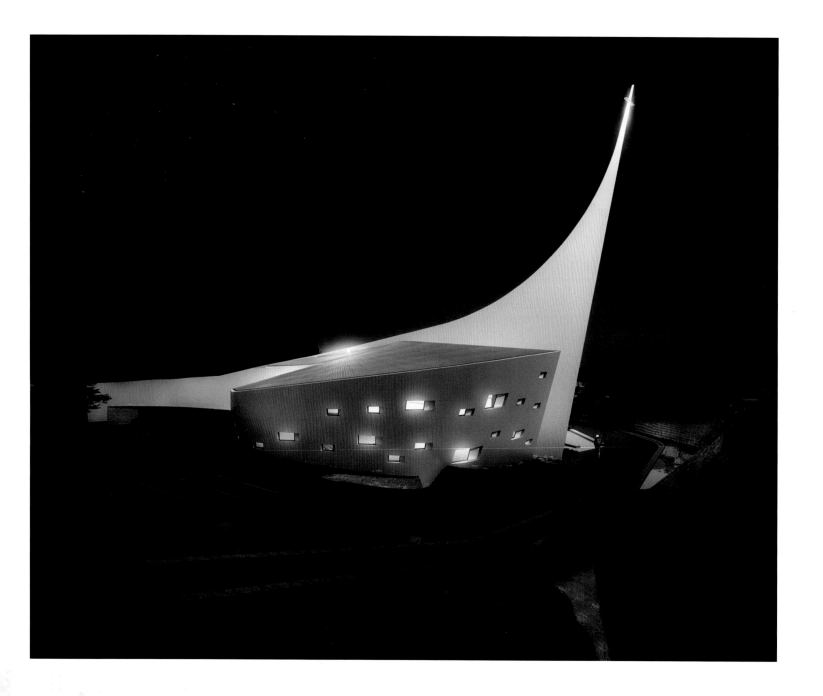

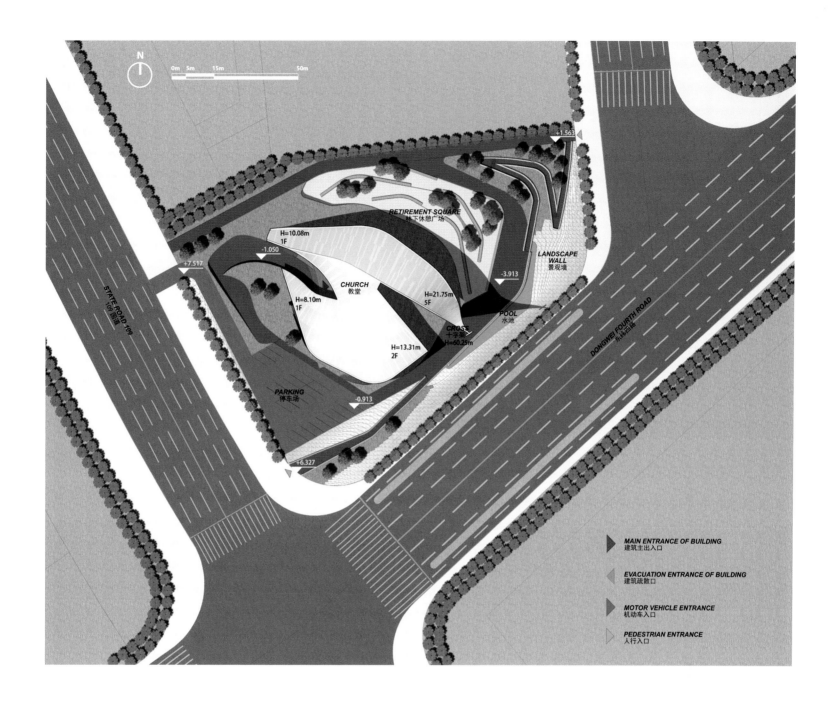

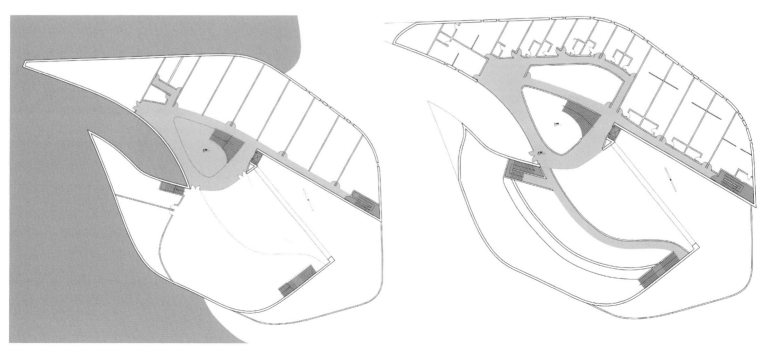

First Floor
一层

Second Floor
二层

Area: 1,142.41m²
面积：1 142.41m²

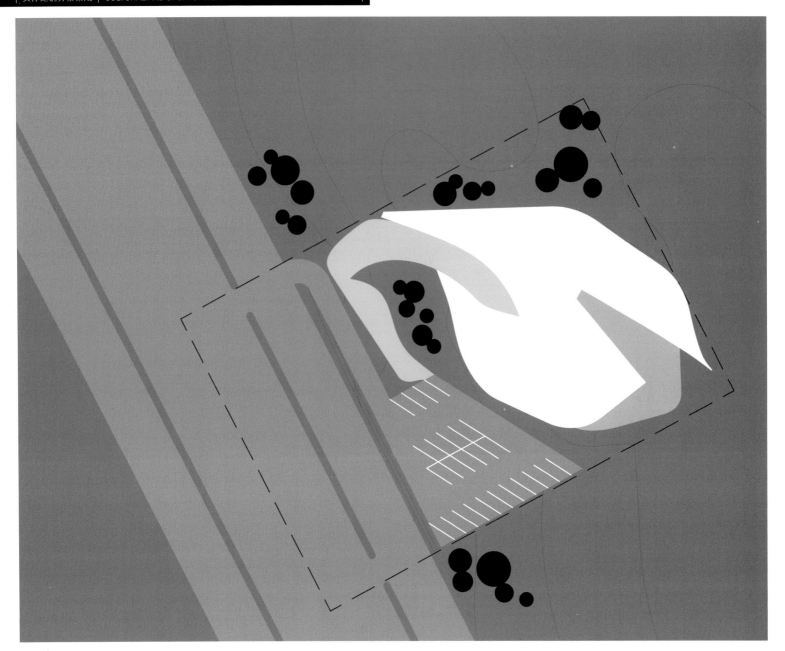
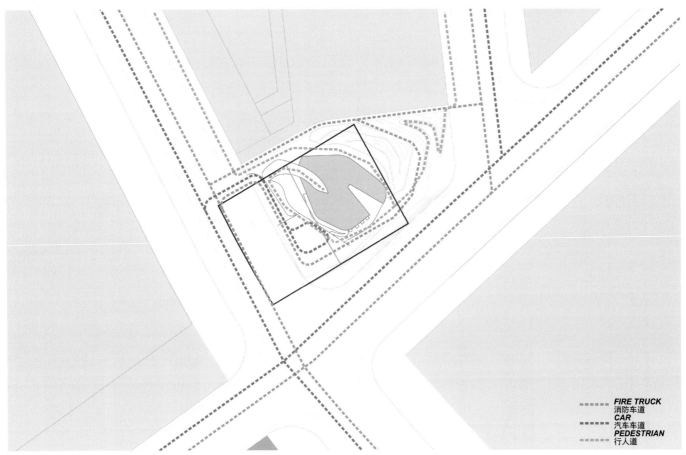

FIRE TRUCK 消防车道
CAR 汽车车道
PEDESTRIAN 行人道

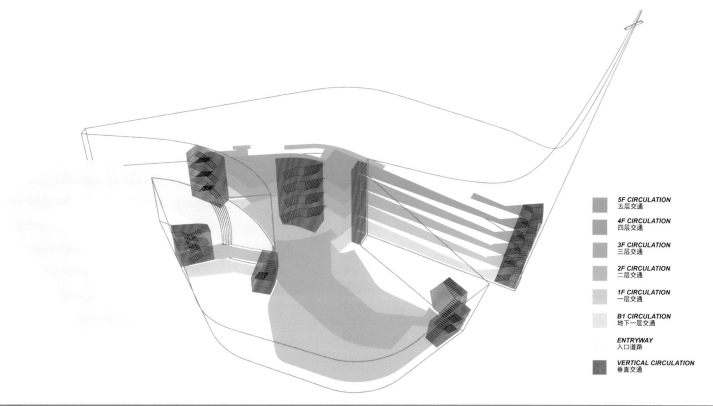

	5F CIRCULATION 五层交通
	4F CIRCULATION 四层交通
	3F CIRCULATION 三层交通
	2F CIRCULATION 二层交通
	1F CIRCULATION 一层交通
	B1 CIRCULATION 地下一层交通
	ENTRYWAY 入口道路
	VERTICAL CIRCULATION 垂直交通

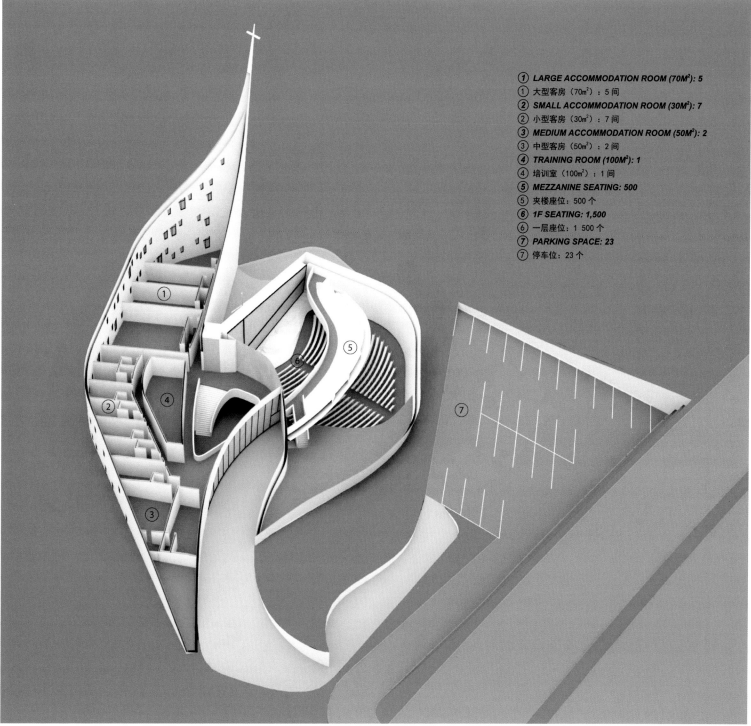

① **LARGE ACCOMMODATION ROOM (70M²): 5**
① 大型客房（70m²）：5 间
② **SMALL ACCOMMODATION ROOM (30M²): 7**
② 小型客房（30m²）：7 间
③ **MEDIUM ACCOMMODATION ROOM (50M²): 2**
③ 中型客房（50m²）：2 间
④ **TRAINING ROOM (100M²): 1**
④ 培训室（100m²）：1 间
⑤ **MEZZANINE SEATING: 500**
⑤ 夹楼座位：500 个
⑥ **1F SEATING: 1,500**
⑥ 一层座位：1 500 个
⑦ **PARKING SPACE: 23**
⑦ 停车位：23 个

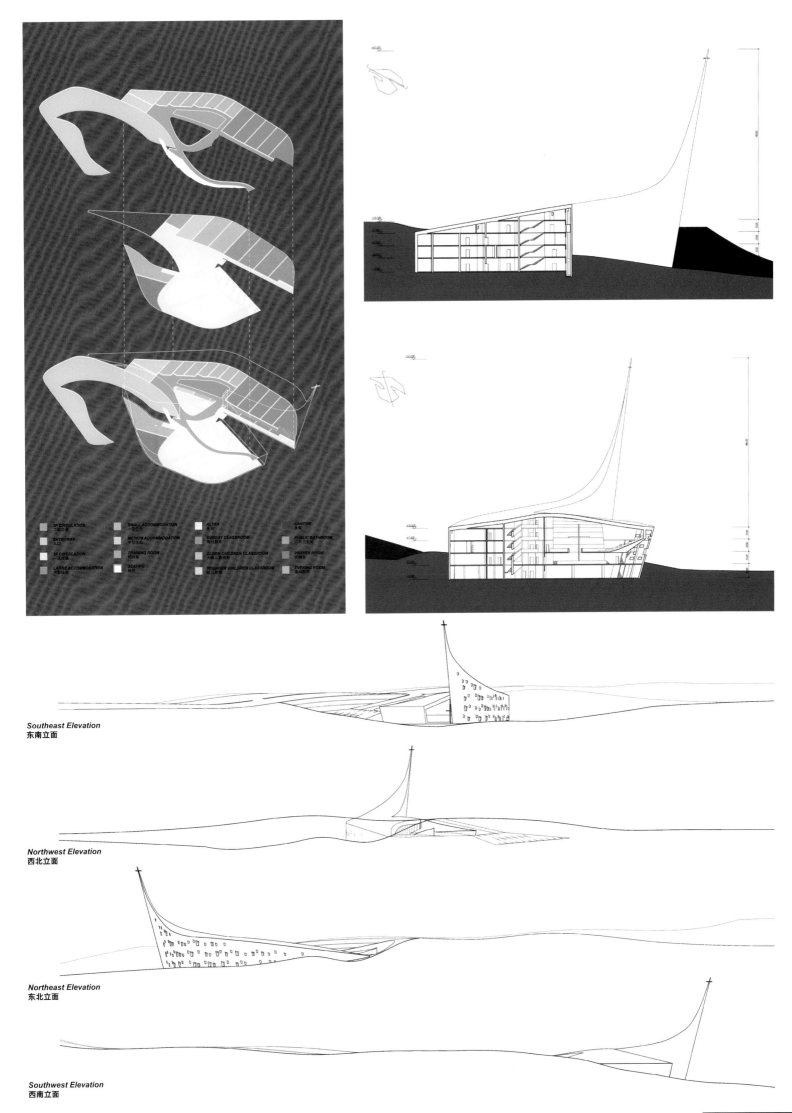

Southeast Elevation
东南立面

Northwest Elevation
西北立面

Northeast Elevation
东北立面

Southwest Elevation
西南立面

| 文体建筑方案集成 | CULTURAL AND SPORTS ARCHITECTURE PROGRAM INTEGRATION |

广西南宁李宁体育园
Lining Sports Park

设计单位：澳大利亚柏涛建筑设计有限公司
委托方：李宁基金会
项目地址：广西省南宁市
项目面积：351 742.31m²
项目获奖：2013年全国优秀工程勘察设计行业奖建筑工程公建一等奖

Designed by: Australia PT Design Consultants Limited
Client: Lining Foundation
Location: Nanning, Guangxi
Area: 351,742.31m²
Award: 2013 National Excellent Engineering Survey and Design Industry Awards Won the First Prize in Public Construction Projects

体育竞技
生态休闲
木工建筑

项目概况

南宁李宁体育园项目是李宁基金会捐赠南宁的第一个体育主题公园工程。项目位于广西南宁凤岭片区，东盟国际商务区南部，用地西侧为南宁天文台。园区主要分综合训练馆、室内游泳馆、员工宿舍、园区辅助用房及配套用房和室外运动设施等。

建筑设计

为了体现对家乡叶落归根的眷念和回报故土的情结，规划设计的立意以根和叶的生长及脉络关系引导整个园区的布局结构。这种生长关系不仅有积极意义的隐喻，也为项目在运营中根据变化拓展功能提供了良好的架构。实践中室外水上娱乐项目的增建和受欢迎的程度证实了这个预判。

项目基地为丘陵山地，按照尊重自然，依山就势，化整为零的设计原则，规划分为含四个标准单元的综合馆及游泳馆两部分，这种布局为使用和管理的独立性带来了较大的便利。

综合馆和游泳馆的入口及配套的商业功能均围绕着中心庭园区域布置，形成积极的空间界面，大量的人流进出和交流活动为这个空间带来了人气聚集的能量场。

进一步的设计把这个场地根据自然高差做成一个露天剧场，剧场的"舞台"为游泳馆的入口区域，而"观众厅"的边界则由综合馆的入口回廊限定而成。在建成后的运营中这个空间获得了巨大的成功，成为了许多演艺及媒体组织争相使用的场所。体育设施的本源目标，群众性、观演性、娱乐性在这里得到了充分的演绎。

生长的概念也体现在构造和造型逻辑上，设计采用混凝土框架与大跨度钢结构屋面相结合的方式。厚重的基座宛如从山体里面生长而出，支撑着屋顶钢结构的有力出挑，淋漓尽致的表现了体育建筑的性格特点。从运营的结果来看，南宁项目得到了各方面的广泛好评，不仅成为许多政府机构的考察对象，更成为南宁市民喜闻乐见的一处场所，一道景观。

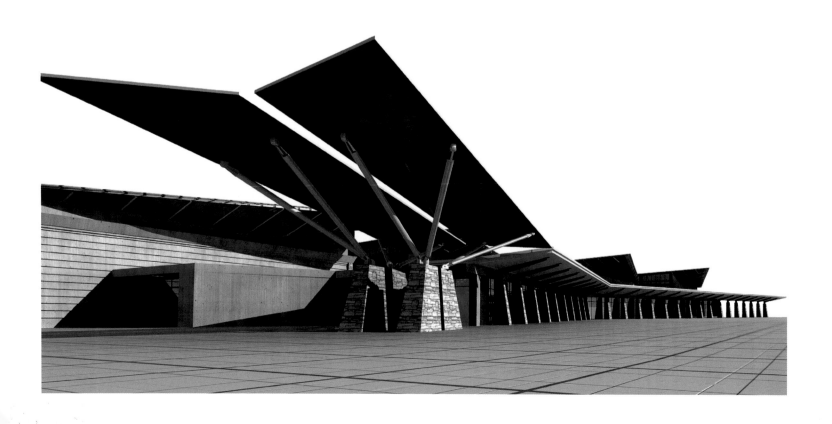

Sports & Competition Ecological & Leisure Woodworking Architecture

Project Overview

The project was commissioned privately for the Malaysian Government. It is a 310 bed tertiary multi specialty, self-contained Hospital on a Greenfield site. The design concept was to break the building mass into 4 blocks, strung together lineally owing to the narrow site constraints. The 4 blocks are separated by large landscaped atria and are arranged to permit light to infiltrate through the building.

Building Design

In order to reflect the missing of back to the home town and the complex of returning the home land, the planning and design conception is to guide the entire park layout structure by the growth of roots and leaves and the context relationships. This growing relationship not only has the positive significance of metaphor, but also provides a good structure for the project to expand functions according to the changes in the process of operation. In practice, the additional construction of outdoor water entertainment project and its popularity has confirmed the anticipation.

The base of the project is hills and mountains. According to the design principles of respecting the nature, being harmonious with the site and breaking up the whole into parts, the planning is divided into two parts including comprehensive hall consisting of four standard units and swimming natatorium, and this layout has brought great convenience for the independence of using and managing. The entrances of the comprehensive hall and swimming natatorium and supporting commercial functions are all designed around the central garden area, developing an active space interface, and the large flow of people and exchange activities bring this space an energy field gathering people here.

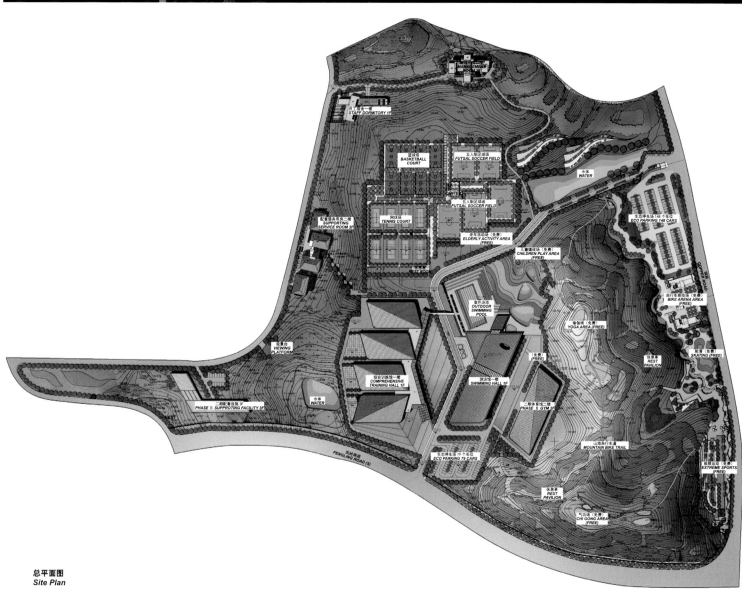

总平面图
Site Plan

游泳馆
Swimming Hall

东立面图
East Elevation

南立面图
South Elevation

西立面图
West Elevation

北立面图
North Elevation

1-1 剖面图
1-1 Section

2-2 剖面图
2-2 Section

管理用房
Management Room

东立面图
East Elevation

西立面图
West Elevation

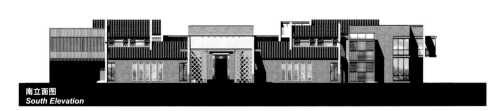

南立面图
South Elevation

北立面图
North Elevation

The site is further designed to be an open-air theater according to the natural elevation difference, and the "stage" of the theater is the entrance area of the swimming natatorium, while the boundary of the "audience hall" is confined by the entrance corridor of the comprehensive hall. After being completed, the space gets a great success in the process of the operation, and becomes a place competed to use by many performing arts and media organizations. Mass, drama performance, entertainment these original targets of sports facilities are fully deduced here.

The growth concept is reflected in the logic of the structure and modeling. The design combines the concrete frame with large-scale steel structural roof. Heavy base as if grows from inside the mountains, supports the strong overhangs of steel structural roof, and incisively and vividly shows the characteristics of sports architecture.

From the point of view of operating results, the project has been obtained wide praises from various aspects. It not only becomes an inspection object of many government agencies, but also becomes a place or a landscape loved by the citizens in Nanning.

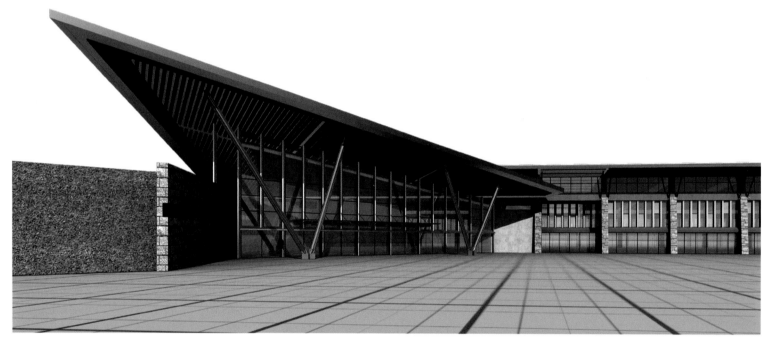

瑞典于默奥艺术馆
Umeå Art Museum

设计单位：Henning Larsen Architects
委托方：Balticgruppen AB
项目地址：瑞典于默奥市
项目面积：15 000m²

Designed by: Henning Larsen Architects
Client: Balticgruppen AB
Location: Umeå, Sweden
Area: 15,000m²

城市风景之窗
新地标建筑

项目概况

艺术馆位于瑞典北部的于默奥大学新艺术学院校园内，新校园还包括了艺术学院、设计学院和建筑学院。一层平面占地 500m²，总建筑面积达 3 500m²。由于建筑位于河岸边，重要的地理区位让艺术馆已然成为该地区新的地标建筑。

建筑设计

于默奥艺术馆由三个叠放的展厅构成，宽敞开阔的方形展厅中仅由四个筒体支撑而不再使用其他承重结构，每个筒体占据了一边。沿着展厅，筒体与外表皮之间是脱离的，形成壁龛一样的窗户，清亮的阳光和自然光线从这里照射进来，与室内温暖的人工照明融合在一起。除了光效之外，壁龛也让人们在参观艺术品之余可以放松一下身心，越往上层就越接近沿河风光和城市图景。四个筒体除了是支撑结构之外还是垂直交通的所在，电梯、楼梯各种管线和垂直通风都设置在这里。

入口和门厅是三层同高的，还包括一个展览馆商店、一个儿童手工作坊和礼堂。三层高的房间将竖向各层空间联通起来。像艺术学院中的其他建筑一样，该艺术馆的外立面以落叶松木的百叶为特色，强调了建筑的垂直线条，仅仅被一些玻璃窗打破整体节奏。于默奥艺术馆希望能吸引国际上的艺术家，而这也加强了该建筑对安全性能和节能方面的标准。另外，建筑的热力系统与地区性的热力管网相连，并且建筑材料都是不需要长期维护的。

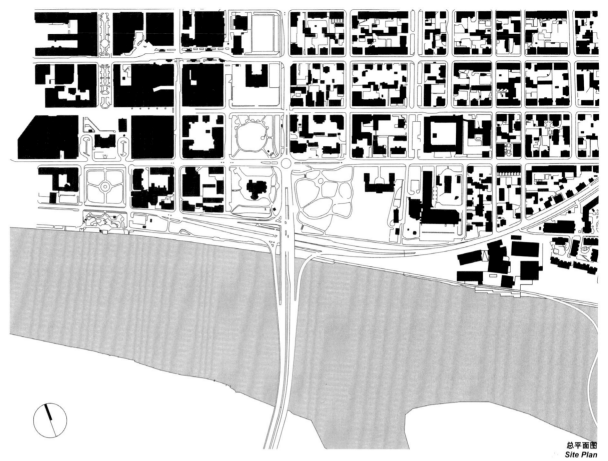

总平面图
Site Plan

| 文体建筑方案集成 | CULTURAL AND SPORTS ARCHITECTURE PROGRAM INTEGRATION |

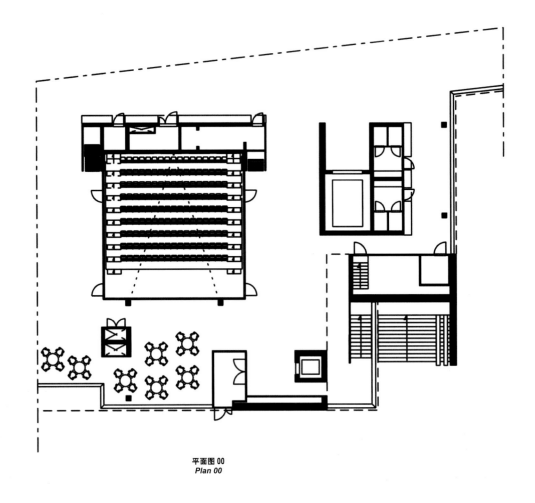

平面图 00
Plan 00

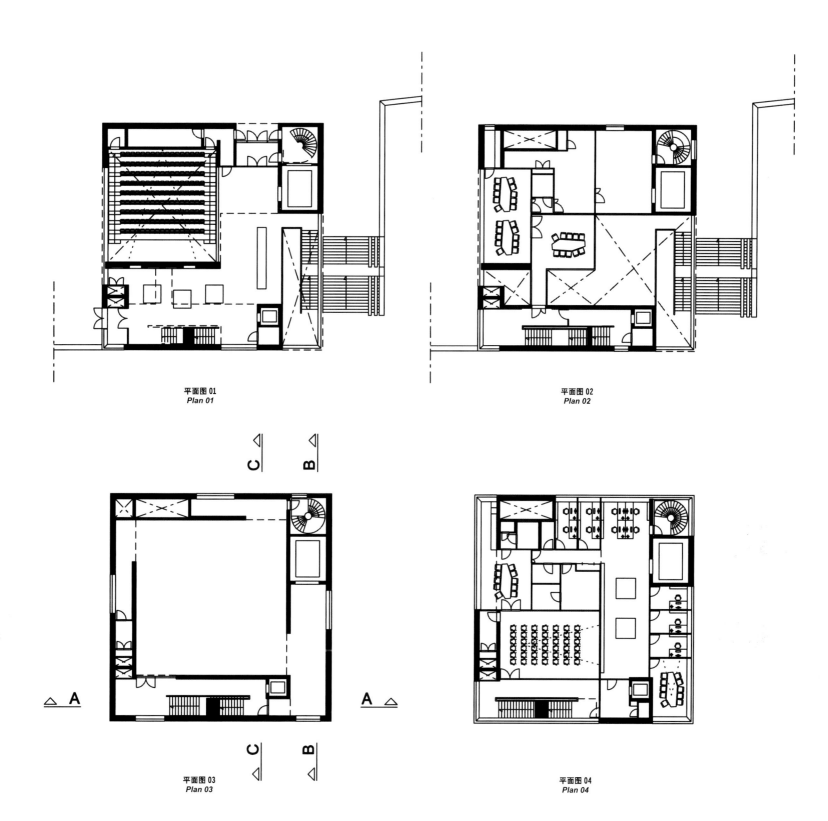

平面图 01
Plan 01

平面图 02
Plan 02

平面图 03
Plan 03

平面图 04
Plan 04

平面图 05
Plan 05

平面图 06
Plan 06

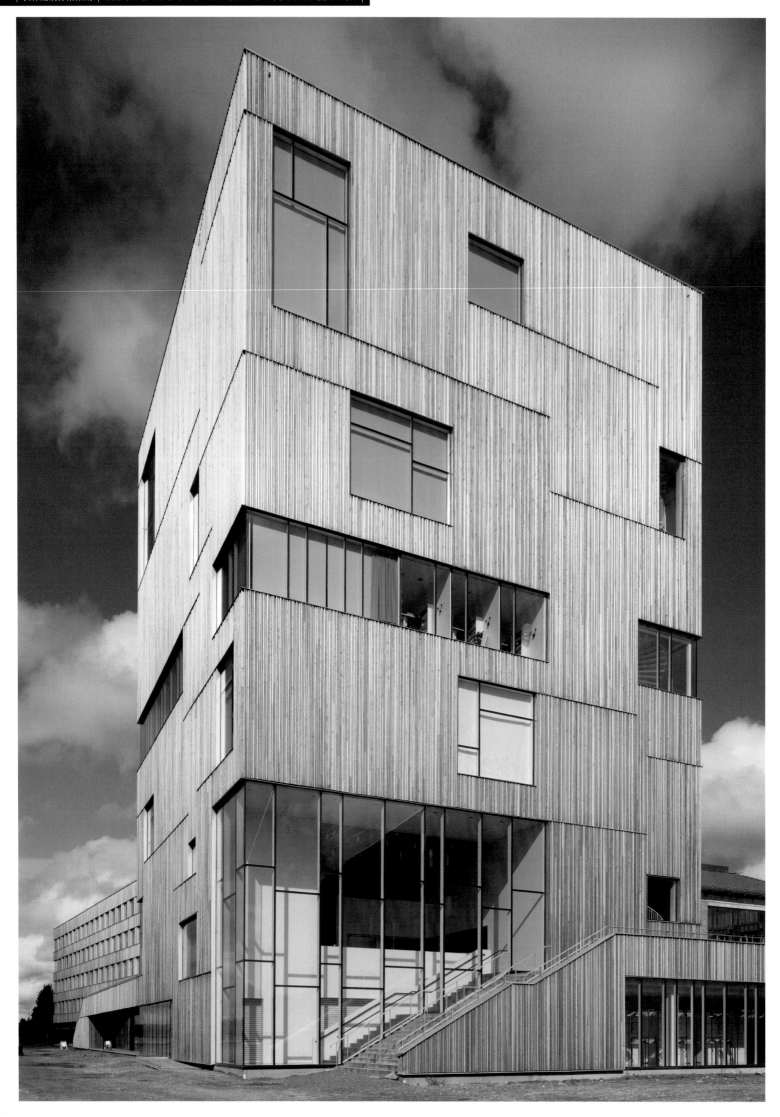

剖面图
Section

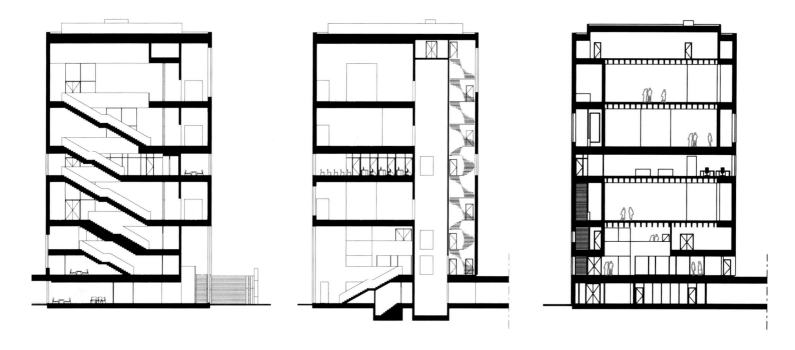

Window of City View
New Landmark Building

Project Overview

The Art Museum, Bildmuseet, at Umeå University is situated at the new Arts Campus by the Umeälven river. The campus also comprises the Academy of Fine Arts, the Institute of Design and the new School of Architecture. The site area of the first floor is 500m^2, and the gross floor area is 3,500m^2. Since it is by the river, so such important location make it a new landmark building of this region.

Building Design

The Umeå Art Museum is consisted of three stacked exhibitions, and the wide open square exhibition is only supported by four barrels, not to use other bearing structure, and each barrel occupies one side. Along the exhibition, the barrel and the out skin are separated, forming the window like niche, from which the bright sunshine and natural light shine in, integrating with the warm interior artificial lighting. Apart from the lighting effect, the niche can also let people relax when visiting the work of art. And you will be more near to the scenery along the river and the city view when you keep on going up. Except for supporting the exhibition, the four barrels are also the place of vertical traffic. The elevator, stairs, other line pipes and vertical ventilation are all arranged here.

The tower comprises three exhibition halls placed upon each other—with inserted floor plans featuring the auditorium, children's workshops and administration. The large, square exhibition halls are free from structural elements and offer ample daylight, let in through niches in the facade. This creates a vibrant and dynamic framework for the various exhibitions. As the other buildings at the Arts Campus, the facade features vertical louvre panelling in Siberian larch, which supports the verticality of the building—only broken by the large windows and the glass floor in the middle. The Art Museum wants to attract international artists, so this also strengthens the standards on safety performance and energy conservation. In addition, the building's thermodynamic system connects with regional heat-supply pipe network, and the building materials do not need to be maintained long-term.

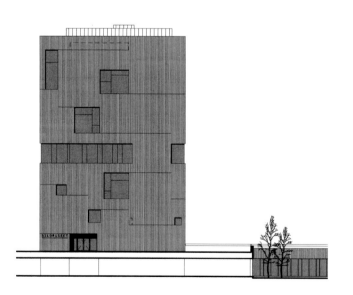

北立面
North Elevation

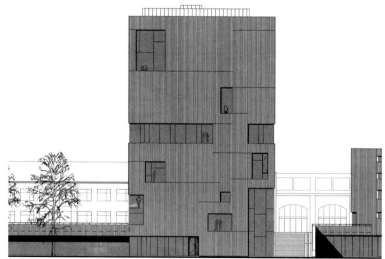

南立面
South Elevation

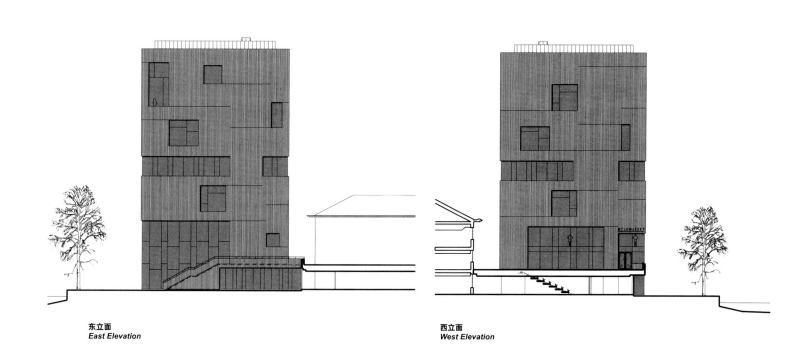

东立面
East Elevation

西立面
West Elevation

广东深圳当代艺术馆与城市规划展览馆
Museum of Contemporary Art and Planning Exhibition

设计单位：Peter Ruge 建筑师事务所
设计团队：Pysall Ruge – Peter Ruge, Kayoko Uchiyama, Matthias Matschewski, Akane Tazawa, Hao Xu
委托方：深圳市文化局与城市规划局
项目地址：广东省深圳市
项目面积：29 688m²

Designed by: Peter Ruge Architekten
Design Team: Pysall Ruge - Peter Ruge, Kayoko Uchiyama, Matthias Matschewski, Akane Tazawa, Hao Xu
Client: Shenzhen Municipal Culture Bureau and Planning Bureau
Location: Shenzhen, Guangdong
Area: 29,688m²

叠加理念
结构灵活

项目概况

深圳当代艺术馆与城市规划展览馆，简称"两馆"，是深圳市中心区最后一个重要大型公共文化建筑。项目有38 500m²的面积用于展示当代艺术品、各种雕塑作品与设计作品；超过19 600m²的面积用于展示建筑和城镇规划的历史和现状；有22 000m²的面积用做休息室、多功能厅、礼堂、会议室、餐厅、书店与管理区。

建筑设计

"两馆"的设计理念是以大型雕塑的形式利用八个椭圆形圆盘以偏置一定位移的方式逐个叠加而成。这种分层方式产生的展区具有能够在各层之间灵活调整永久展览与临时展览的比例的功能。各种空间空隙以及楼梯连接为观众提供了良好的室内视觉定位。各层之间的交错重叠不仅为露天展览和餐厅提供了阳台位置，而且还在夜晚将向下的照明灯光投射到前院。

当代艺术馆以当代艺术的展示、收藏、研究、推广、教育为主要功能的大型公益性艺术机构，内容包括雕塑、摄影、绘画、影像、多媒体等现代艺术和工业设计、建筑设计、服装设计、平面设计、动漫设计等设计艺术领域。城市规划展览馆，将运用声光电三维模型、四维动态模拟场景体验等各种新型展示技术和先进设施来展现深圳市规划成就和城市发展历程，采用智能多媒体查询、动手规划创作等新技术激发观众积极参与，是展示城市规划和发展历程的窗口，也是公众参与城市规划和家园教育的重要场所。

外观详图
Facade Detail

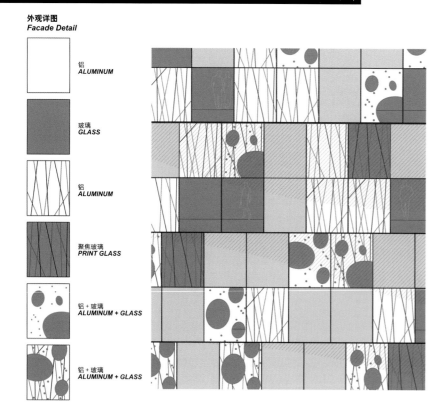

- 铝 ALUMINUM
- 玻璃 GLASS
- 铝 ALUMINUM
- 聚焦玻璃 PRINT GLASS
- 铝 + 玻璃 ALUMINUM + GLASS
- 铝 + 玻璃 ALUMINUM + GLASS

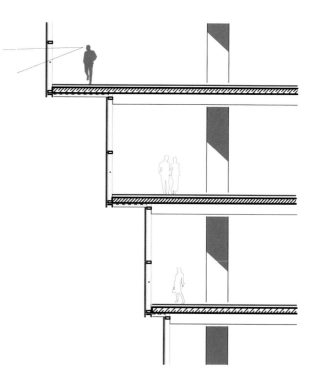

Superposition Concept
Flexible Structure

Project Overview

The Museum of Contemporary Art and Planning Exhibition represents the last component of the cultural building ensemble at the centre of Shenzhen. Pieces of contemporary art, sculptures and design objects are shown on 38,500 square meters, and the history and presence of architecture and town planning are displayed over 19,600 square meters. The foyer, multi-functional rooms, auditorium, conference hall, restaurants, bookshop and administration areas are placed together in 22,000 square meters.

Building Design

The large-scale sculpture that comprises the project is generated out of eight one-upon-the-other offset elliptical discs. The layering forms exhibition areas and functions that allow a flexible combination of the permanent collection and temporary exhibitions floor by floor. Spatial voids and access stair connections provide good visual orientation in the interior. The twisting of the layers produces not only floor terraces for open-air exhibitions and restaurants, but also projections whose illuminated undersides light the forecourt aesthetically at night. The Museum of Contemporary Art is a large public art institution with main functions of contemporary art exhibition, collection, research, promotion and education, including sculpture, photography, painting, image, multimedia and other modern art and industrial design, architecture design, costume design, graphic design, animation design, and other areas of the design art. Urban planning exhibition museum will use all sorts of new display technology and advanced facilities such as sound and light three-dimensional model and four-dimensional dynamic simulation scenario experience to show the Shenzhen's planning achievements and the urban development process, use the new technologies such as intelligent multimedia query and beginning planning creation to inspire the audience to take an active part in, which is the window to show the city planning and development, and also the important place of the public participating in urban planning and home education.

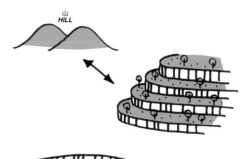

山 HILL

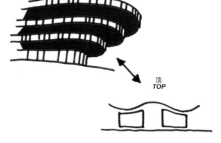

顶 TOP

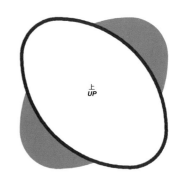

上 UP

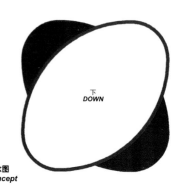

下 DOWN

概念图 / Concept

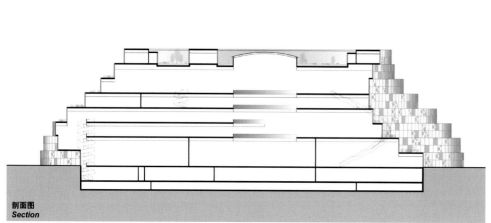

剖面图 / Section

总平面图
Site Plan

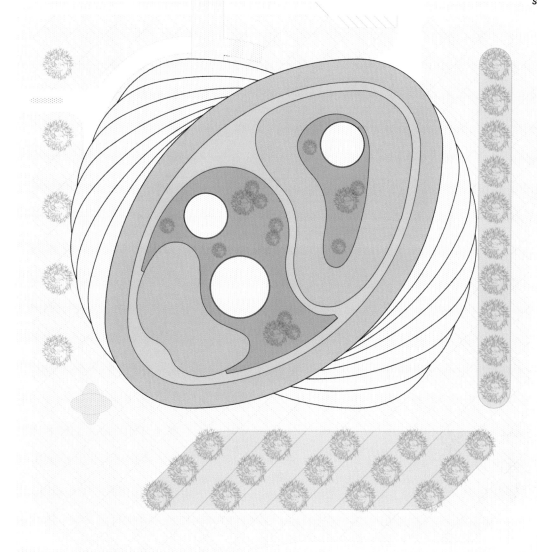

西班牙毕尔巴鄂新圣马梅斯体育场
New San Mames Stadium

设计单位：IDOM – ACXT	Designed by: IDOM - ACXT
委托方：Alberto Tijero	Client: Alberto Tijero
项目地址：西班牙毕尔巴鄂市	Location: Bilbao, Spain
项目面积：114 500m²	Area: 114,500m²

宏大体量
玻璃幕墙

项目概况

新圣马梅斯体育场位于毕尔巴鄂市延伸出的"扩充区"，俯瞰着河流、Olabeaga 地区和佐罗扎里半岛。体育场设计有3 000个座位，能同时容纳53 500名观众欣赏精彩赛事。另外，场馆还设有俱乐部、博物馆、官方商店、餐厅、咖啡厅、会议区、运动中心、游泳池、高性能中心等。

建筑设计

设计考虑到体育场作为一种特殊的建筑要素，强调其拥有雄伟而稳健的体量立面与外观，同时也体现出对城市周边其他建筑的敏锐感受。因此，该建筑体量作为一种城市结构，要与其他建筑协调融合，又不能被人们单纯地视作一种体育设施而存在。同时，设计也重视并突出体育场馆建筑中一直被人们所忽视的部分空间的价值——位于体育馆周边区域和顶棚看台后部之间的交通空间。为了突出这些空间的价值，设计策略不仅包括营造高品质的空间，也包括加强建筑物与城市及周边环境之间的紧密联系。因此，为新圣马梅斯体育场设计的玻璃幕墙立面，也是令该建筑独具特色的一个必要因素。体育场的灯光设计又进一步强调了这些概念，并清晰地形成了四个独立"体育馆"的空间感官效果——无论是在白天，还是在夜晚；无论是在比赛当日，还是在射球入门的时刻。

整个看台设计为三层。其中，第一层离球场最近，位于街道平面以下。这样既减少了体育场对城市的干扰，也从街道一侧提供了直接方便的看台入口。中层看台的一条条过道延伸至街道，因此，也可从室外轻松抵达。最后，在看台顶层，沿周边布置了数量充足的楼梯出入口，以确保体育场能安全有效地疏散整座建筑的人流。

| 文体建筑方案集成 | CULTURAL AND SPORTS ARCHITECTURE PROGRAM INTEGRATION |

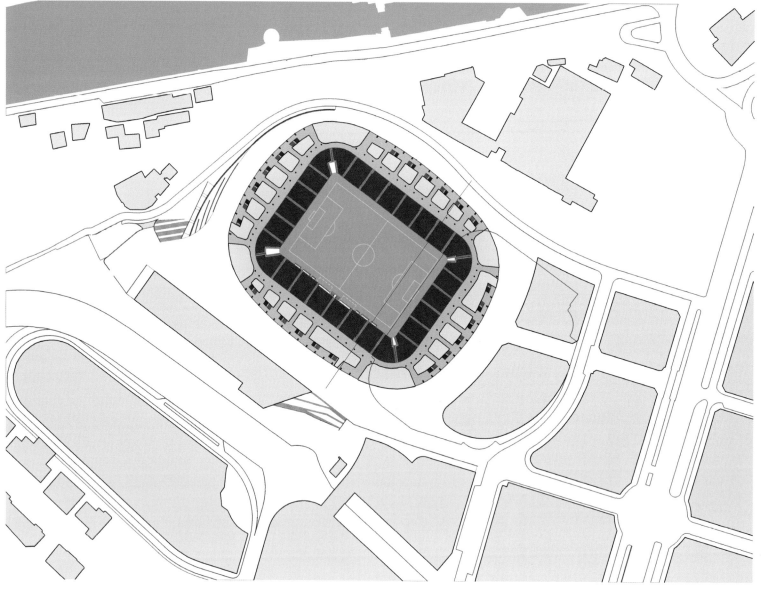

场地位置
Site Location

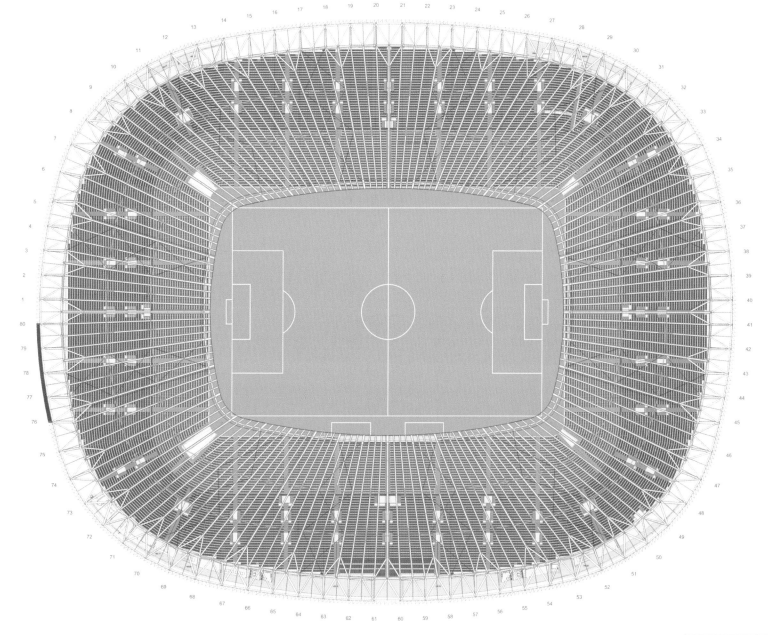

文体建筑方案集成 | CULTURAL AND SPORTS ARCHITECTURE PROGRAM INTEGRATION

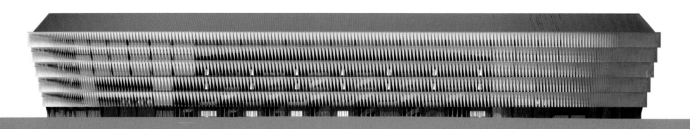

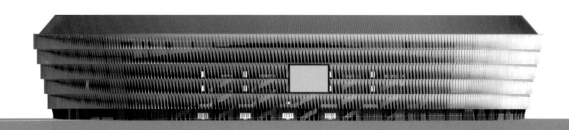

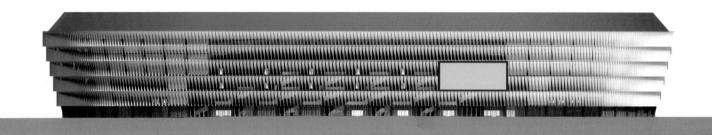

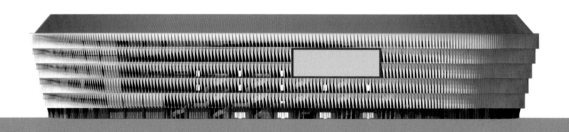

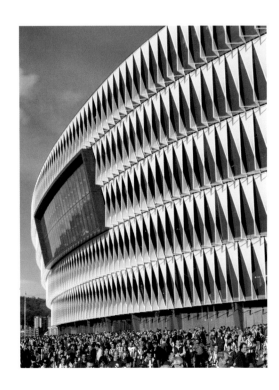

Grand Volume Glass Curtain Wall

Project Overview

The location of the new stadium, at the end of the urban mesh of the expansion district of Bilbao, peeping over the estuary with privilege, turns the building into a piece of architecture that must be introduced categorically and with force, but at the same time, respecting the rest of the buildings that make up that area of the city. The stadium has ample hospitality areas, with VIP boxes, premium seating and its leisure and meeting areas, restaurants, cafes, the Club's Museum, the Official Shop and areas for meetings, as well as a sports centre open to the general public under one of its stands. Its capacity will exceed 53,500 spectators, and there are 3,000 seats.

侧面
Alzado Lateral

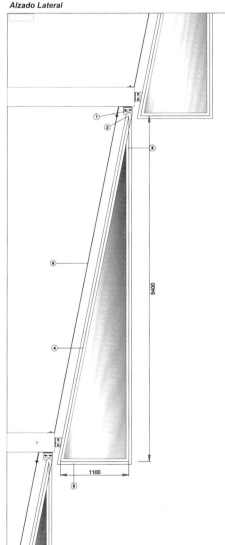

正面 A
Alzado Frontal "A"

正面 B
Alzado Frontal "B"

底部
Planta

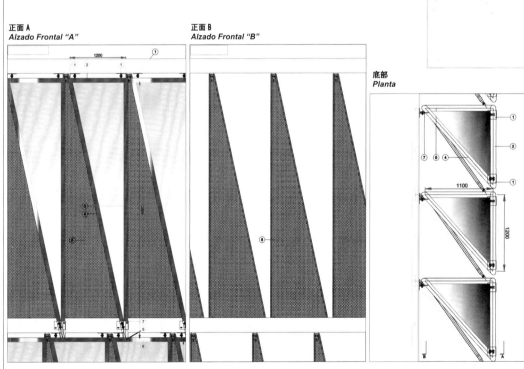

1. 上级
1. SUJECTIÓN SUPERIOR.
2. 管直径 Ø90×5（上边）
2. TUBO Ø90×5 (LADO SUPERIOR).
3. 管直径 Ø90×5（垂直长边）
3. TUBO Ø90×5 (LADO LARGO APROX. VERTICAL).
4. 管直径 Ø90×5（对角长边）
4. TUBO Ø90×5 (LADO LARGO DIAGONAL).
5. 管直径 Ø90×5（下边）
5. TUBO Ø90×5 (LADO INFERIOR).
6. 压缩带
6. TIRANTE TRACCIÓR-COMPRESIÓN.
7. 下级
7. SUJECTIÓN INFERIOR.
8. 穿孔板 =2MM
8. CHAPA PERFORADA E=2mm

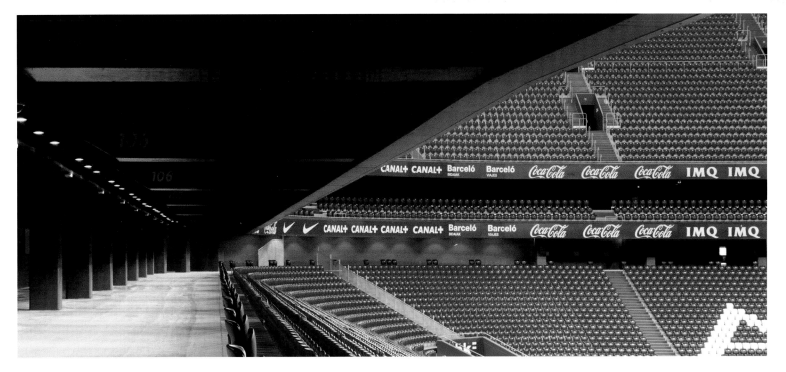

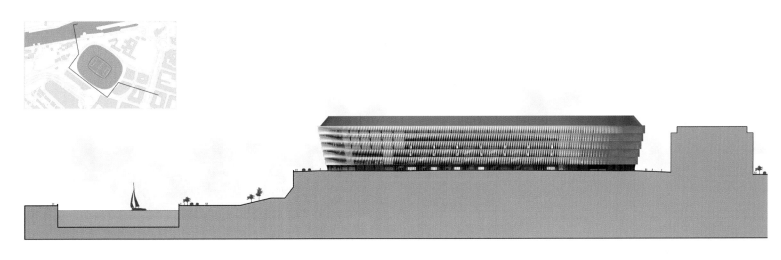

| 文体建筑方案集成 | CULTURAL AND SPORTS ARCHITECTURE PROGRAM INTEGRATION |

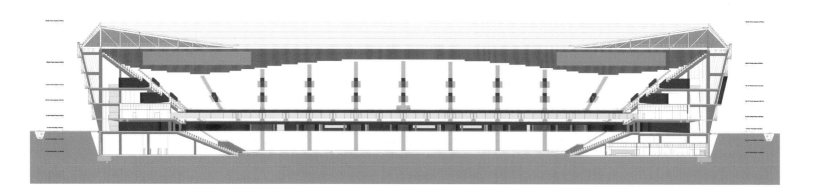

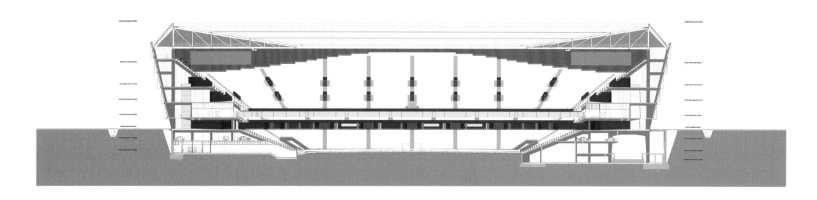

Building Design

The location of the new stadium turns the building into a piece of architecture that must be introduced categorically and with force, but at the same time, respecting the rest of the buildings that make up that area of the city. From this reflection comes one of the first aspects borne in mind for its design. That is, the perception of the erected construction as an urban building, in relation to the others and not just as simple sports facilities. It was intended for those stadium areas that are traditionally worthless to become valuable. These are located between the stadium's perimeter and the rear part of the stands and constitute the circulation areas through which you can access and exit the stands, which are, after all, the main part of the whole football stadium. In order to give these areas an added value, the strategy of the project consisted of, not only giving them spatial features, but also making sure that they had a very intense connection with the city and the surroundings. For this purpose, a basic element that will surely give character to the New San Mames stadium is put into play on the façade. This is, the repetition of a twisted ETFE element, giving the elevation energy and unity. This element will be illuminated at night, thus creating an urban landmark over the estuary, projecting a new image of Bilbao from within, thanks to one of the most advanced dynamic lighting systems in the world. The roof, formed by powerful radial metal trusses orientated towards the centre of the pitch, is covered with white ETFE cushions, covering the entire stands. The set-up of the stands is totally focused on the field, maximizing the pressure that the fans exert on the game, just like in the old San Mames, known the world over for being like a pressure cooker where the public would be on top of the players.

The entire stands are designed as three layers. The first layer is the nearest to the court, below the street plane. This can not only reduce the disturbance made by the stadium to the city, but also offer a direct and convenient stand entry from one side of the street. The aisles of middle layer stand extend to the street, so the spectators can also easily arrive in the stadium from outside. At last, on the top layer of the stand, around which we arrange sufficient stair entries to ensure that people can be evacuated safe and effective.

芬兰赫尔辛基古根海姆博物馆
Helsinki Guggenheim Museum

设计单位：绿舍都会
委托方：所罗门—古根海姆博物馆及基金会
项目地址：芬兰赫尔辛基市
项目面积：18 000m²

Designed by: SURE Architecture
Client: Solomon R Guggenheim Museum and Foundation
Location: Helsinki, Finland
Area: 18,000m²

时尚前卫 空间美感

项目概况

本案设计独特新颖，充分尊重并且增加了赫尔辛基这座古老城市的优雅与魅力，使得当地的公共、私人空间更加时尚前卫而且美轮美奂。

建筑设计

项目从不同方面重新诠释了赫尔辛基历史悠久的城市形态与构造。广阔的林荫道将市场与公园连接起来，与坡道、庭院、屋顶和桥梁一起勾勒出博物馆的轮廓。设计师在这里搭建起壮观的露天平台，将来作为展示区使用，从这里可以看到海港的迷人风景。

博物馆内外有两条路（公用及博物馆专用），交叉却不重叠，营造出空间感，又创造了视觉上的动态认知体验。

项目设计时有人质疑和否定"艺术空间"这一说法，因而建筑边界不再像原来那么清楚。就像是在埃舍尔的画中一样，博物馆的空间存在于上部的同时又体现在其下部，是内在的同时又是外在的，在上升的同时又在下降。这些不同的体验将激发和调动游客的感觉和情感的共鸣，使他们与周围环境融为一体，进行思维碰撞，留下永恒的记忆。

| 文体建筑方案集成 | CULTURAL AND SPORTS ARCHITECTURE PROGRAM INTEGRATION |

总平面图
Site Plan

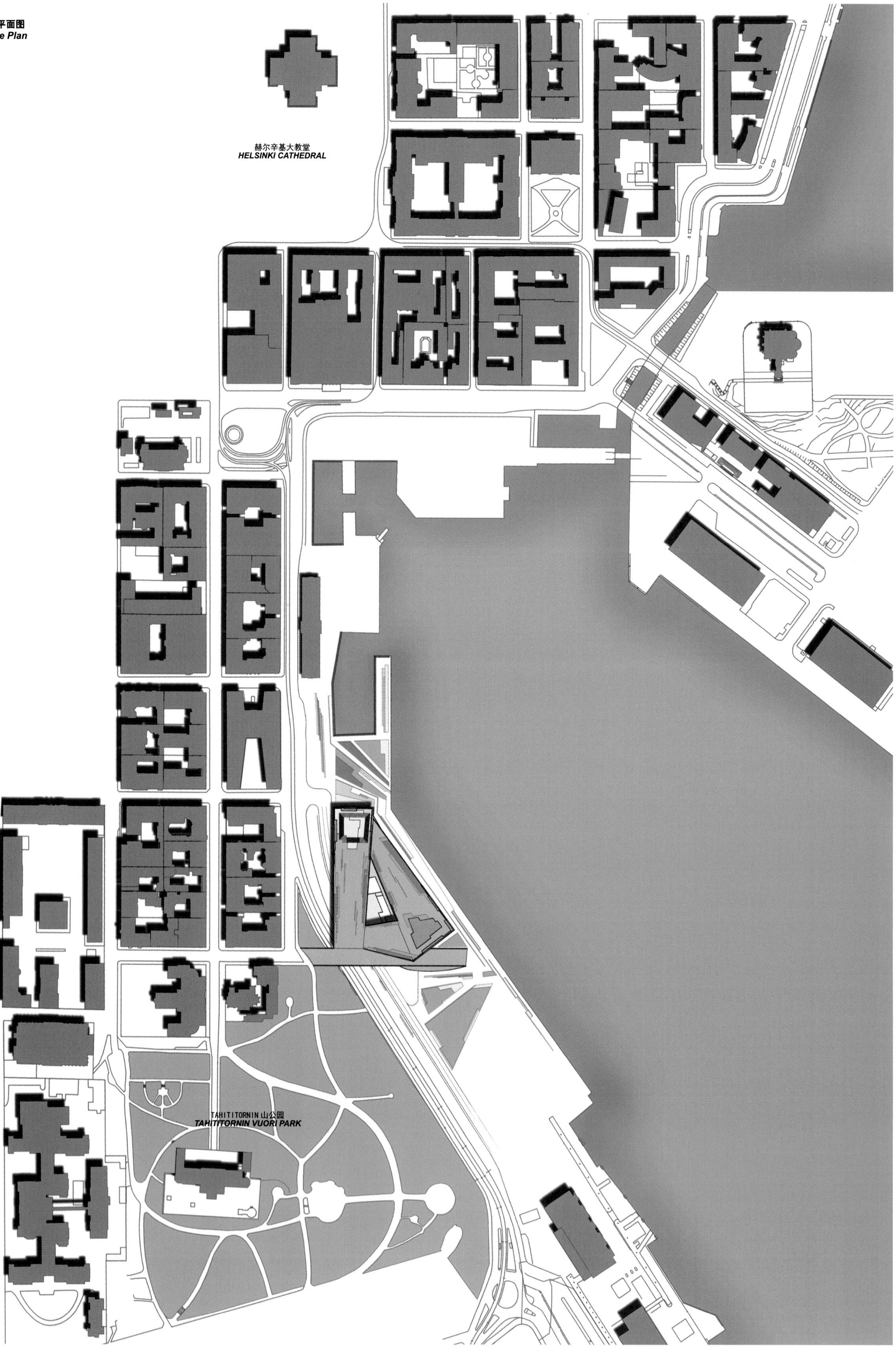

赫尔辛基大教堂
HELSINKI CATHEDRAL

TAHITITORNIN 山公园
TAHITITORNIN VUORI PARK

赫尔辛基大教堂

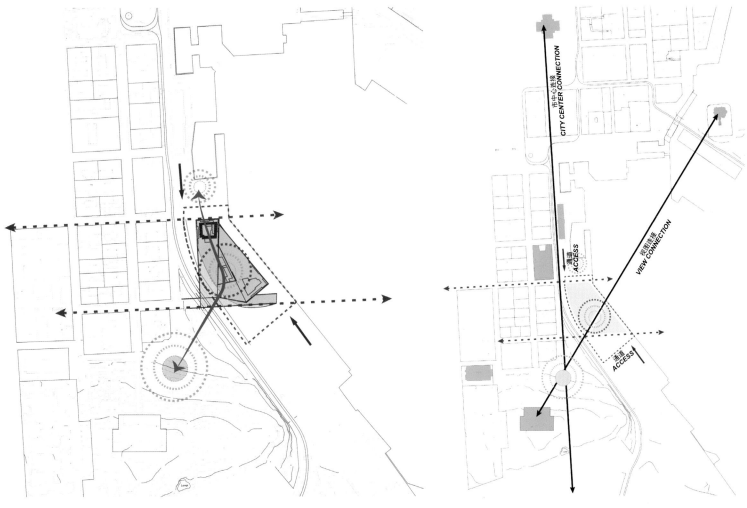

景观关系图 1
Landscape Relations 1

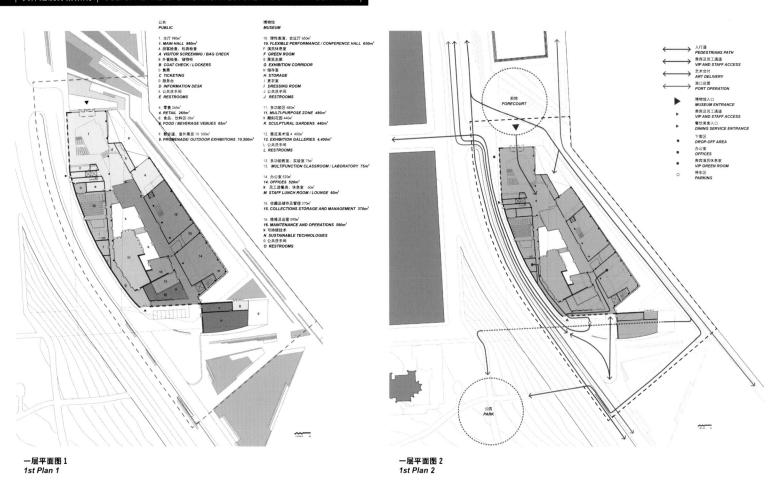

一层平面图 1
1st Plan 1

一层平面图 2
1st Plan 2

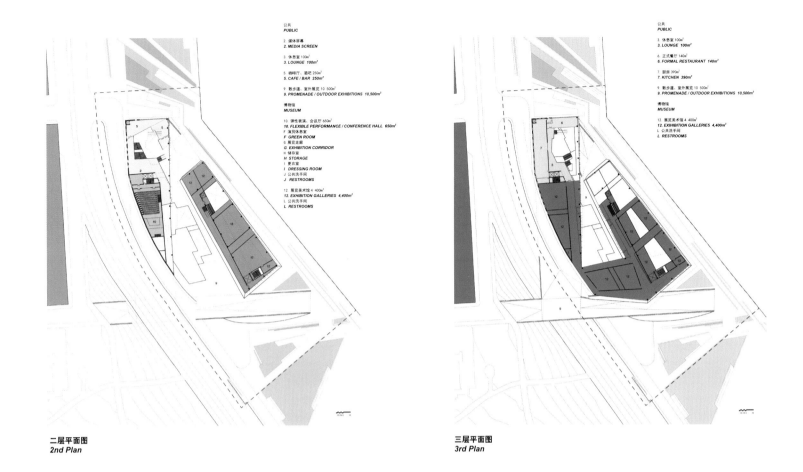

二层平面图
2nd Plan

三层平面图
3rd Plan

公共
PUBLIC
2. 媒体屏幕
2. MEDIA SCREEN
3. 休息室 100m²
3. LOUNGE 100m²
5. 咖啡厅、酒吧 250m²
5. CAFE / BAR 250m²
9. 散步道、室外展览 10,500m²
9. PROMENADE / OUTDOOR EXHIBITIONS 10,500m²
博物馆
MUSEUM
10. 弹性表演、会议厅 650m²
10. FLEXIBLE PERFORMANCE / CONFERENCE HALL 650m²
F. 演员休息室
F. GREEN ROOM
G. 展览走廊
G. EXHIBITION CORRIDOR
H. 储存室
H. STORAGE
I. 更衣室
I. DRESSING ROOM
J. 公共洗手间
J. RESTROOMS
12. 展览美术馆 4,400m²
12. EXHIBITION GALLERIES 4,400m²
L. 公共洗手间
L. RESTROOMS

公共
PUBLIC
3. 休息室 100m²
3. LOUNGE 100m²
6. 正式餐厅 140m²
6. FORMAL RESTAURANT 140m²
7. 服务 390m²
7. KITCHEN 390m²
9. 散步道、室外展览 10,500m²
9. PROMENADE / OUTDOOR EXHIBITIONS 10,500m²
博物馆
MUSEUM
12. 展览美术馆 4,400m²
12. EXHIBITION GALLERIES 4,400m²
L. 公共洗手间
L. RESTROOMS

| 266-267 |

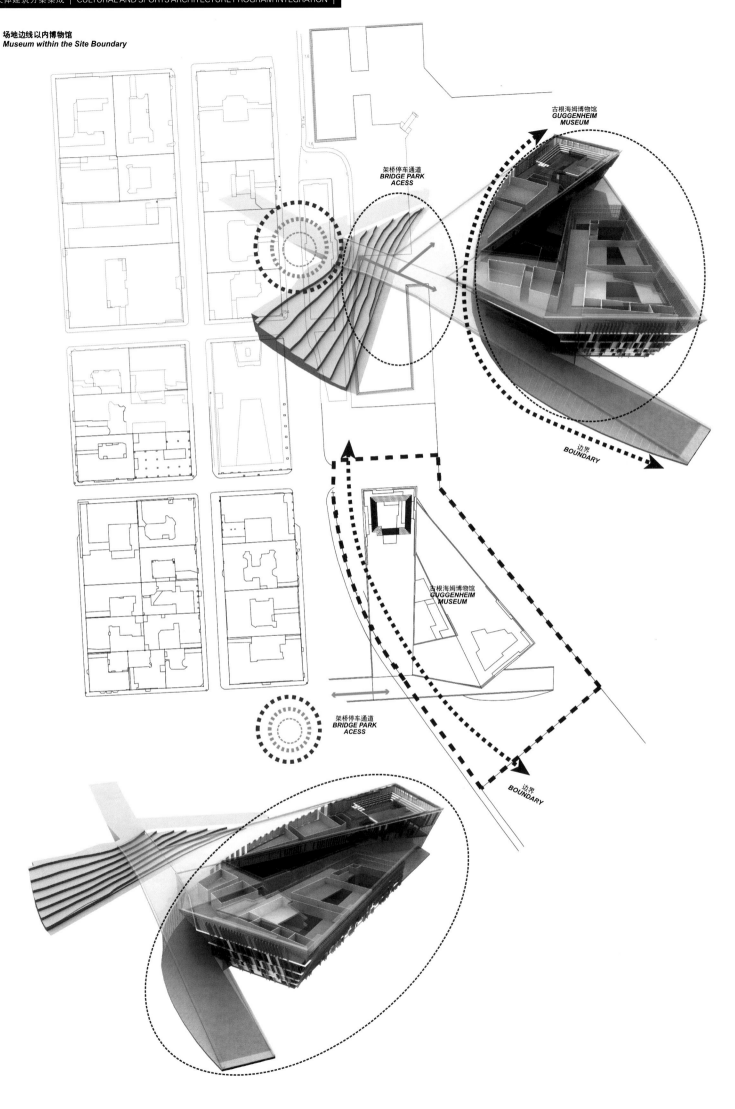

Fashion Bold Space Beauty

Project Overview
The Helsinki Guggenheim Museum is an innovative design which respects and enhances the elegance of the old city, and a bold yet beautiful addition to the local precinct which will offer a mix of public and private spaces.

Building Design
The historic Helsinki Urban morphology and structure has been reinterpreted at different scales of our project. The expansive public boulevards that give shape to the museum connect the market with the park by a game of ramps, courtyards, rooftops and bridges; culminating in a spectacular open air terrace which will become an exhibition area with views of the harbor.

Two distinctive paths (public / museum) interweave without crossing each other inside and outside the museum, these two paths are designed to provoke a visual, spatial, kinesthetic and cognitive experience.

As a Space of Art statements are questioned and challenged. Boundaries are not as clear as usual. Like in an Escher drawing, spaces of the museum are at the same time above and below, inside and outside, rising and descending. This variety of experiences evokes feeling and emotion and will cause the visitor to interact with their surroundings, challenging the mind and thought, whilst creating everlasting memories.

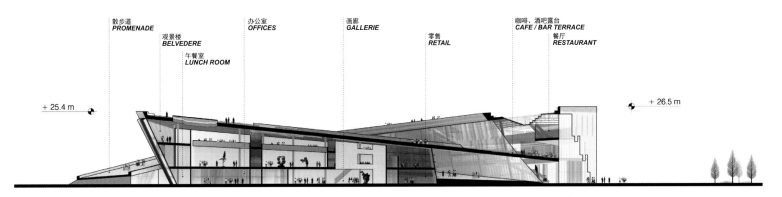

西剖面图
West Section

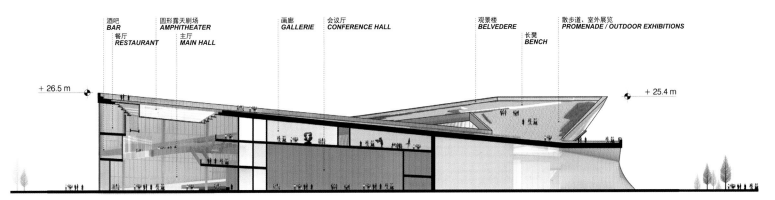

东剖面图
East Section

| 文体建筑方案集成 | CULTURAL AND SPORTS ARCHITECTURE PROGRAM INTEGRATION |

模糊玻璃 / *BLUR GLASS*　　木质网格 / *WOOD MESH*　　蓝色玻璃 / *BLUE COLOR GLASS*　　木 / *WOOD*　　白色石头 / *WHITE STONE*

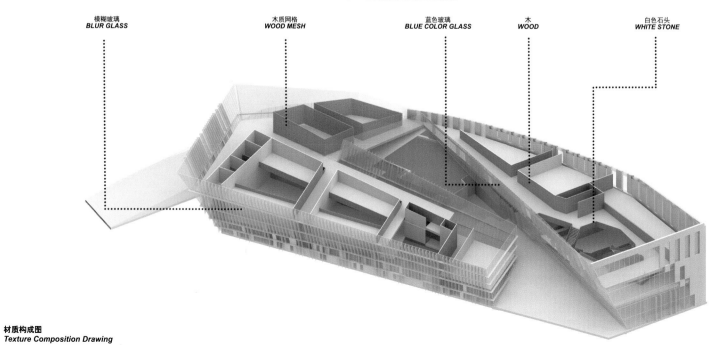

材质构成图
Texture Composition Drawing

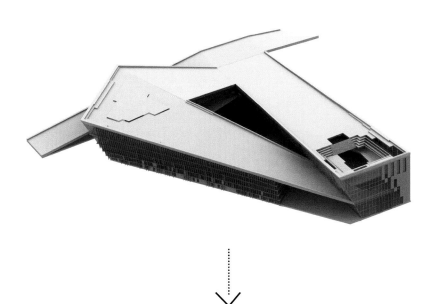

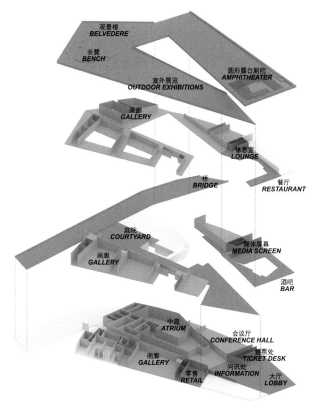

交织—博物馆路径及公共路径
Interweaving - Museum Path and Public Path

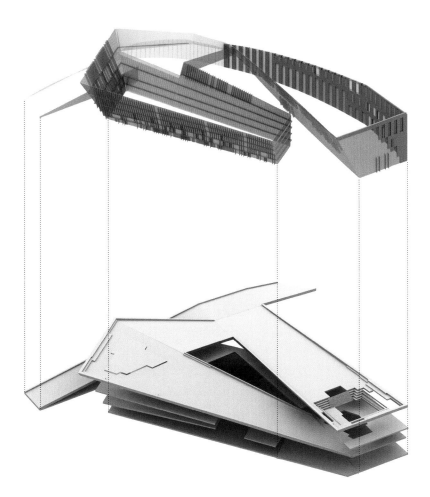

体量外观 + 平板
Volume Facade + Slabs

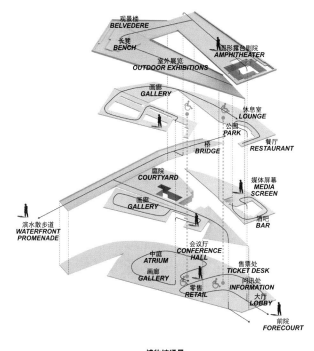

博物馆通量
Museum Flux

艺术空间
Space of Art

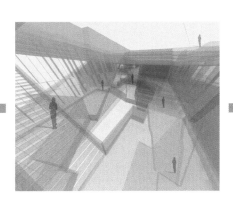

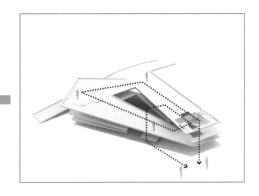

| 270-271 |

| 文体建筑方案集成 | CULTURAL AND SPORTS ARCHITECTURE PROGRAM INTEGRATION |

主厅流通图
Main Hall's Circulation

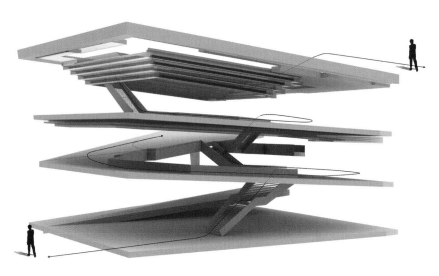

立面分析图——重新解释
Elevation Analysis – Reinterpretation

直线元素城市连接点
LINEAL ELEMENT CITY CONNECTION

窗户周围连接点
WINDOW SURROUNDING CONNECTION

景观关系图 2
Landscape Relations 2

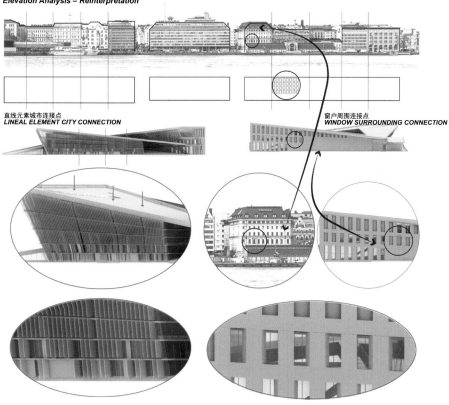

可持续性
Sustainability

环境策略
ENVIRONMENTAL STRATEGY
精选材料
CAREFUL CHOICE OF THE MATERIALS

- FSC 认证芬兰木材 / FSC CERTIFIED FINNISH WOOD
- 低挥发性涂料及地毯 / LOW VOC PAINTS AND CARPET
- 双重抽水马桶 / DUAL FLUSH TOILET
- 低流动性固定装置 / LOW FLOW FIXTURES
- 节能照明设备 / ENERGY EFFICIENT LIGHTING
- 能源之星电气用具 / ENERGY STAR APPLIANCES

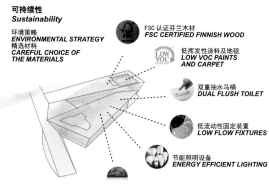

太阳能板
SOLAR PANEL
产能
TO PRODUCE ENERGY

自然光及空气循环
NATURAL LIGHT AND AIR CIRCULATION

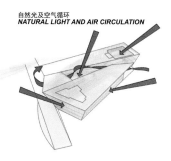

景观及绿色屋顶
LANDSCAPE AND GREEN ROOFTOP
本地物种
WITH NATIVE SPECIES
+
白色石头
WHITE COLOR STONE
降低热岛效应
REDUCE HEAT ISLAND EFFECT

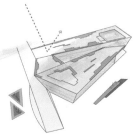

雨水及污水收集，生物处理及再利用
RAIN WATER AND GREY WATER HARVESTING, BIO-TREATMENT AND RE-USE

过滤玻璃面板
FILTERING GLASS PANELS
+
三倍上釉双层立面
DOUBLE SKIN FACADE OF TRIPLE GLAZING

横截面—夏季
Transversal Section—Summer

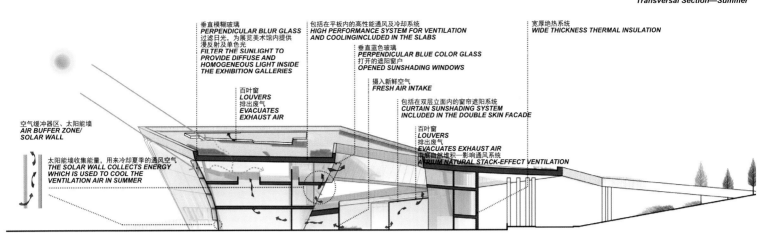

横截面—冬季
Transversal Section—Winter

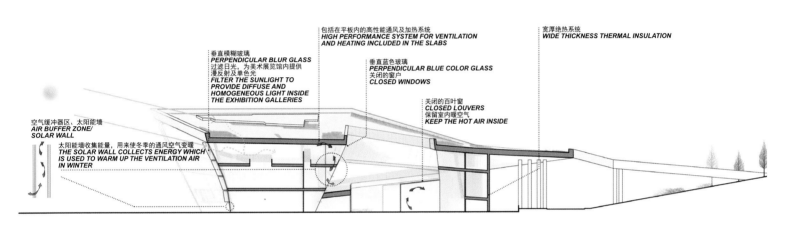

匈牙利布达佩斯 Pancho 体育场
Pancho Arena

设计单位：Tamás dobrosi, Doparum Architects Ltd.	Designed by: Tamás dobrosi, Doparum Architects Ltd.
委托方：The Puskás Academy Foundation	Client: The Puskás Academy Foundation
项目地址：匈牙利布达佩斯 Felcsút	Location: Felcsút, Budapest, Hungary
项目面积：12 000m²	Area: 12,000m²

结构非凡
舒适球场

项目概况

Pancho 体育场位于匈牙利首都布达佩斯以西 40km 的 Felcsút 社区。自 2004 年开始,当地一所全国最大的足球教育学院为匈牙利培养足球新星。Ferenc Puskás 是世界上最伟大的足球运动员之一,曾获得欧洲世界杯冠军和亚军。足球学院以 Ferenc Puskás 命名,Pancho 则为其昵称。

建筑设计

学院管理层决定建造一所欧洲足球协会联盟三级足球场,以举办匈牙利足球联赛、青年锦标赛,甚至是欧洲联赛及欧洲冠军联赛等各类国际赛事。

考虑到体育场的主要功能是该寄宿体育学院的主运动场,学院管理层经由欧洲足协代表的同意,决定将坐席缩减至 3 400 个,同时提高座位的舒适度。每排座椅之间距离为 100cm,相邻座椅的距离为 55cm,达到了西欧体育场的商务级座椅级别。此外还设计了更为惬意舒适拥有 420 个席位的 VIP 区域。西翼二层为主要赞助商准备了 7 间舒适的的包间,还设有电视和电台解说区、容纳 50 人的新闻发布室、设有 70 个席位的媒体看台以及配备先进设备的媒体直播间。

尽管体育场覆盖面积多达 12 000m² (安装了加热装置),但安装上所有必要设施仍显得空间紧凑。除了南北看台下方的外露空间之外,其他所有空间均被隐藏起来。体育馆充分利用了阶梯式地形特点,每一层的入口通道都与外部地面直接相通。人造草皮覆盖的运动员更衣室及旁边的热身室严格遵照欧洲足协的规定进行设计,内部配备了教练员、裁判员和兴奋剂检查官员所用的各项设施。地面层设有 12 间衣帽间、一间健身房和一间健康中心,可供年轻球员们自由使用,由东面的独立监控门可以进入这些区域。观众可从地下层南北两侧进入体育馆,从不同入口经过具有清晰指向的开放式覆顶空间通向不同的看台区。

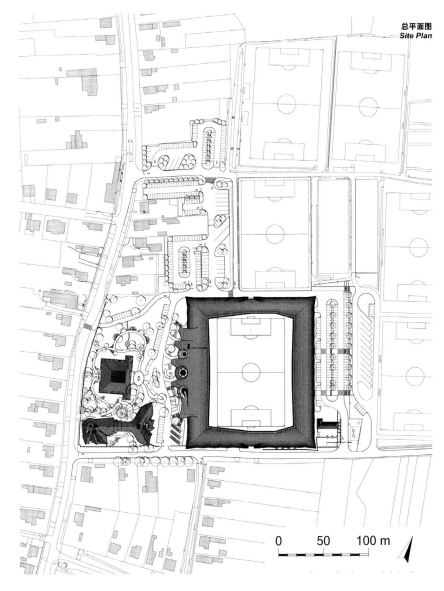

总平面图
Site Plan

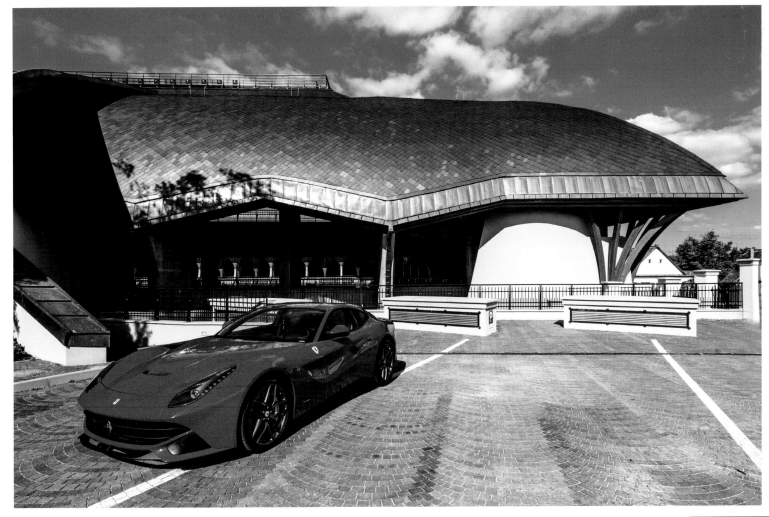

已建建筑和球场除了具有指示方位的基本功能外，更重要的是决定了 Pancho 体育场的建筑场地，此外 8m 的斜坡也是建筑师面临的另一挑战。球场设在地势较低的区域，这样在西面形成了较低的立面。相对狭窄的空间被凸起的拱形屋顶结构进一步分割成更小的区域。VIP 区域、媒体室以及球员的独立入口通道均设在西面，锁扣式的穹顶将这些独立的入口连接起来。格子图案的木屋顶凸起于钢筋混凝土梁柱之上，如同覆盖空地的树冠般笼罩着整个看台区域。其设计理念是打造一个具有自由自在的建筑逻辑和清晰形态的独特而非凡的结构，以创新的方式将当代匈牙利体育馆的建筑潮流与复杂的轮廓和结构结合起来。由于地处乡村，无法从远距离欣赏该建筑，因此建筑的外观相对单调。设计将重点放在入口上方散开的屋顶结构和多样的内部空间上。从正面墙壁的开口处可以看到体育场内部的情景，借此一睹赛场上激烈的比赛场景。

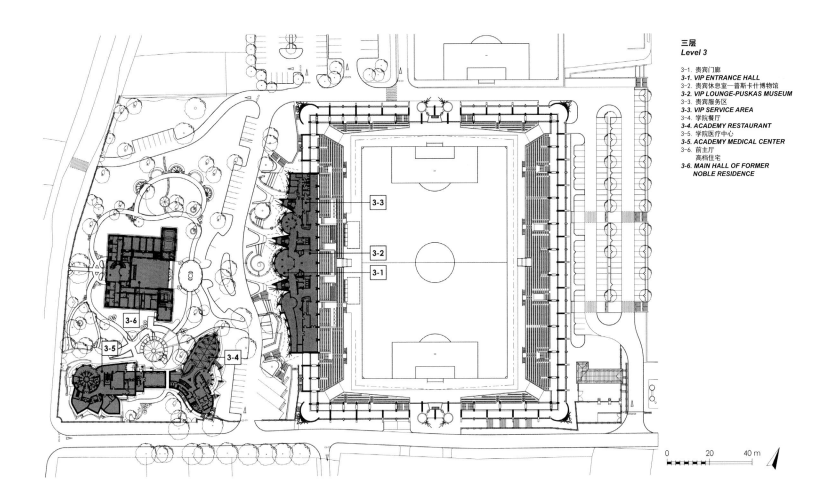

三层
Level 3

3-1. 贵宾门廊
3-1. VIP ENTRANCE HALL
3-2. 贵宾休息室—普斯卡什博物馆
3-2. VIP LOUNGE-PUSKAS MUSEUM
3-3. 贵宾服务区
3-3. VIP SERVICE AREA
3-4. 学院餐厅
3-4. ACADEMY RESTAURANT
3-5. 学院医疗中心
3-5. ACADEMY MEDICAL CENTER
3-6. 前主厅
高档住宅
3-6. MAIN HALL OF FORMER NOBLE RESIDENCE

| 文体建筑方案集成 | CULTURAL AND SPORTS ARCHITECTURE PROGRAM INTEGRATION |

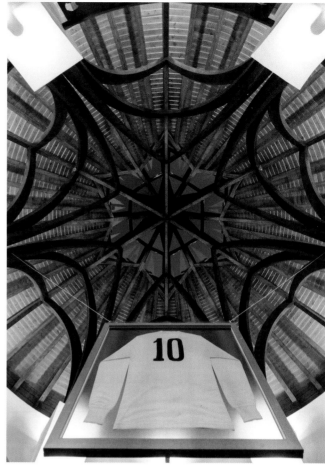
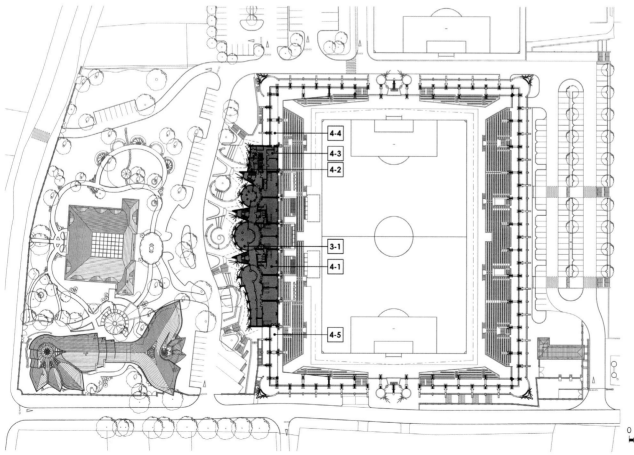

四层
Level 4

4-1. 天空盒
4-1. SKYBOXES
4-2. 服务区—厨房
4-2. SERVICE AREA-KITCHEN
4-3. 体育场控制室
4-3. STADIUM CONTROL ROOM
4-4. 媒体平台
4-4. PITCHVIEW MEDIA TERRACE
4-5. 天空盒露台
4-5. SKYBOX TERRACE

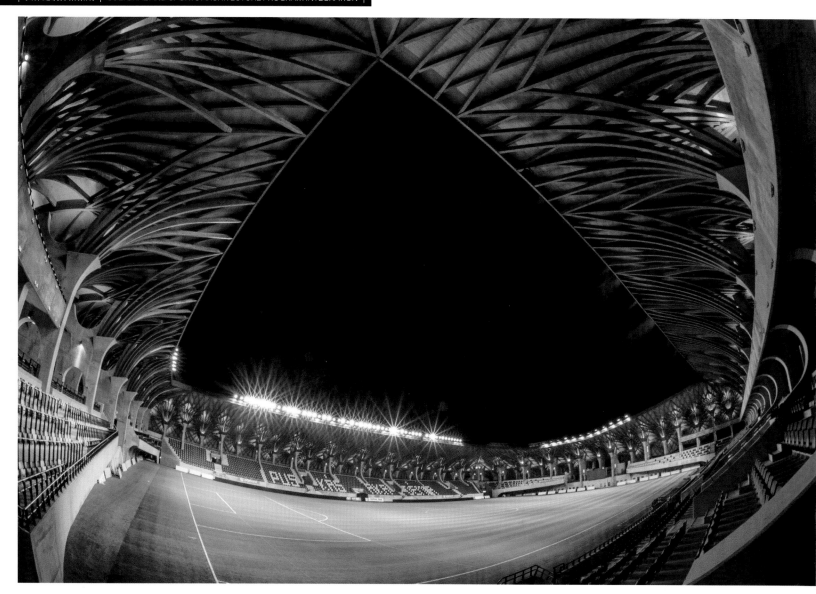

Unique Structure
Comfortable Court

Project Overview

Felcsút is a community of 1,800 souls in the Váli Valley, 40 kilometers west of Budapest, the capital of Hungary. Since 2004, the locality hosts the largest education center for aspiring young footballers in the country; an institution named in 2006 after one of the greatest football legends in history, Olympic champion, European Cup winner and World Cup silver medalist Ferenc Puskás.

Building Design

The management of the Academy decided to build a UEFA category III football stadium suitable for hosting Hungarian league matches and junior tournaments as well as any kind of international competition up to the second qualifying round of the Europa League and the Champions League.

Considering that serving a large urban agglomeration was not among the priorities for the future arena, because it was to be used as the main pitch of a boarding sports academy, the management decided in agreement with the UEFA delegates that the number of seats would remain relatively low at 3,400 while offering greater than usual comfort. With a 100 cm distance between rows and 55 cm between seats, the stadium matches the level of comfort of business class sections in Western European stadiums. Reserved for special guests, the 420-seat VIP section echoes the more intimate and cozy atmosphere of the Academy's existing buildings. Seven lodges on the second floor of the west wing offer full comfort for the main sponsors while press correspondents are served by five TV and radio commentary positions, a press conference room capable of hosting 50 people, a 70-seat media tribune and several studios equipped for high quality live broadcast of any sporting event.

Although the arena occupies more than 12,000 square meters, all of which

西立面
West Elevation

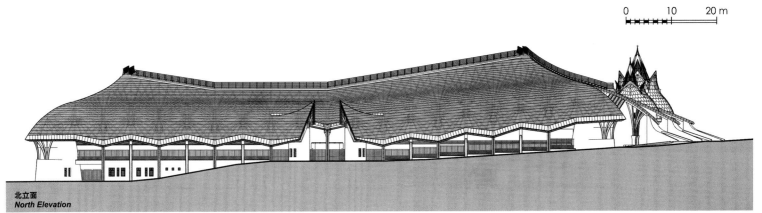

北立面
North Elevation

横截面
Cross Section

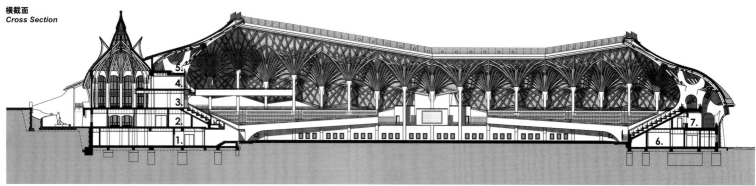

1. 团队更衣室，新闻发布会室
1. TEAM DRESSING ROOMS, PRESS CONFERENCE ROOM
2. 新闻看台，工作室，解说位置
2. PRESS STANDS, STUDIOS, COMMENTARY POSITIONS
3. 贵宾休息室
3. VIP LOUNGE
4. 天空盒
4. SKYBOXES
5. 摄影机位置
5. CAMERA POSITIONS
6. 学院更衣室
6. ACADEMY DRESSING ROOMS
7. 公共设施（食品，饮料，厕所，急救）
7. PUBLIC FACILITIES (FOOD, BEVERAGES, TOILETS, FIRST AID)

纵截面
Longitudinal Section

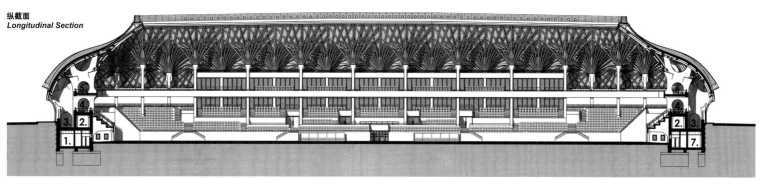

1. 办公室，学院医疗中心
1. OFFICES, ACADEMY MEDICAL CENTER
2. 公共设施（食品，饮料，厕所，急救）
2. PUBLIC FACILITIES (FOOD, BEVERAGES, TOILETS, FIRST AID)
3. 观众区的下部走廊
3. SPECTATORS' AREA-LOWER CORRIDOR
4. 观众区的上部走廊
4. SPECTATORS' AREA-UPPER CORRIDOR
5. 天空盒露台
5. SKYBOX TERRACE
6. 媒体平台
6. PITCHVIEW MEDIA TERRACE
7. 学院更衣室
7. ACADEMY DRESSING ROOMS

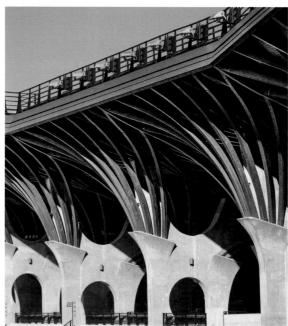

fitted with heating, the space still proved to be rather tight to host all necessary facilities. Except for the area below the northern and southern tribunes protruding over the pitch, all the space is built in. Making good use of the properties of the terrain, the building is served by exits communicating directly with the outside world on three different floors. The changing rooms of both the home side and the visitors, complete with their own adjacent warm-up rooms covered in artificial grass, were designed in full accordance with UEFA provisions, containing facilities for trainers, referees and doping control officers as well. On the basement level, further 12 locker rooms, a gym and a health center are at the disposal of young footballers. These can be approached from east through separate and monitored gates. Fans are admitted from the north and south on the subground floor, with separate gates leading to different sectors, through clearly indicated open and roofed spaces.

Beyond general requirements for orientation, the existing buildings and pitches had a defining role in designating the exact site of construction. The terrain, with its 8-meter inclination, was yet another challenge to be overcome. The lowered pitch made it possible to build a lower facade on the western side. Here, the relatively narrow space is further divided into smaller chambers by elements of the protruding roof structure which are running into the earth between the domes of the foyers. Separate gates to the VIP and media rooms as well as the players' entrance are located on this side, connected by the interlocking structure of rooftop domes. Sprouting from pillars of reinforced concrete, the lattice wood roof is spread out over the tribunes like the canopy of the tree line around a clearing. The underlying concept was to build a structure which is unique and exceptional due to its own inner logic and clean static system, integrating the rational tendency of contemporary Hungarian stadium architecture with more complex forms and structures in an innovative context.

下

龙志伟 编著
Edited by Long Zhiwei

| 规矩偏不落窠臼 | 智欲圆而行欲方 |
| Unique Style | Cautious Ideas |

广西师范大学出版社
·桂林·

目录 Contents

规矩偏不落窠臼
Unique Style ... 6

重庆图书馆
Chongqing Library ... 8

甘肃兰州省图书馆新建项目
New Project in Gansu Provincial Library ... 16

美国西雅图中央图书馆
Seattle Central Library ... 24

斯洛文尼亚卢布尔雅那国家大学图书馆 II
NUKII, National and University Library II Ljubljana ... 28

芬兰赫尔辛基中央图书馆
Helsinki Central Library ... 36

韩国大邱 Gosan 公共图书馆（方案四）
Daegu Gosan Public Library ... 42

韩国大邱 Gosan 公共图书馆（方案五）
Daegu Gosan Public Library ... 50

澳大利亚阿德莱德新文化中心
Adelaide Culture Center ... 58

法国圣日耳曼 -lès- 阿尔帕容新文化中心
St Germain-Lès-Arpajon Cultural Center ... 64

| 比利时亨克城市中心 | 70 |
| GENK | |

| 荷兰合作银行 Westelijke Mijnstreek 咨询中心 | 78 |
| Rabobank Westelijke Mijnstreek Advice Centre —Sittard | |

| 湖南长沙中联重科新总部展览中心 | 88 |
| Zoomlion Headquarters Exhibition Center | |

| 上海国际设计一场 | 94 |
| Shanghai International Design No. 1 | |

| 塞浦路斯尼科西亚 RAUF RAIF DENKTAS 纪念馆和博物馆 | 102 |
| Rauf Raif Denktas Memorial and Museum | |

| 美国路易斯安那州博物馆与名人堂 | 108 |
| Louisiana State Museum and Sports Hall of Fame | |

| 德国汉诺威史普格尔博物馆扩建 | 118 |
| Sprengel Museum Extension | |

| 德国拜罗伊特理查德·瓦格纳博物馆扩建 | 122 |
| Richard Wagner Museum Extension | |

| 奥地利韦尔斯 Welios | 126 |
| Wels Welios | |

| 斯洛文尼亚马里博尔美术馆 | 134 |
| Slovenia Maribor Art Gallery | |

| 四川成都安仁桥馆 | 140 |
| Museum-Bridge in Anren | |

俄罗斯彼尔姆芭蕾舞馆
Opera Ballet Theater, Perm, Russia — 146

挪威 Våler 新教堂
New Church of Våler — 156

丹麦奈斯特韦兹竞技场
Nåstved Arena — 162

俄罗斯莫斯科卢日尼基奥体中心游泳池
Swimming Pool of Luzhniki Olympic Complex — 166

智欲圆而行欲方
Cautious Ideas — 174

阿富汗喀布尔国家博物馆
Afghanistan National Museum — 176

委内瑞拉库马纳会展中心
Cumana Convention Center — 184

北京国家美术馆
National Art Museum of China — 192

广东广州三馆一场
Guangzhou Three Museums One Square — 198

江苏镇江科技馆
Zhenjiang Science and Technology Museum — 206

塞尔维亚贝尔格莱德科学促进中心	214
Center for Promotion of Science Belgrade, Republic of Serbia	
荷兰鹿特丹伊拉斯姆斯大学生活动中心	220
Erasmus Pavilion	
意大利佛罗伦萨音乐文化公园	226
Florence Music and Culture Park	
俄罗斯圣彼得堡阿里纳斯田径场综合楼	230
Complex of Track and Field Arenas	
刚果共和国布拉柴维尔体育场	236
Brazzaville Stadium	
哥斯达黎加瓜纳卡斯特 Miravalles 教堂	240
Miravalles Church	
韩国大邱 Gosan 公共图书馆	244
Daegu Gosan Public Library	
挪威克里斯蒂安松歌剧院与文化中心	250
Opera and Culture House—Kristiansund, Norway	
瑞典哥德堡世界文化博物馆	258
Museum of World Culture—Gothenburg, Sweden	
丹麦奥尔堡音乐厅	268
House of Music	

规矩偏不落窠臼

Unique Style

重庆图书馆
Chongqing Library

设计单位：Perkins Eastman	**Designed by: Perkins Eastman**
委托方：重庆市政府	**Client: Chongqing Municipal Government**
项目地址：中国重庆市	**Location: Chongqing, China**
建筑面积：50 000m²	**Building Area: 50,000m²**
项目摄影：ZhiHui Gu 版权	**Photography: Copyright ZhiHui Gu**

城市综合体
传统而活跃
视觉连接

项目概况

重庆图书馆成立于1947年，是中国五个国家图书馆之一，新馆被设计为城市综合体。图书馆的计划体现了从书本房子改造成为一个文化中心，其中包括一个公共剧场、展览厅、开放电脑学习设施、古代档案和阅览室。该中心是学习的主要目的地，为过夜的学者提供酒店客房，还有会议中心、专题讨论会、餐厅和咖啡厅。

建筑设计

图书馆的建筑形式是基于中国庭院的建筑传统与西方图书馆多层内部的中庭大厅。室外庭院空间为繁华城市提供了一个独特的绿洲，它从街道可视面层下沉到底层，实现人文活动与城市街景的分离。3个主要设计空间——开放的公共接入区、安全的公共区和私人安全区，都围绕着一个中心庭院而建。

内部阅览室被设想为森林之中的浮动平台，读者可以从一个平台眺望另一个平台。设计给读者提供了多种小型阅读区，既倍感温馨，同时也获得了视觉效果。"Y"形混凝土柱支撑基准面上的轻质檐篷，稳固而美观。石头，作为一种内外饰材料，兼具丰富图案和多种色彩，并与城市的老建筑和街道相匹配。

设计采用玻璃和石材阐明了图书馆的公共和私人领域。面对丰田大道的公共区域，以玻璃覆盖，对于使用者及行人意味着透明度和知名度。而西边的书架和服务区都覆盖着石头以达到建筑安全性与图书馆馆藏的牢固性。

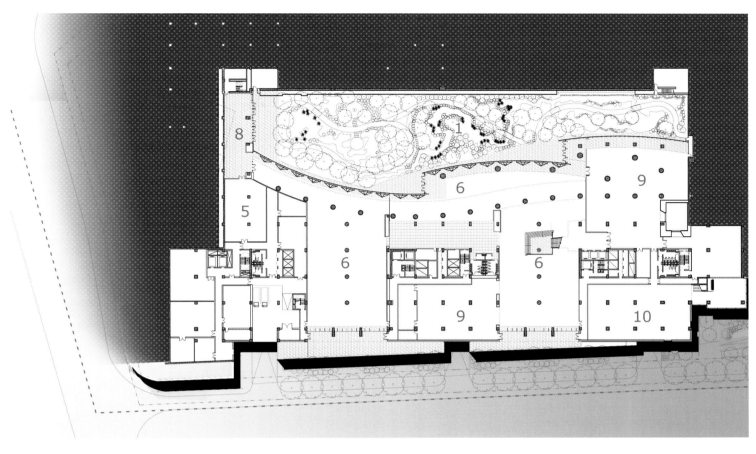

Lowered Courtyard - First Floor
一层庭院

← N 0 30ft / 9M

1. **COURTYARD** / 庭院
2. **REFLECTING POOL** / 倒影池
3. **MAIN ENTRY** / 主入口
4. **LIBRARY RECEPTION** / 图书馆接待处
5. **ADMINISTRATION** / 管理处
6. **READING ROOM** / 阅览室
7. **READING TERRACE** / 阅览室露台
8. **MEETING ROOM** / 会议室
9. **COLLECTION STACKS** / 收藏书库
10. **CHILDREN'S BOOKS** / 儿童读物区

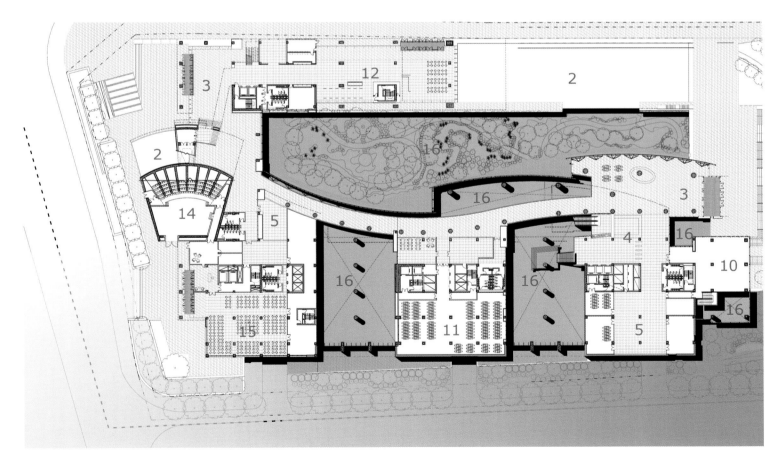

Entrance / Street Level – Second Floor
二层入口 / 街道水平

← N 0 30ft / 9M

1. **COURTYARD** / 庭院
2. **REFLECTING POOL** / 倒影池
3. **MAIN ENTRY** / 主入口
4. **LIBRARY RECEPTION** / 图书馆接待处
5. **ADMINISTRATION** / 管理处
6. **READING ROOM** / 阅览室
7. **READING TERRACE** / 阅览室露台
8. **MEETING ROOM** / 会议室
9. **COLLECTION STACKS** / 收藏书库
10. **CHILDREN'S BOOKS** / 儿童读物区
11. **MULTIMEDIA CENTER** / 多媒体中心
12. **EXHIBITION HALL** / 展览厅
13. **CLASSROOMS** / 教室
14. **AUDITORIUM** / 观众席
15. **RESTAURANT** / 餐厅
16. **OPEN TO BELOW** / 朝下打开

| 10–11 |

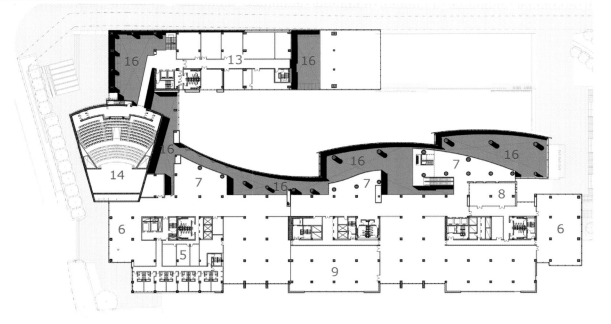

Stacks / Auditorium / Classrooms – Third Floor
三层书库 / 观众席 / 教室

← N 0 30ft / 9M

1. *COURTYARD*
  1. 庭院
2. *REFLECTING POOL*
  2. 倒影池
3. *MAIN ENTRY*
  3. 主入口
4. *LIBRARY RECEPTION*
  4. 图书馆接待处
5. *ADMINISTRATION*
  5. 管理处
6. *READING ROOM*
  6. 阅览室
7. *READING TERRACE*
  7. 阅览室露台
8. *MEETING ROOM*
  8. 会议室
9. *COLLECTION STACKS*
  9. 收藏书库
10. *CHILDREN'S BOOKS*
  10. 儿童读物区
11. *MULTIMEDIA CENTER*
  11. 多媒体中心
12. *EXHIBITION HALL*
  12. 展览厅
13. *CLASSROOMS*
  13. 教室
14. *AUDITORIUM*
  14. 观众席
15. *RESTAURANT*
  15. 餐厅
16. *OPEN TO BELOW*
  16. 朝下打开

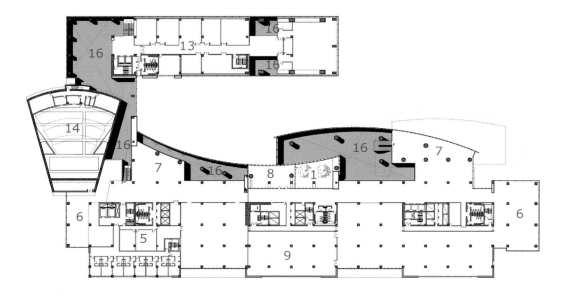

Stacks / Classrooms – Fourth Floor
四层书库 / 教室

← N 0 30ft / 9M

1. *COURTYARD*
  1. 庭院
2. *REFLECTING POOL*
  2. 倒影池
3. *MAIN ENTRY*
  3. 主入口
4. *LIBRARY RECEPTION*
  4. 图书馆接待处
5. *ADMINISTRATION*
  5. 管理处
6. *READING ROOM*
  6. 阅览室
7. *READING TERRACE*
  7. 阅览室露台
8. *MEETING ROOM*
  8. 会议室
9. *COLLECTION STACKS*
  9. 收藏书库
10. *CHILDREN'S BOOKS*
  10. 儿童读物区
11. *MULTIMEDIA CENTER*
  11. 多媒体中心
12. *EXHIBITION HALL*
  12. 展览厅
13. *CLASSROOMS*
  13. 教室
14. *AUDITORIUM*
  14. 观众席
15. *RESTAURANT*
  15. 餐厅
16. *OPEN TO BELOW*
  16. 朝下打开

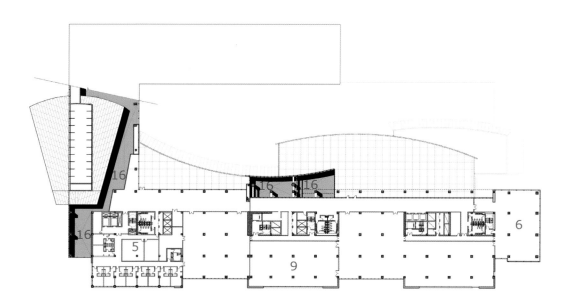

Stacks / Roof Deck – Fifth Floor
五层书库 / 屋顶平台

← N 0 30ft / 9M

1. *COURTYARD*
  1. 庭院
2. *REFLECTING POOL*
  2. 倒影池
3. *MAIN ENTRY*
  3. 主入口
4. *LIBRARY RECEPTION*
  4. 图书馆接待处
5. *ADMINISTRATION*
  5. 管理处
6. *READING ROOM*
  6. 阅览室
7. *READING TERRACE*
  7. 阅览室露台
8. *MEETING ROOM*
  8. 会议室
9. *COLLECTION STACKS*
  9. 收藏书库
10. *CHILDREN'S BOOKS*
  10. 儿童读物区
11. *MULTIMEDIA CENTER*
  11. 多媒体中心
12. *EXHIBITION HALL*
  12. 展览厅
13. *CLASSROOMS*
  13. 教室
14. *AUDITORIUM*
  14. 观众席
15. *RESTAURANT*
  15. 餐厅
16. *OPEN TO BELOW*
  16. 朝下打开

Urban Complex
Traditional and Active
Visual Connection

Project Overview

The previous Chongqing Library was set up in 1947 as one of five National Chinese Libraries. The program of the Chongqing Library exemplifies a transformation from a house for books into a cultural center which includes a public theater, exhibition hall, open computer learning facilities, ancient archives, and reading rooms. The center is a major destination for learning—offering hotel rooms for scholar's overnight stays, a conferencing center for symposia, a restaurant, and a café.

Building Design

The building's form is based on both the Chinese architectural tradition of the courtyard and the interior multi-story atrium halls of traditional western libraries. The outdoor courtyard spaces provide a unique oasis within the bustling city, visible from the street level but sunken below grade to achieve separation from the active urban streetscape. Three main programmatic elements-the open public access zone, secure public zone, and the private secure zone-are built around a center courtyard garden.

The building's interior reading rooms are envisioned as floating platforms among the forest. Readers can view from platform to platform, which provides a variety of smaller, intimate reading areas while allowing for visual interest. "Y" shaped concrete columns support light canopies over reference areas. Stone is used both as an exterior and interior material, with patterns and colors matching the older buildings and streets of the city.

The design uses glass and stone to articulate public and private areas of the library. The public areas facing Feng Tian Avenue are sheathed in glass to connote transparency and visibility for users and pedestrians. The book stacks and service areas to the west are clad in stone to represent security and the solidity of the library's collection.

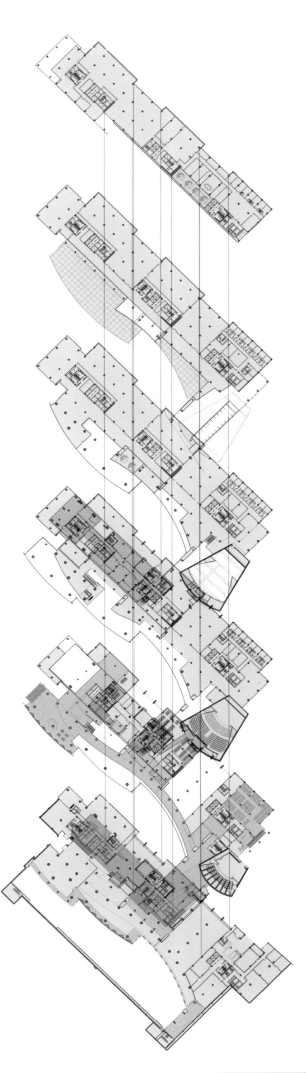

甘肃兰州省图书馆新建项目
New Project in Gansu Provincial Library

设计单位：北京殊舍建筑设计有限公司
项目地址：甘肃省兰州市
建筑面积：25 000m²

Design by: Beijing Shushe Architectural Design Co., Ltd.
Location: Lanzhou, Gansu
Building Area: 25,000m²

纯净外边
三角形中庭
改造广场

项目概况

2013年4月，北京殊舍建筑参与甘肃省省图书馆公开竞标，以第一名的身份中标。方案延续殊舍建筑的一贯风格，分析基地和功能要求，提出需要解决的问题，问题的一步一步解决就是设计的过程，问题的答案就是最终建筑。

此图书馆为了与老图书馆平衡，最大限度的减少自己的重量。素雅简单的外表皮下藏着丰富灵活的内部空间，透露出无限阳光和绿色的气息。

建筑设计

建筑外表看起来虽然简单，但是复杂实用的空间隐藏在建筑内部，透过纯净的表皮可以隐约感知内部空间。南楼和北楼连接处做了一个封闭退台组成的中庭，它既是交通空间，也是休息空间。接近三角形的中庭布置，化东西向为南北向，尽量减少西面面积，增加南向采光面，让北楼有更长的南向采光空间。西面不利于阅读的空间作为庭院和行走通道等休闲和交通空间。

此外，为了打造全民参与、开放的图书馆，在不破坏原有绿化和交通便利的原则下，该项目还设计了一个改造广场。广场的形状是在现有树木的位置和大小以及从主道路到图书馆的各种功能流线综合分析上得来的。树的位置就是广场上洞口的位置，广场下面可以通过洞口来采光，广场上面可以用这些洞口作为休息座椅。树木、休息座椅、台阶、行走通道构成了一个空间丰富、生态实用、开放互动的市民广场。该改造广场，有利于增加图书馆与道路的联系，引导道路和公园广场人流进入馆内。市民可以在广场上休息、交流、互动，图书馆广场不仅作为交通疏散空间，而且将成为甘肃人民日常生活服务的市民广场。对于图书馆本身，也可以利用这个广场下面的空间作为临时停车场，解决地面临时停车问题，且也不会影响整个馆区的交通和视觉。

Pure Façade Triangle Atrium Modified Square

Project Overview

April 2013, Beijing Shushe Architecture participated in Gansu Provincial Library public bidding to the identity of the first bid. The program is continuous the consistent style of Shushe Architecture to analysis the base and functional requirements, putting forward to the problem need to be solved. Therefore, to solve the problem step by step is our design process and the answer is our final construction.

In order to balance with the old library, the new library in the design process as much as possible reduced the massing. Simple but elegant appearance hides rich and flexible interior space, revealing the infinite sunshine and green atmosphere.

Building Design

The building may look simple, but complicated practical space is hidden inside the building, which can be vaguely perceived the interior space through pure façade. The connection between the south and north building is made an atrium composed by close back sets, and it is both the transportation space and the rest space. Nearly triangular atrium layout changes the east-west into the north-south to minimize the west area and increase the south lighting, so the north building has a longer south lighting space. The west space not conducive to reading will be as patio and walkways and other recreational and transportation space.

Furthermore, in order to build public participation and open library, without destroying the original principles of green and convenient transportation, the project has also designed a renovation square. The shape of the square is made come from a comprehensive analysis of existing trees' location and size and from the main road to the various functional flow lines of the library. The location of the trees is on the square holes' location, and the square underside can be light through the holes, as well as the square upside can use these holes as seats to rest. Trees, rest seats, stairs and walkways constitute a public square with rich space, practical ecology and open interaction. The modified square is conducive to increasing connection between the library and roads, guiding the flow of people from roads and park into the library. The public can rest, communicate and interact in the square. The library square is not only as the space for transportation evacuation, but will become the public square for Gansu People's daily life services. For the library itself, you can also use this space under the square as a temporary parking to solve the temporary parking problem on the ground, which will not affect the traffic and the vision of the whole library area.

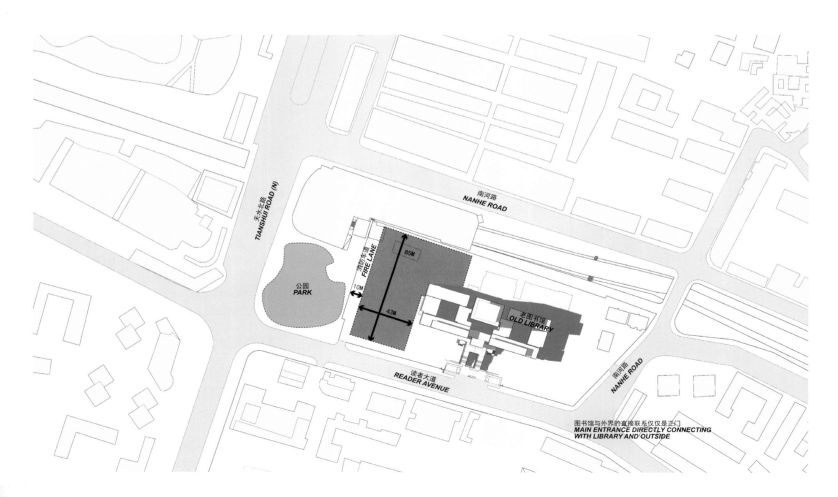

MAIN ENTRANCE DIRECTLY CONNECTING WITH LIBRARY AND OUTSIDE

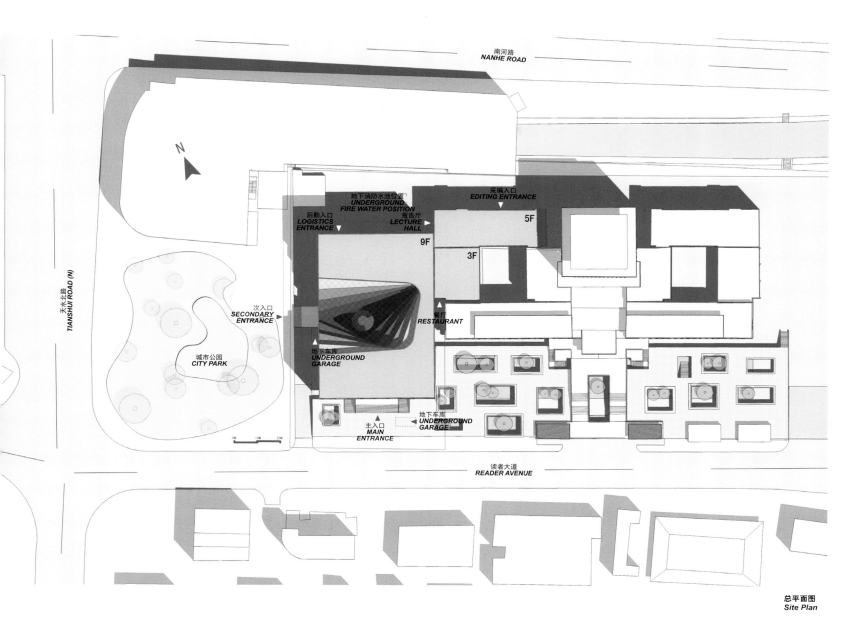

总平面图
Site Plan

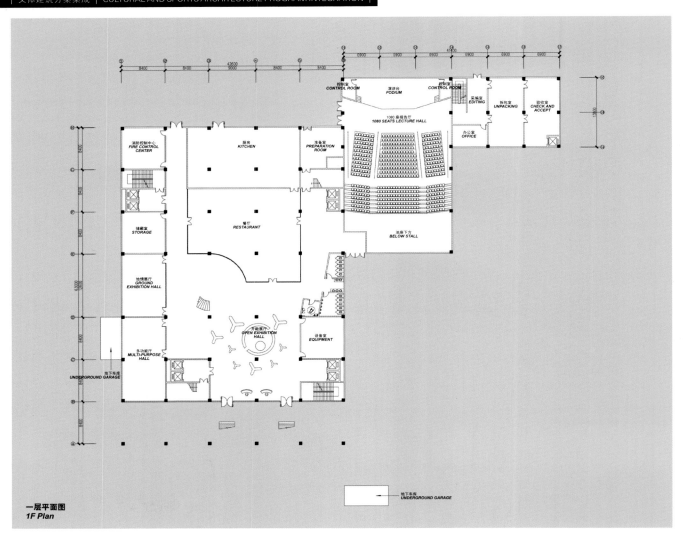

一层平面图
1F Plan

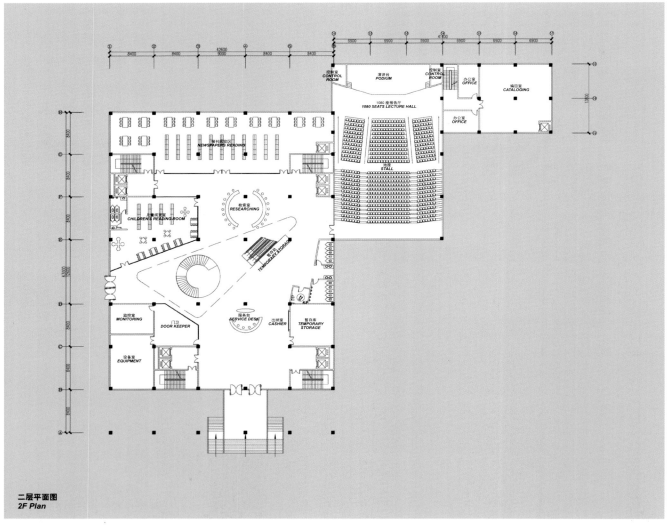

二层平面图
2F Plan

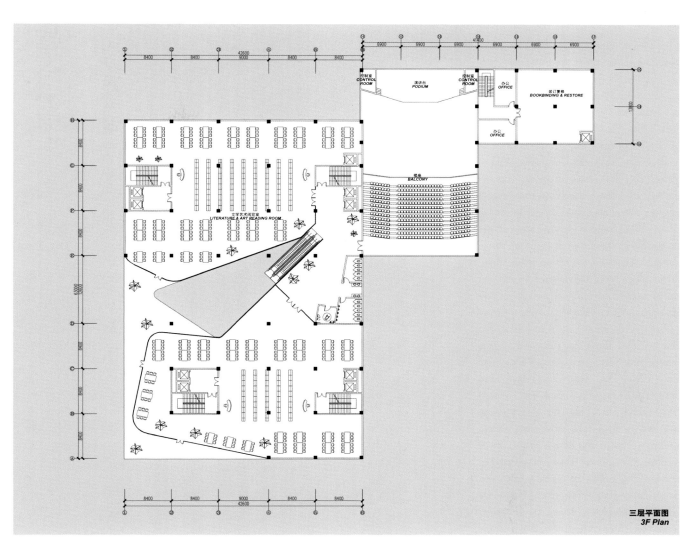

三层平面图
3F Plan

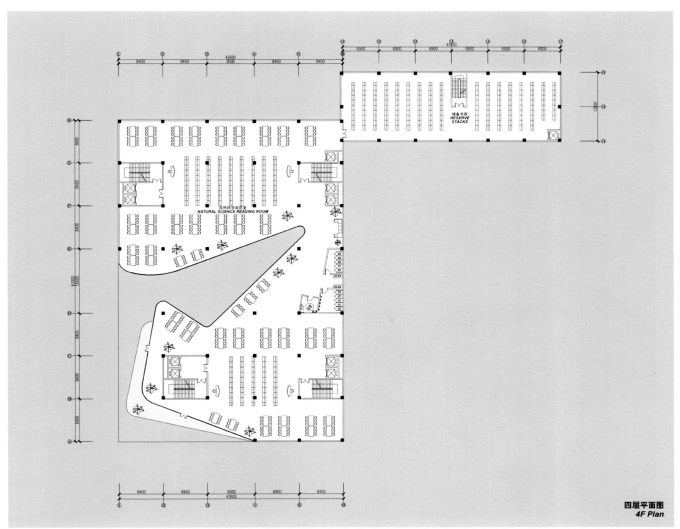

四层平面图
4F Plan

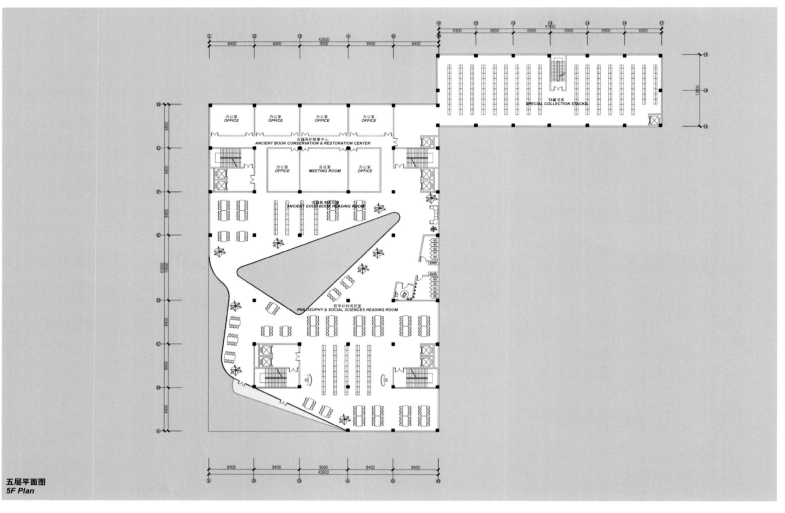

五层平面图
5F Plan

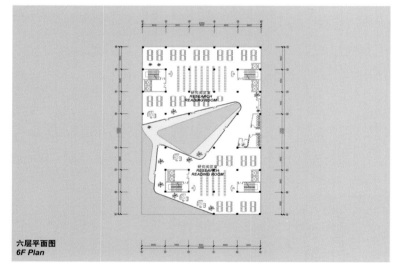

六层平面图
6F Plan

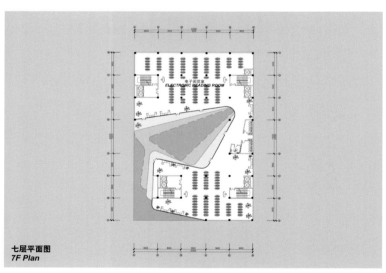

七层平面图
7F Plan

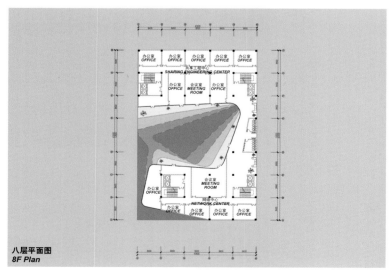

八层平面图
8F Plan

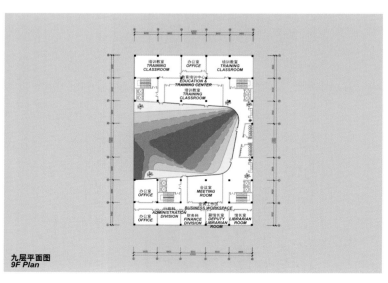

九层平面图
9F Plan

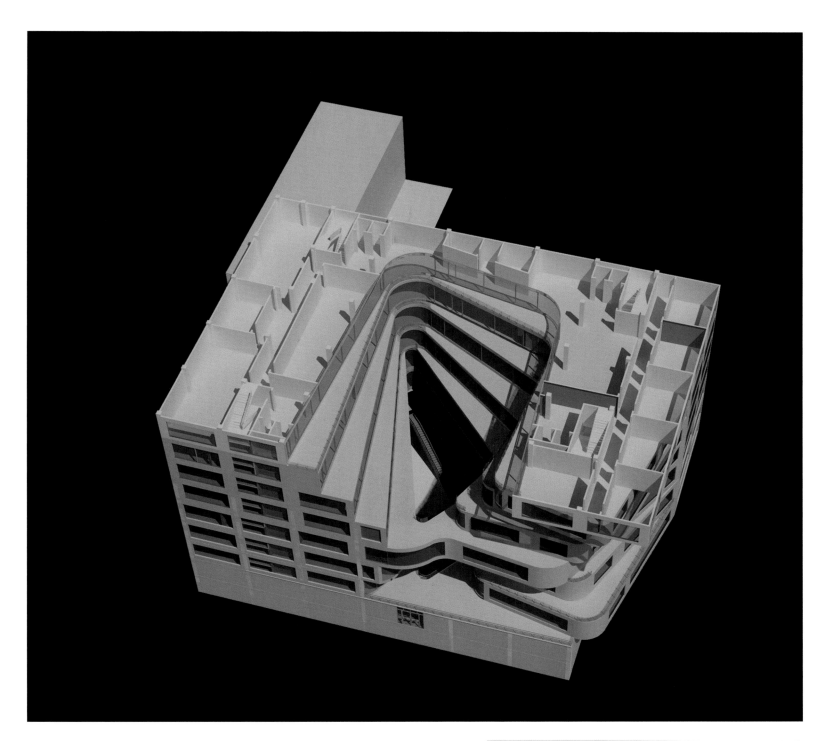

西立面图
West Elevation

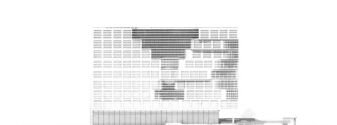

南立面图
South Elevation

1-1 剖面
1-1 Section

2-2 剖面
2-2 Section

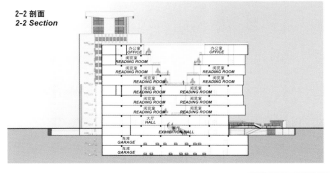

美国西雅图中央图书馆
Seattle Central Library

设计单位：大都会建筑事务所
委托方：西雅图公共图书馆
项目地址：美国西雅图市
项目面积：38 300m²

Designed by: OMA
Client: The Seattle Public Library
Location: Seattle, USA
Area: 38,300m²

旗舰馆
蜘蛛网外观
浮动平台

项目概况

西雅图中央图书馆是美国西雅图公共图书馆系统的旗舰馆。它位于市中心，是一幢由11层、高56m的玻璃和钢铁组成的建筑。这个巨大的图书馆容纳了约145万册书籍和其他资料，其中包括超过400台电脑向公众开放，另设有一个地下公共停车场。图书馆有一个独特且突出的外观——构成若干分立"浮动平台"，就像置身于一个大的蜘蛛网。

建筑设计

西雅图中央图书馆把图书馆重新定义为一个不再只专注于书本的机构，还可作为一个信息存储源，在那里一切有效的新旧媒体形式都可以清晰地呈现。

中央图书馆网络使得传统的以收藏图书为主的图书馆模式发生了变化，交流无限制，图书馆的所有空间也有都了交流的特质。灵活布置也使图书馆打破了传统的单一大空间。

项目设计了五个平台模式：办公、书籍及相关资料、交互交流区、商业区、公园地带。这五个平台从上到下依次排布，最终形成一个综合体。每个平台是从结构上定义，并配备强大、专用性能的程序化集群。它们的大小、弹性、循环、调色板、结构和MEP都是不同的。平台之间的空间就像交易区，不同的平台交互界面被组织起来，这些空间或用于工作，或用于交流，或用于阅读，有一种特别的空间交融感觉。建筑形体随着平台面积和位置的变化形成新奇的多角结构，有新现代主义的某些特征。

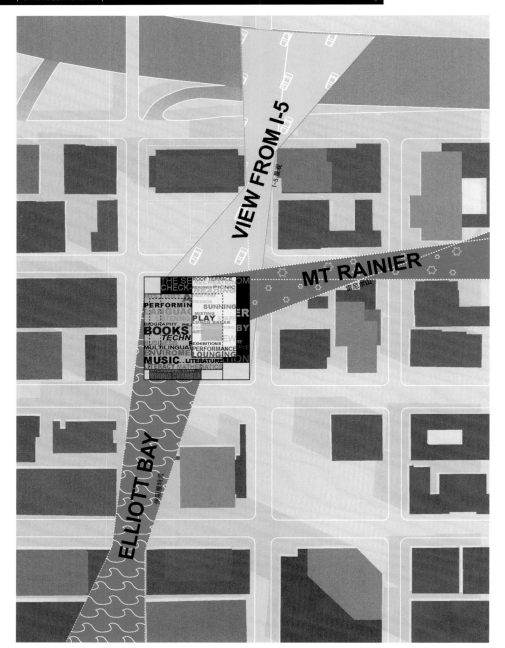

Flagship Library Cobweb Appearance Floating Platform

Project Overview

Seattle Central Library is the flagship library for Seattle Public Library system. It is located in the city center, and composed by 11 layers and glass and steel with 56m height. This huge library accommodates about 1.45 million books and other details, including more than 400 computers open to the public, and an underground public parking lot. The library has a unique and outstanding appearance to constitute several discrete "floating platforms", just like placing itself in a big cobweb.

Building Design

The Seattle Central Library redefines the library as an institution no longer exclusively dedicated to the book, but as an information store where all potent forms of media—new and old—are presented equally and legibly.

Central Library network has made a change for the traditional library with the pattern of collection books for the basis so that communication is unlimited, and all the space in the library also has communicating qualities. The flexible layout makes the library break the traditional and single large space.

The project is designed 5 platforms: offices, books and related materials, interactive exchanges area, business area and park zone. This five platforms are arranged in order from top to bottom, ultimately forming a complex. Each platform is a programmatic cluster that is architecturally defined and equipped for maximum, dedicated performance. Because each platform is designed for a unique purpose, their size, flexibility, circulation, palette, structure, and MEP vary. The spaces in between the platforms function as trading floors where librarians inform and stimulate, where the interface between the different platforms is organized—spaces for work, interaction, and play.

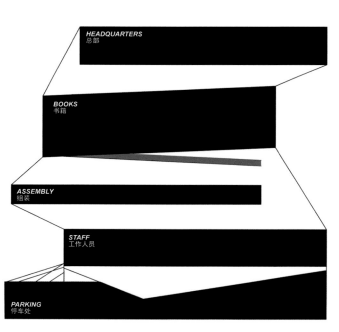

| 26–27 |

| 文体建筑方案集成 | CULTURAL AND SPORTS ARCHITECTURE PROGRAM INTEGRATION |

斯洛文尼亚卢布尔雅那国家大学图书馆 II
NUKII, National and University Library II Ljubljana

设计单位：NL Architects
委托方：斯洛文尼亚高等教育、科技部
项目地址：斯洛维尼亚卢布尔雅那
项目面积：4 220m²

Designed by: NL Architects
Client: Ministry of Higher Education, Science and Technology, Republika Slovenija
Location: Ljubljiana, Slovenia
Building Area: 4,220m²

"微广场"
步行区
楼梯椅子

项目概况

卢布尔雅那,一个真正独特的地方。在某种程度上,可以直接从卡尔维诺到达这个看不见的城市。而在前罗马镇Emona的遗迹之上建立这样一个新的国家大学图书馆,这不是一个反直觉的叠加,而是一个难得的、有趣的机会。

该建筑从上空俯视呈现在眼中的是一个十字型的高层建筑,但是它高度的增加依靠的是横向的拓展和纵向的加深,它与周围的建筑物保持着一定的距离,给人一种舒适的感觉。

建筑设计

在建筑设计上,十字型的交叉处,产生了一个"微广场",它的创建是根据设计给定的站点边界挖掘的,这个干净的"剪切"有助于阐明给建筑的"随意性"。为了留下原有的废墟,该建筑被抬升。一个"拱形"的空间在正上方的前交叉处应运而生,因此该建筑被类似于立交桥的"桥梁"支撑顶起,形成该建筑的另一道风景。也由于此原因,坡型的广场减少了汽车的流通,形成一个步行区,一个连接两个图书馆的广场也就应运而生。

建筑内部的公众地板是由同等宽度的楼梯相连,楼梯兼作椅子,它们形成了一个连续的但有区别的内部空间,以满足使用者放松、阅读、学习、创造、沟通和思考。公共职能部门被设置得相对靠近地面,可直接通往街道。壮丽的楼梯通向入口大厅、信息台、夜间阅读室和一个小商店。另一个公共空间位于建筑顶层,它涵盖了咖啡馆、餐厅、露天吧台等,在这里可以看到整个城市的壮丽景色,能够真正的吸引游客。

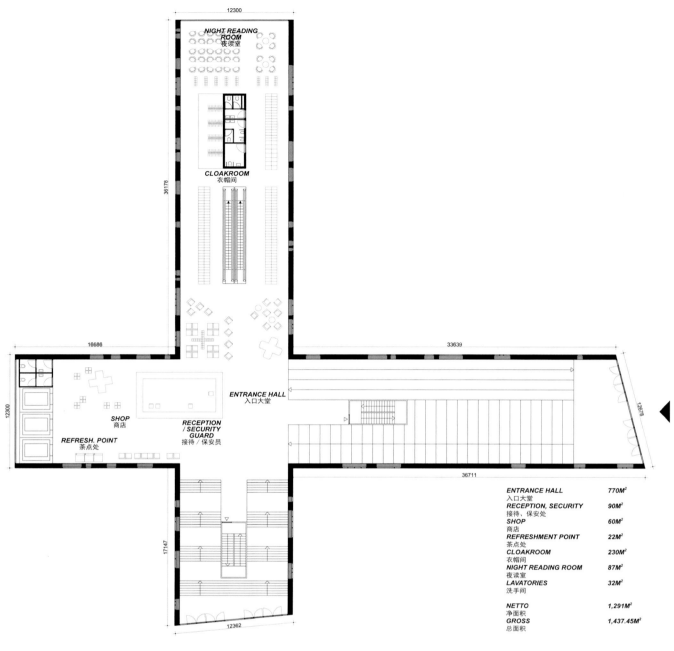

ENTRANCE HALL 入口大堂	770M²
RECEPTION, SECURITY 接待、保安处	90M²
SHOP 商店	60M²
REFRESHMENT POINT 茶点处	22M²
CLOAKROOM 衣帽间	230M²
NIGHT READING ROOM 夜读室	87M²
LAVATORIES 洗手间	32M²
NETTO 净面积	1,291M²
GROSS 总面积	1,437.45M²

Entrance + Public 1:200
入口 + 公共 1:200

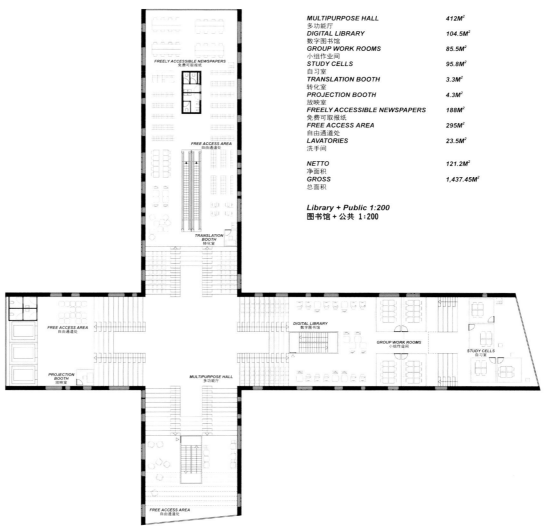

MULTIPURPOSE HALL	*412M²*
多功能厅	
DIGITAL LIBRARY	*104.5M²*
数字图书馆	
GROUP WORK ROOMS	*85.5M²*
小组作业间	
STUDY CELLS	*95.8M²*
自习室	
TRANSLATION BOOTH	*3.3M²*
转化室	
PROJECTION BOOTH	*4.3M²*
放映室	
FREELY ACCESSIBLE NEWSPAPERS	*188M²*
免费可取报纸	
FREE ACCESS AREA	*295M²*
自由通道处	
LAVATORIES	*23.5M²*
洗手间	
NETTO	*121.2M²*
净面积	
GROSS	*1,437.45M²*
总面积	

Library + Public 1:200
图书馆 + 公共 1:200

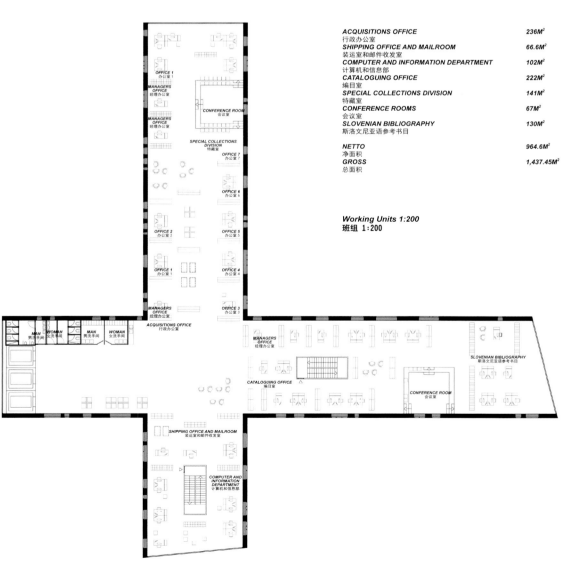

ACQUISITIONS OFFICE	**236M²**
行政办公室	
SHIPPING OFFICE AND MAILROOM	**66.6M²**
装运室和邮件收发室	
COMPUTER AND INFORMATION DEPARTMENT	**102M²**
计算机和信息部	
CATALOGUING OFFICE	**222M²**
编目室	
SPECIAL COLLECTIONS DIVISION	**141M²**
特藏室	
CONFERENCE ROOMS	**67M²**
会议室	
SLOVENIAN BIBLIOGRAPHY	**130M²**
斯洛文尼亚语参考书目	
NETTO	**964.6M²**
净面积	
GROSS	**1,437.45M²**
总面积	

Working Units 1:200
班组 1:200

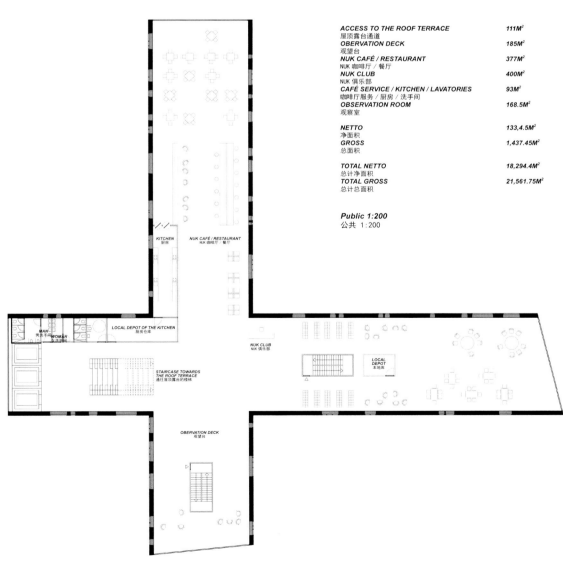

ACCESS TO THE ROOF TERRACE 屋顶露台通道	111M²
OBERVATION DECK 观望台	185M²
NUK CAFÉ / RESTAURANT NUK 咖啡厅 / 餐厅	377M²
NUK CLUB NUK 俱乐部	400M²
CAFÉ SERVICE / KITCHEN / LAVATORIES 咖啡厅服务 / 厨房 / 洗手间	93M²
OBSERVATION ROOM 观察室	168.5M²
NETTO 净面积	133,4.5M²
GROSS 总面积	1,437.45M²
TOTAL NETTO 总计净面积	18,294.4M²
TOTAL GROSS 总计总面积	21,561.75M²

Public 1:200
公共 1:200

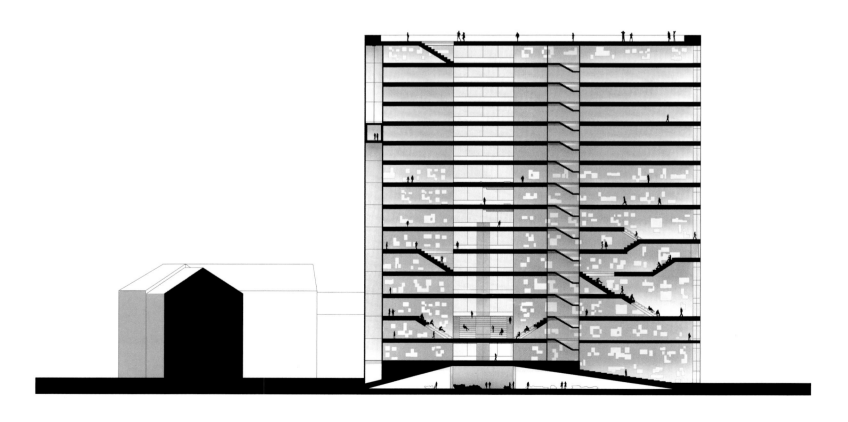

"Micro Square"
Pedestrian Zone
Stair-like Chair

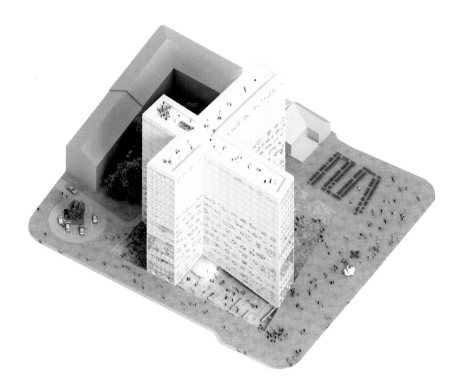

Project Overview

Ljubljana has formulated an intriguing opportunity by aiming to build the new library on top of the remains of the former Roman town Emona. The location is beautiful: a convenient position in the city and next door to the exquisite existing library by Plečnik. But how to construct a building on top of such valuable remains?

The further constraints are demanding as well: an enormous program has to be placed within tight limitations. It seems that eliminating one of the restrictions is the only way out. What if we skip the maximum height? Would it be possible to be tall and modest at the same time?

Building Design

Starting point of the proposal is the ancient Roman crossing. A small fragment of the public space of former Emona, framed by the site boundary, has been extruded to accommodate the entire program. What used to be void now becomes mass and vice versa: a Rachel Whiteread-like "urban inversion". The former street becomes building, but a building with the character of public space…

The public floors are connected by stairs over the full width. The stairs double as chairs. They form a continuous but differentiated interior "landscape", to meet, relax, read, study, create, communicate, to think. The floors are articulated in the facade, an X-Ray of the internal organization.

Since NUKII is cross-shaped, the increased height is compensated by relatively modest width and depth, the building keeps a distance: instead of inserting a bulky mass, a slender tower emerges. The solid cross produces "micro squares" in its armpits, creating surprisingly precise relationships with the immediate surroundings.

In order to leave the ruins untouched the building is lifted. An "arched" space comes into being right above the former crossing. The building as such forms an intricate "bridge" with four landings: strangely familiar…

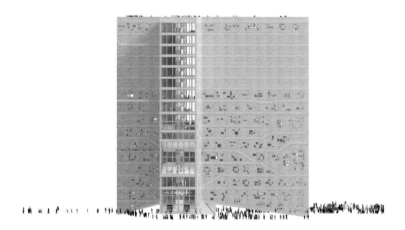
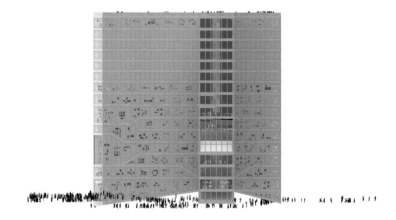
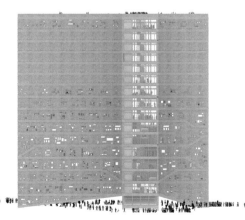
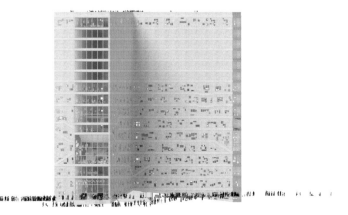

芬兰赫尔辛基中央图书馆
Helsinki Central Library

设计单位： Urban Office Architecture
项目地址： 芬兰赫尔辛基
建筑面积： 16 000m²

Designed by: Urban Office Architecture
Location: Helsinki, Finland
Building Area: 16,000m²

大脑概念
功能分区
中央花园

项目概况

新的赫尔辛基中央图书馆是作为周边一个自然公园的扩展。因为该公园东西两侧的城市街区"围困"，发展受限，设计师们试图通过建造一个新的图书馆，使这块地面重新发挥功效——因为园区的水平力将建筑划分成东西两翼，使其变身成一个室内阅读花园，衔接其中，共同发挥作用。

建筑设计

新赫尔辛基中央图书馆的概念是基于人类大脑设计的，就像人的大脑，各方面管理不同的功能，图书馆也设置了不同的区域和功能，共同发挥作用。

类似大脑硬脑膜，图书馆外部携带的信息都会向用户和它周围的环境反应。图书馆内部空间是多种多样的、复杂的和令人兴奋的。围绕中心花园的大堂组织，可以设计成一个密闭的冬季花园，图书馆的两个主翼作为中央花园核心，类似于围绕树木创建的美丽而宁静的地区。通过广场的坡道可直接到达图书馆的屋顶。顶楼餐厅的爵士酒吧和桑拿浴的两个大窗房间，坐北朝南面向亚历山大和芬兰议会，并分别与国会众议院和亚历山大图书馆对齐。这些景观激活了对图书馆过去、现在和未来的连接，并欢迎游客以一个前瞻性的思维层面重新体验。

此外，该建筑的设计是使用建筑围护结构作为主要的热和电发电机和调节器，并使用集中供热、制冷和网格电力作为高效的备份。

| 文体建筑方案集成 | CULTURAL AND SPORTS ARCHITECTURE PROGRAM INTEGRATION |

Brain Concept
Functional Partition
Central Garden

Project Overview

The new Helsinki Central Library serves as an extension of the natural park surrounding it. Because the east and west sides of the city block became "trapped", the designers tried to build a new library to make the piece of the ground to be effective again—the horizontal force of the park divides the building into two wings, North and East and transforms itself into an interior reading garden. They are converged, which together play a role.

Building Design

The concept for the New Central Library of Helsinki is based on the Human Brain to design, and like the Human Brain, the library is organized in very different zones and functions to play a role together.

Similar to the Dura Mater of the brain, the Library envelope carries information both to its users and as a response to the environment around it. Inside the Library spaces are diverse, complex and exciting. Organized around the central garden lobby, which can be designed as an enclosable winter garden, the two main wings of the library act as extension branches from the central garden core, similarly to the beautiful and serene areas created around the trees. It is accessible on the library's roof via a ramp of the square.

The two large window-rooms of the top floor restaurant, Jazz bar and Sauna face south towards Alexandria and the Finnish Parliament, aligning respectively with the Parliament and House, and the Alexandria Library. These views activate a connection to the past, present and future of the library and welcome visitors to re-experience it in a forward thinking dimension.

This building is designed to use the building envelope as the primary thermal and electrical generator and regulator and uses the district heating, cooling and grid electricity as efficient back-up.

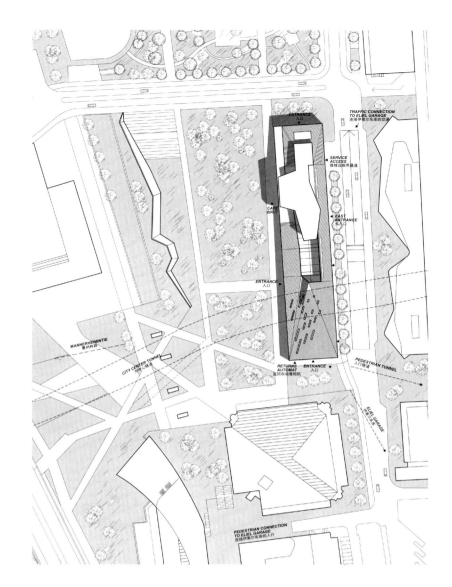

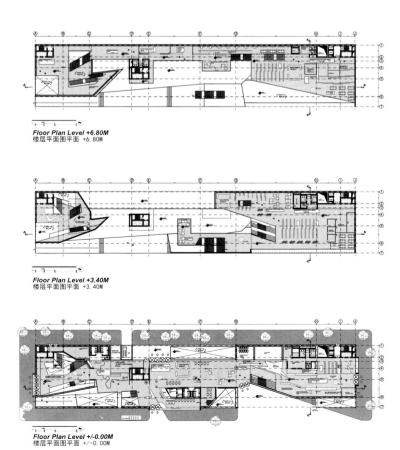

Floor Plan Level +6.80M
楼层平面图平面 +6.80M

Floor Plan Level +3.40M
楼层平面图平面 +3.40M

Floor Plan Level +/-0.00M
楼层平面图平面 +/-0.00M

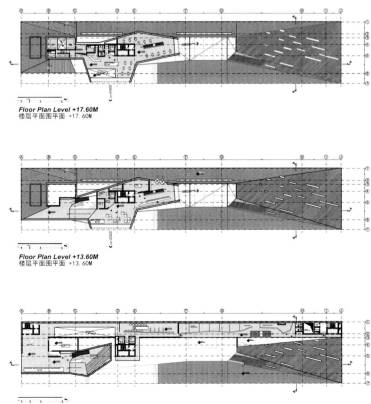

Floor Plan Level +17.60M
楼层平面图平面 +17.60M

Floor Plan Level +13.60M
楼层平面图平面 +13.60M

Floor Plan Level +10.20M
楼层平面图平面 +10.20M

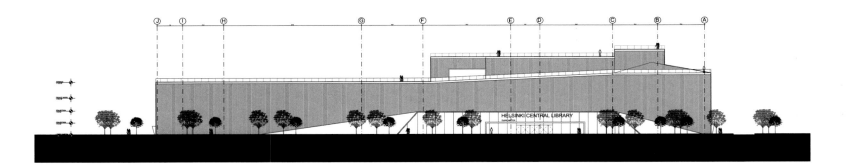

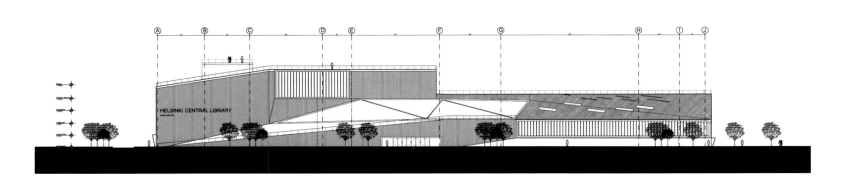

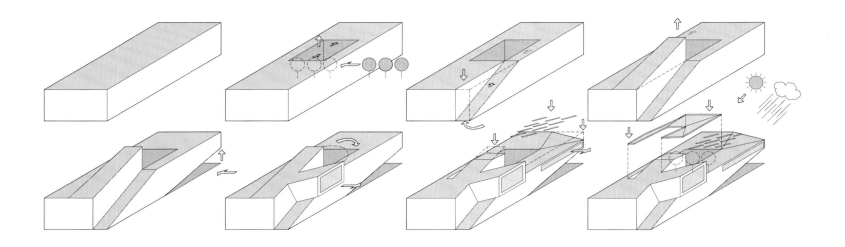

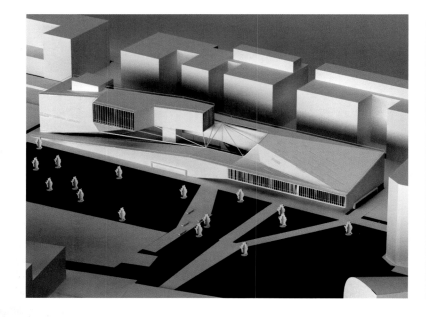
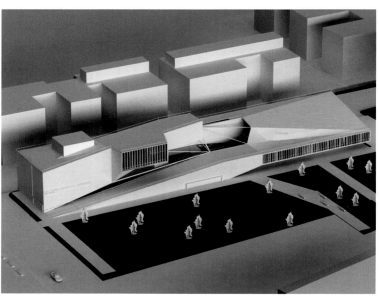

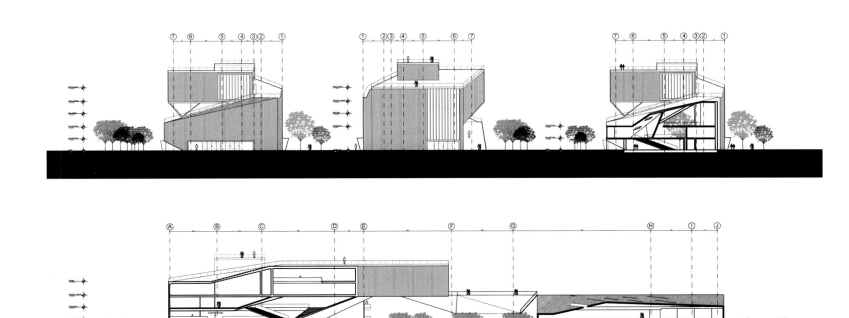

韩国大邱 Gosan 公共图书馆（方案四）
Daegu Gosan Public Library

设计单位：筱崎弘之建筑设计事务所
委托方：大邱广域市寿城区政府
项目地址：韩国大邱市
占地面积：2 080m²
建筑面积：3 231.96m²

Designed by: Hiroyuki Shinozaki Architects
Client: Daegu Metropolitan City Suseong-gu Office
Location: Daegu, South Korea
Site Area: 2,080m²
Building Area: 3,231.96m²

**自然时尚
低层建筑
优化阅读**

项目概况

本案将图书馆作为当地社区的文化和社会空间。项目的重点是获得构成对邻里的日常生活的重要组成部分的功能，吸引力和舒适的设施。项目周边城市网格由中小型住宅建筑组成。新图书馆计划的规模与环境有直接的关系。

这是一个全新的图书馆类型，它结合了自然时尚的各种要素，提供了必要的灵活性，以顺利地适应未来的变化。

建筑设计

就像书籍根据各自的流派被放置在不同的货架上，不同的空间在图书馆的布置根据使用目的以全面的方式被划分。为了使优化构建的房间顺利使用，地板跨度将允许和图书馆地面一样宽。就像一个灵活的书架，地板空间作为一个平台容纳各种设施。一楼被认为是作为一个文化空间，第二层和第三层将分别用作一般的收集空间和儿童空间。地下楼层提供机房和预留室。为不干扰沿图书馆北部延伸的绿化带，图书馆将以一个低矮的方式来构建，此外，在楼层中为了保持与其他建筑的和谐将提供一个阅读空间，让人联想到一个公园。

为了提升公园以及巴士站方便的通道，北侧和东侧将作为人流入口。同样，南侧和西侧将保留给乘车到来的游客。北侧和东侧用作行人通道，使到达图书馆像漫步公园一样。

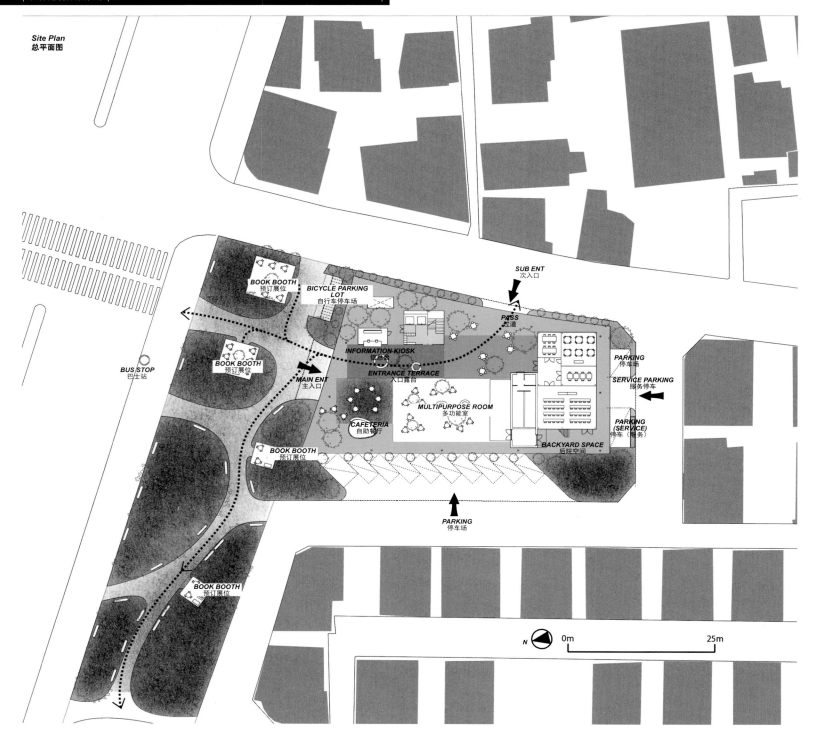

Site Plan
总平面图

Natural Fashion Low-rise Building Optimization of Reading

Project Overview

The case made the Library as a cultural and social space for the local communities. The focus of the project is to get to the important part of the daily life of the neighborhood function composition, attractive and comfortable facilities. The project grid around the city is composed by the medium-sized small residential buildings. The size of the new library program has a direct relationship with the environment.

This will be a new type of library, and it is combined various elements in a natural fashion and provides the degree of flexibility necessary to adapt smoothly to future changes.

Building Design

Just like books are arranged into shelves according to their respective genre, the different spaces furnished in the library are divided in a comprehensive manner according to purpose of use. In order for rooms of sizes optimal for their purpose of use to be constructed smoothly, the floor span will be as wide as allowed by the library grounds. Just like a flexible bookshelf, the floor space serves as a platform housing a range of facilities. The first floor is thought of as a Culture Space; while the second and third floor will serve as General Collection Space and Children Space respectively. The underground floors provide a Machine Room and a Preservation Room. As not to interfere with the green belt stretching out along the north of the library grounds, the library will be constructed in a low-rise fashion. Moreover, in order to preserve harmony with other buildings in the floors will provide a reading space reminiscent of a park.

To promote easy access from the park as well as the bus stop, the north and east sides will serve as entrances for people. Similarly, the south and west sides will be reserved for visitors arriving by car. With the north and east sides functioning as pedestrian passages, accessing the library will be similar to taking a stroll through the park.

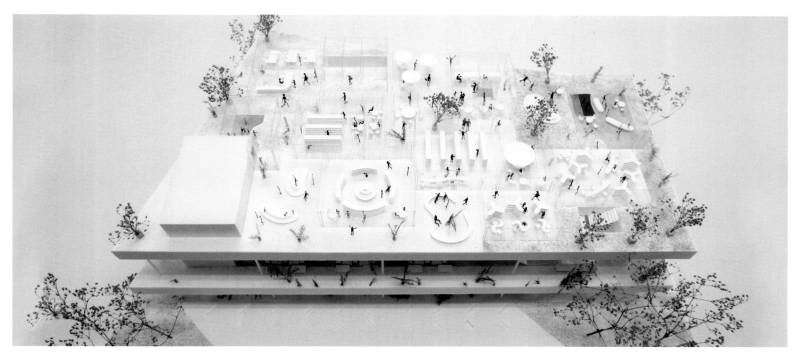

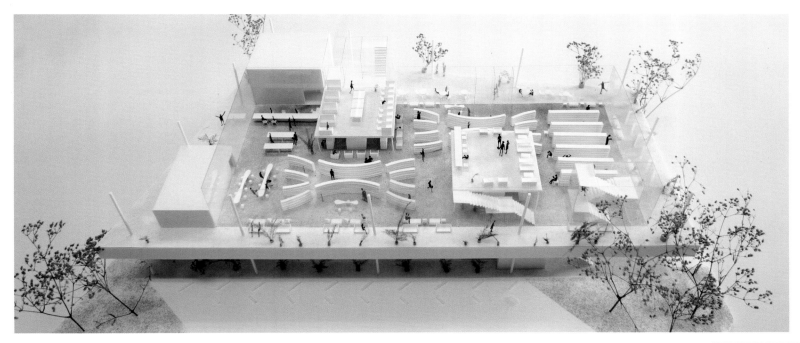

| 文体建筑方案集成 | CULTURAL AND SPORTS ARCHITECTURE PROGRAM INTEGRATION |

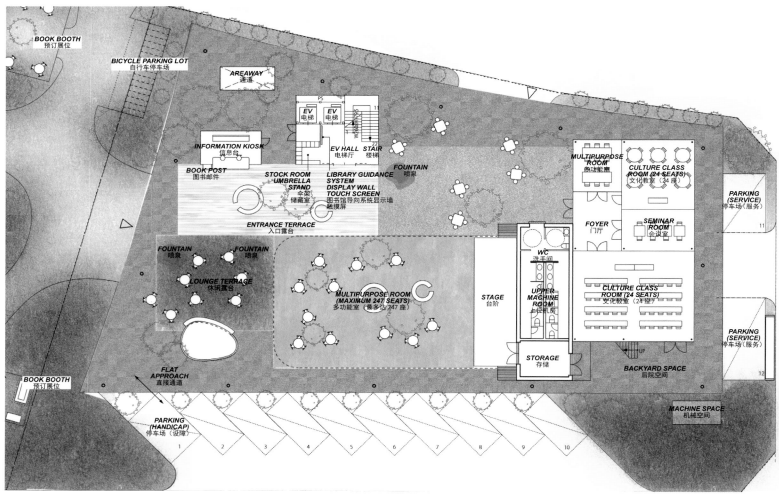

1F Plan
一层平面图

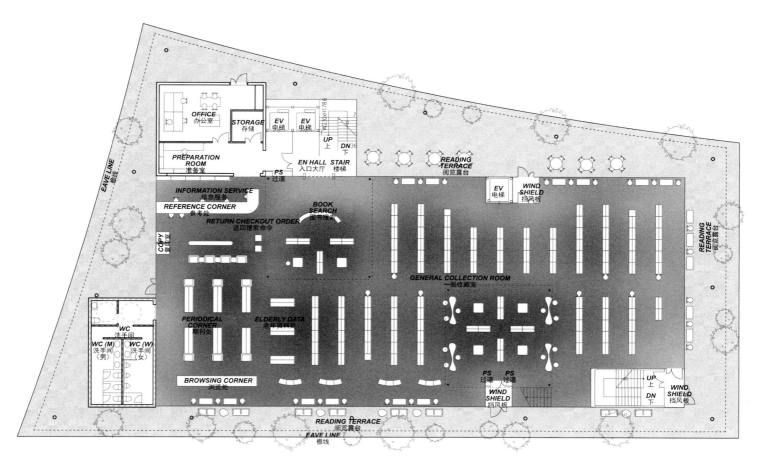

2F Plan
二层平面图

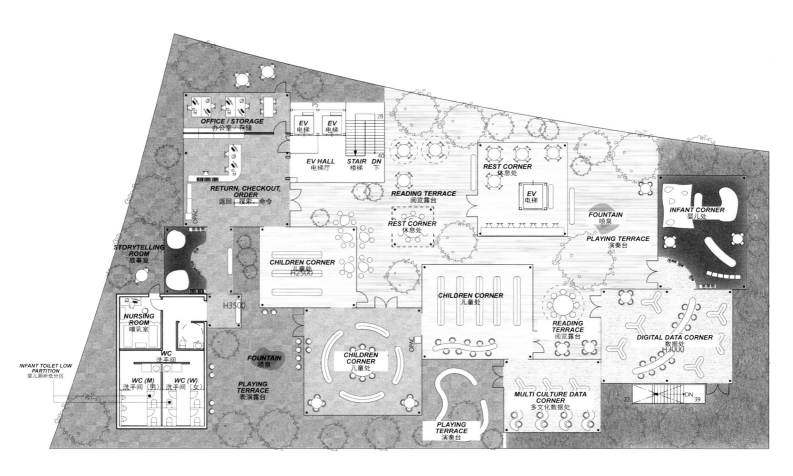

3F Plan
三层平面图

| 文体建筑方案集成 | CULTURAL AND SPORTS ARCHITECTURE PROGRAM INTEGRATION |

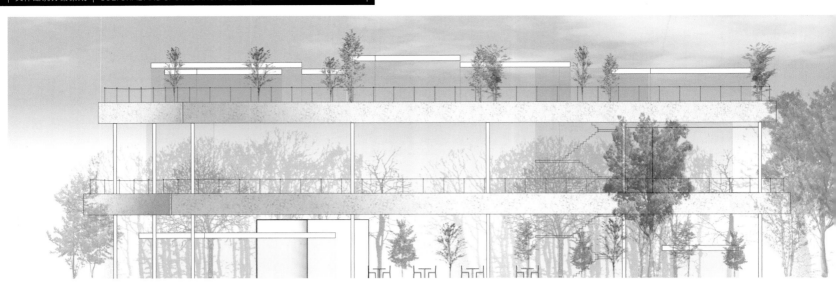

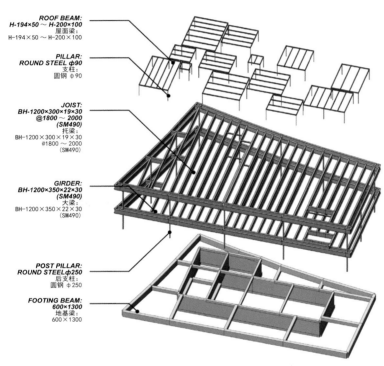

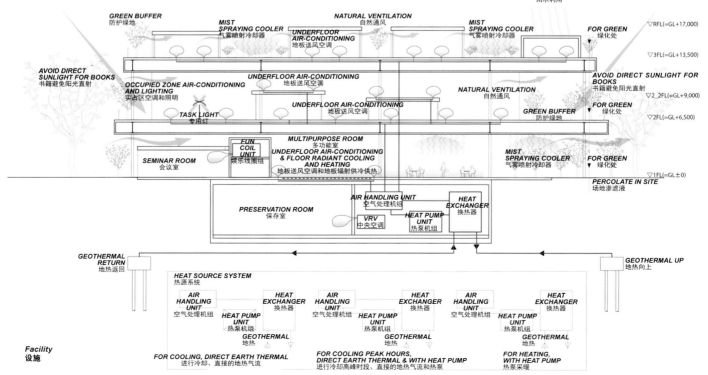

Facility
设施

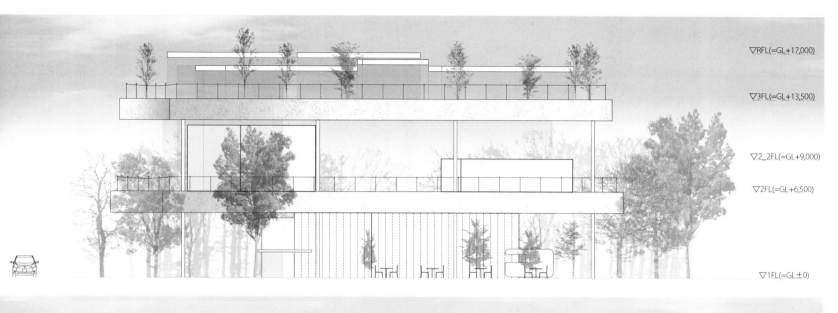
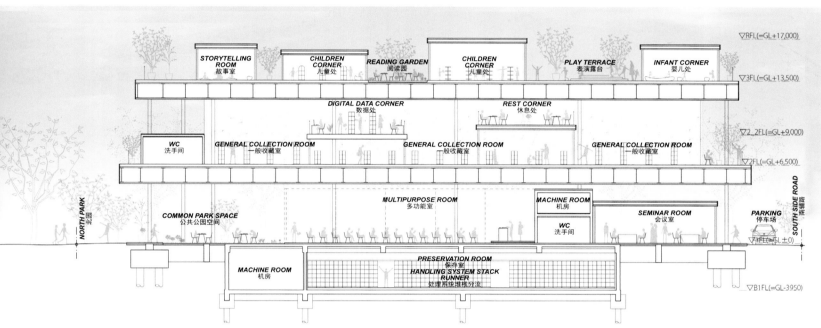
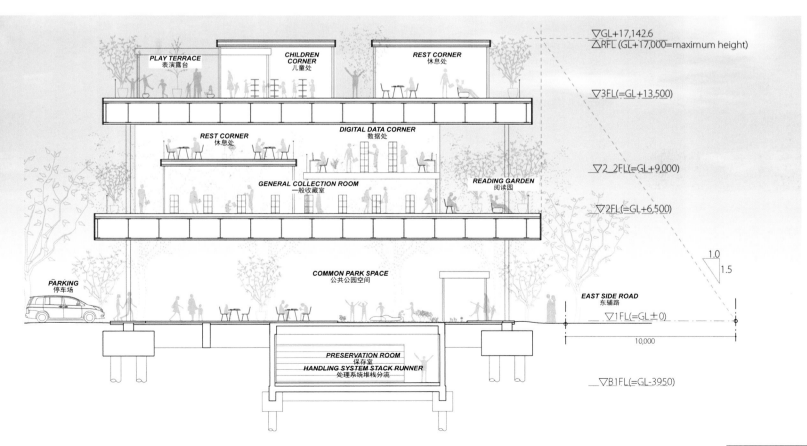

韩国大邱 Gosan 公共图书馆（方案五）
Daegu Gosan Public Library

设计单位：MSB Arquitectos	*Designed by: MSB Arquitectos*
委托方：大邱广域市寿城区政府	*Client: Daegu Metropolitan City Suseong-gu Office*
项目地址：韩国大邱市	*Location: Daegu, South Korea*
项目面积：2 897.60m²	*Area: 2,897.60m²*

绿色建筑
文化交互

项目概况

项目位于韩国大邱市。保持绿色公共空间是设计师考虑的重点,每一层均设计有露台,空中花园加强了内部和外部空间的联系,提高舒适性,更加开放,便于人与人之间的交流和互动。设计方案将图书馆的灵活性、多功能性,以及适应各类社会活动的开放性——展现出来。

建筑设计

本案依据给定场地后的解释和假设,明确两个强大的思想作为指导:一个是争取对现有园林的维护,因为这个空间作为会议、交互和人与人之间的关系等多学科知识的空间,有运作良好的可能性,相同的功能是为了保持和强调拟议的新图书馆;而且这个新的空间的物理发展,并不是作为一个图书馆,而是作为一个"知识剧院"、多学科的空间,灵活而牢固地向社区呈现和开放,增强社会的人际关系。

大楼的部署占用现有花园的整个区域,满足所需的空间和创造公共人行道和停车位。该体量代表建筑的第二阶段,在整个项目开发中为四层楼。第三阶段是移除这个大体量内部的基体部分,形成悬浮的花园,增强内外部空间之间的关系。在每个相互作用的楼层进行连接并创建不同的庭院,以提高舒适度。开放壁龛的创建,将促使其用户探索建筑和结识新朋友。最后,第四个阶段是在这些不同庭院重生的花园中进行维护,创造一个"绿色"建筑。

| 文体建筑方案集成 | CULTURAL AND SPORTS ARCHITECTURE PROGRAM INTEGRATION |

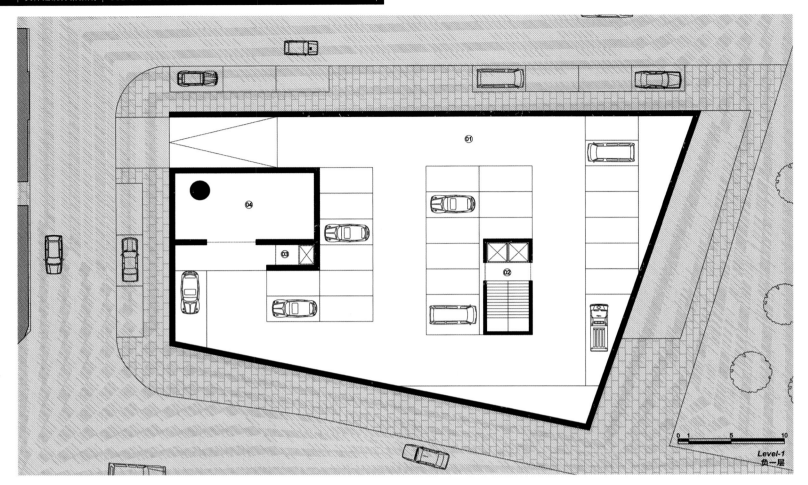

Level-1
负一层

01. GARAGE	**08. RECEPTION**	**15. BIKES PARKING**	**22. CHILDREN'S TOILETS**	**29. INFORMATION DESK**	**36. PRESERVATION ROOM**	**43. DRESSING ROOM**
01. 车库	08. 接待处	15. 自行车停放处	22. 儿童洗手间	29. 服务台	36. 保存室	43. 更衣室
02. MAIN ACCESS	**09. PERIODICALS / COMPUTERS**	**16. MULTIPURPOSE RECEPTION**	**23. NURSERY ROOM**	**30. CIRCULATION DESK**	**37. CULTURE CLASSROOM**	**44. ORDER DEPARTMENT**
02. 主要通道	09. 期刊 / 电脑室	16. 多目标接待	23. 育婴室	30. 图书借阅台	37. 文化教室	44. 签发订单部门办公室
03. STAFF ACCESS	**10. LOUNGE**	**17. MULTIPURPOSE LOBBY**	**24. INFANT CORNER**	**31. DIGITAL DATA CORNER**	**38. SEMINAR ROOM**	**45. PREPARATING ROOM**
03. 员工通道	10. 休息室	17. 多用途大厅	24. 婴儿处	31. 数字数据处	38. 会议室	45. 准备室
04. MACHINE ROOM	**11. STOCK ROOM**	**18. MULTIPURPOSE ROOM**	**25. STORY TELLING**	**32. COLLECTIONS**	**39. LANGUAGE LAB**	**46. STAFF ROOM**
04. 机房	11. 储藏室	18. 多功能室	25. 讲故事室	32. 收发处	39. 语音室	46. 教研室
05. 24H LOBBY	**12. STAFF ENTRY**	**19. MULTIPURPOSE STORAGE**	**26. DATA ROOM**	**33. BALCONY**	**40. OFFICE**	**47. VOLUNTEER ROOM**
05. 24 小时大堂	12. 员工入口	19. 多用途储存室	26. 资料室	33. 阳台	40. 办公室	47. 志愿者室
06. BOOK RETURN	**13. TOILETS**	**20. PARENTS ROOM**	**27. MULTIPURPOSE ROOM**	**34. TOILETS**	**41. MEETING ROOM**	
06. 书籍归还处	13. 洗手间	20. 父母房间	27. 多功能室	34. 洗手间	41. 会议室	
07. MAIN LOBBY	**14. EXTERIOR GARDEN**	**21. TOILETS**	**28. CHILDREN CORNER**	**35. READING SPACE**	**42. DIRECTOR OFFICE**	
07. 主大厅	14. 外园	21. 洗手间	28. 儿童处	35. 阅读空间	42. 主任办公室	

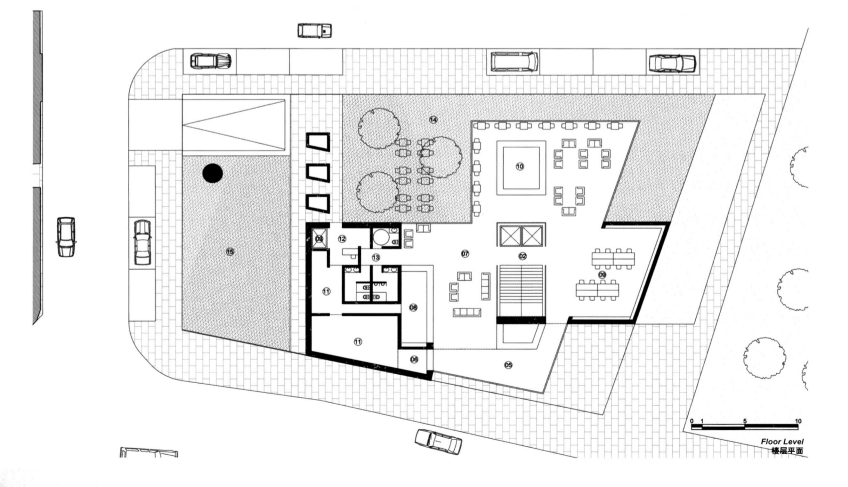

Floor Level
楼层平面

Level 1
一层

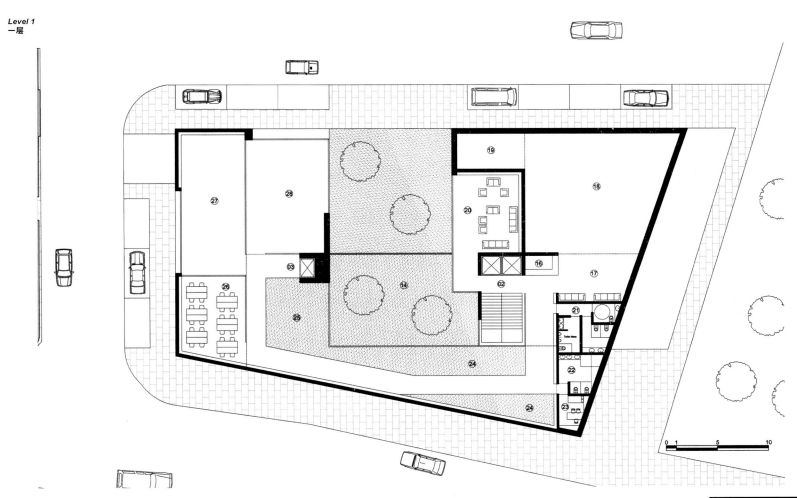

Level 2
二层

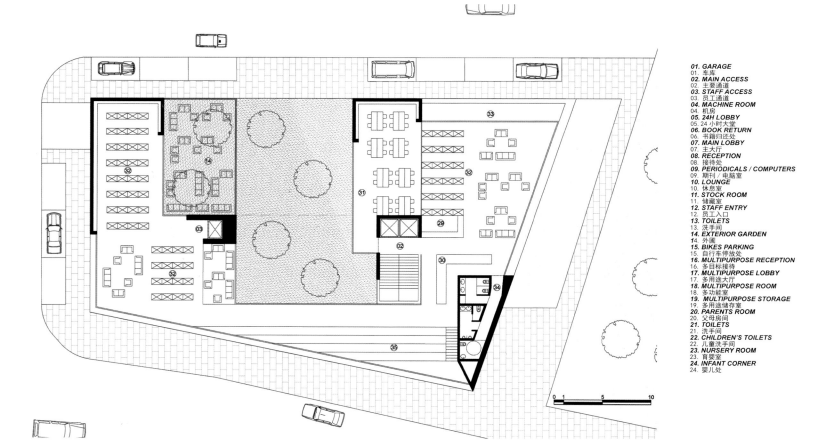

01. **GARAGE**
01. 车库
02. **MAIN ACCESS**
02. 主要通道
03. **STAFF ACCESS**
03. 员工通道
04. **MACHINE ROOM**
04. 机房
05. **24H LOBBY**
05. 24小时大堂
06. **BOOK RETURN**
06. 书籍归还处
07. **MAIN LOBBY**
07. 主大厅
08. **RECEPTION**
08. 接待处
09. **PERIODICALS / COMPUTERS**
09. 期刊 / 电脑室
10. **LOUNGE**
10. 休息室
11. **STOCK ROOM**
11. 储藏室
12. **STAFF ENTRY**
12. 员工入口
13. **TOILETS**
13. 洗手间
14. **EXTERIOR GARDEN**
14. 外园
15. **BIKES PARKING**
15. 自行车停放处
16. **MULTIPURPOSE RECEPTION**
16. 多目标接待
17. **MULTIPURPOSE LOBBY**
17. 多用途大厅
18. **MULTIPURPOSE ROOM**
18. 多功能室
19. **MULTIPURPOSE STORAGE**
19. 多用途储存室
20. **PARENTS ROOM**
20. 父母房间
21. **TOILETS**
21. 洗手间
22. **CHILDREN'S TOILETS**
22. 儿童洗手间
23. **NURSERY ROOM**
23. 育婴室
24. **INFANT CORNER**
24. 婴儿处

Level 3
三层

25. STORY TELLING
25. 讲故事室
26. DATA ROOM
26. 资料室
27. MULTIPURPOSE ROOM
27. 多功能室
28. CHILDREN CORNER
28. 儿童处
29. INFORMATION DESK
29. 服务台
30. CIRCULATION DESK
30. 图书借还台
31. DIGITAL DATA CORNER
31. 数字数据处
32. COLLECTIONS
32. 收发处
33. BALCONY
33. 阳台
34. TOILETS
34. 洗手间
35. READING SPACE
35. 阅读空间
36. PRESERVATION ROOM
36. 保存室
37. CULTURE CLASSROOM
37. 文化教室
38. SEMINAR ROOM
38. 会议室
39. LANGUAGE LAB
39. 语音室
40. OFFICE
40. 办公室
41. MEETING ROOM
41. 会议室
42. DIRECTOR OFFICE
42. 主任办公室
43. DRESSING ROOM
43. 更衣室
44. ORDER DEPARTMENT
44. 签发订单部门办公室
45. PREPARATING ROOM
45. 准备室
46. STAFF ROOM
46. 教研室
47. VOLUNTEER ROOM
47. 志愿者室

| 文体建筑方案集成 | CULTURAL AND SPORTS ARCHITECTURE PROGRAM INTEGRATION |

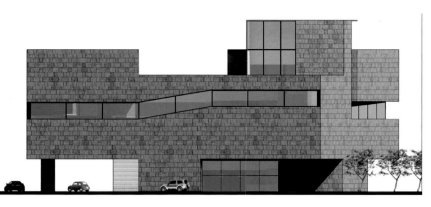
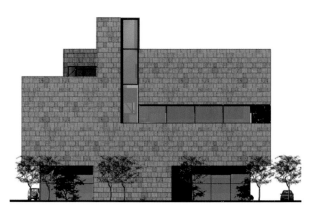
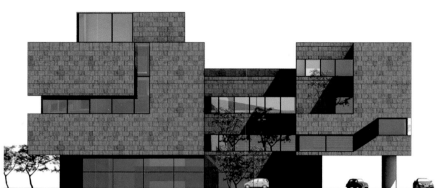
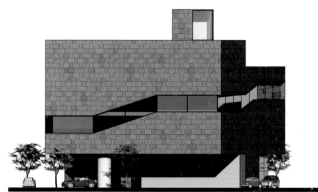

Green Building Cultural Interaction

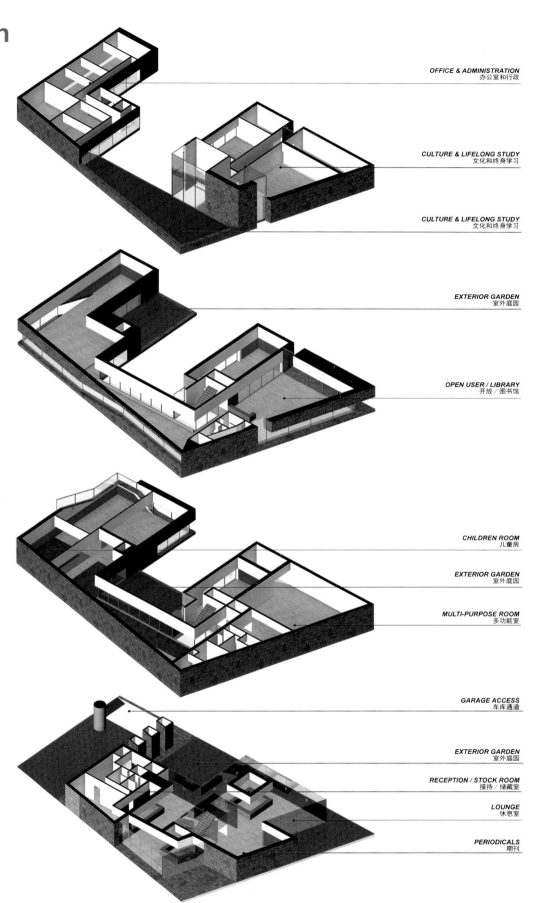

Project Overview

The project is located in the city of Daegu, South Korea. To maintain green public space is the focus of designers. Each layer is designed with terraces, and the hanging gardens strengthen the space link between the interior and the exterior and improve comfort, with more open space to facilitate communication and interaction between people. The library's design flexibility, versatility and openness adapting to all kinds of social activities are showed one by one.

Building Design

After the interpretation of the site and the assumptions given by the program basis, we defined two strong ideas as guidelines: one that is fighting for the maintenance of the existing garden, because this space has the potential to work well as a space for meeting, interaction and multidisciplinary knowledge of human relations, the same features that are intended to maintain and emphasize the proposed new library; but also the physical development of this new space, not as a library but as a "Theatre of Knowledge", a multidisciplinary space, flexible, with strong presence, open to the community that enhances human relationships.

The deployment of the building occupied the entire area of the existing garden, meeting the required clearances and creating public sidewalk and parking spaces. The volume represents the second phase of construction that develops on four floors, with the entire program. The third phase is to remove the part of the base inside this large volume, creating the suspended gardens and reinforcing the relation between interior and exterior space, connecting and creating different patios in each floor, that always allows this interaction, and enhancing the creation of comfortable but open niches that will motivate its users to explore the building and meet new people. Finally, the fourth phase is the maintenance of the garden, now reborn in these different courtyards, which are also fundamental in creating a "green" building.

澳大利亚阿德莱德新文化中心
Adelaide Culture Center

设计单位：绿舍都会
委托方：澳大利亚政府
项目地址：澳大利亚阿德莱德市

Designed by: Sure Architecture
Client: Australian Government
Location: Adelaide, Australia

城市地标
文化符号
节能环保

项目概况

项目拟定建于阿德莱德市中心的黄金地段。综合楼计划为覆盖多种类型表演，既混合日夜开放，又对所有年龄人群都具有吸引力、迎合当地居民和游客的多功能文化中心。项目完成后将成为澳大利亚阿德莱德市地标建筑。

建筑设计

项目设计为一个可以演习、讨论、教授、学习和表演的地方；一个为剧院、音乐和舞蹈演出与教育的最好平台；一个对于文化生活至关重要且能代表阿德莱德和整个澳大利亚的文化符号。设计旨在建设一个在各方面都有魅力又极具代表性的大方开放性的建筑。

综合楼反映了当地特有的城市规划特色，并表现在突出的中心区。建筑设计体现阿德莱德市民和音乐的生动表达。交流、开放、集中和冷静一起笼罩在音乐和舞蹈的顶端下，成为设计的引导性主题并指引阿德莱德市新一代的灵感空间创意。主楼被设想作为一个大胆、有活力的几何形体。两个纵向的建筑物令人感觉像两个舞者。光线被平滑、光透的曲面通过多种方式反射，所以使建筑物外观在一天的变化中同引人注目的光线变化相互依存。

项目可区分为连通的公共空间序列和娱乐空间，为会话创造一个生动、讨论自由的环境。场地中的活动、观点和关系清楚地表明其自身作为发展建筑生产的基本参数。建筑物的入口具有空间集聚和视觉直接关联的特色。

通过对地形和自然环境的分析，为了减少风力的影响，建筑物的形状被设计成一个空间多面体。为了减少楼房的能源消耗，建筑物的结构、外观和位置高效地利用了自然光线和空气流通。另外，设计也采用了被动节能手段，比如说保温外墙，中空低辐射玻璃和自动阴影系统等。

| 文体建筑方案集成 | CULTURAL AND SPORTS ARCHITECTURE PROGRAM INTEGRATION |

Concept
理念

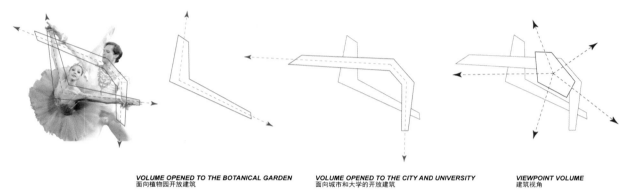

VOLUME OPENED TO THE BOTANICAL GARDEN
面向植物园开放建筑

VOLUME OPENED TO THE CITY AND UNIVERSITY
面向城市和大学的开放建筑

VIEWPOINT VOLUME
建筑视角

Phases
阶段

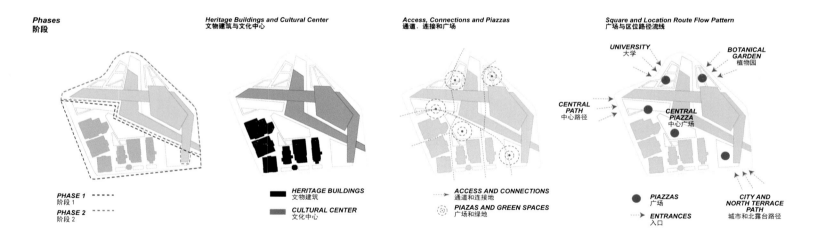

PHASE 1 阶段 1
PHASE 2 阶段 2

Heritage Buildings and Cultural Center
文物建筑与文化中心

■ HERITAGE BUILDINGS 文物建筑
■ CULTURAL CENTER 文化中心

Access, Connections and Piazzas
通道、连接和广场

ACCESS AND CONNECTIONS 通道和连接地
PIAZAS AND GREEN SPACES 广场和绿地

Square and Location Route Flow Pattern
广场与区位路径流线

UNIVERSITY 大学
BOTANICAL GARDEN 植物园
CENTRAL PATH 中心路径
CENTRAL PIAZZA 中心广场
● PIAZZAS 广场
→ ENTRANCES 入口
CITY AND NORTH TERRACE PATH 城市和北露台路径

City Landmark
Cultural Symbol
Energy Saving

Project Overview

Proposed at a prime location in the heart of the City of Adelaide, the complex is planned to house multiples types of performing including a mix of day and night attractions for all ages; catering for both local residents and tourists. After completion, it will become a landmark and an exciting jewel in Adelaide's crown!

Building Design

Conceived as a place to rehearse, discuss, teach, study and perform, the complex is to become the premier venue for theatre, music and dance performances and education—a vital element of the cultural life and identity of Adelaide and of all Australia. The goal of the design is to conceive an open building that is effective as an inviting and yet powerful symbol in all directions while being permeated by generosity and openness.

Being bold, yet not ostentatious, the complex reflects the very specific urban planning features of the site and bundles these into a striking Centre-piece. As one with its surroundings, the complex will invite historical buildings into an exciting contemporary new Adelaide Precinct. With direct access and flow, public thoroughfares from the Botanical Garden, the University, The CBD and North Terrace Cultural Precinct give life to the re-birth of this location, and will become an Iconic Cultural Adelaide site, becoming a new milestone for the country. The building embodies a living statement for the music and people of Adelaide. Communication and openness, concentration and calmness are united beneath the "roof of music and dance", becoming guiding themes for designs, and leading to the creation of an inspiring place for the new generation in Adelaide.

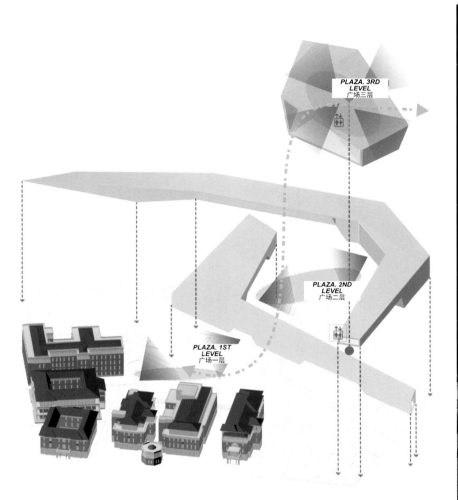

| 文体建筑方案集成 | CULTURAL AND SPORTS ARCHITECTURE PROGRAM INTEGRATION |

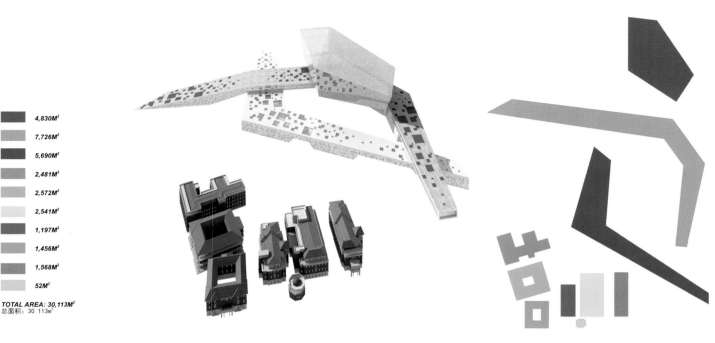

- 4,830M²
- 7,726M²
- 5,690M²
- 2,481M²
- 2,572M²
- 2,541M²
- 1,197M²
- 1,456M²
- 1,568M²
- 52M²

TOTAL AREA: 30,113M²
总面积：30 113m²

AUDITORIUM
会堂
LOBBY
大堂
THEATER HALL (800 SEATS)
戏剧厅（800个座位）
STAGE
舞台
CONTROL CABINS
控制室
DRESSING ROOMS
化妆间
REHEARSING AND WARM UP ROOMS
排练及热身室
BATHROOMS
浴室
OFFICES, STORAGE ROOMS
办公室、储藏室

CENTER FOR INVESTIGATION AND DOCUMENTATION OF DANCE
舞蹈研究和编制中心
LOBBY / RECEPTION
大堂／接待区
LIBRARY / ARCHIVE
图书馆／档案馆
VIDEO / AUDIO LIBRARIES
视频／音频库
DEPOSIT
存储室
CONFERENCE AND AUDIOVISUAL ROOM
会议及视听室
RESEARCH ROOM
研究室
RESTING HALL
休息大堂
AUDIOVISUAL LAB
视听实验室
OFFICES AND ADMINISTRATION, STORAGE ROOM
办公行政、储藏室

SCHOOL OF DANCE & ART
舞蹈与艺术学校
LOBBY
大堂
4 LARGE CLASSROOMS (WOOD FLOORING, SOUNDPROOFED, MIRRORS)
四个大教室（木地板、隔音、后视镜）
4 MEDIUM-SIZED CLASSROOMS
四个中型教室
STUDY ROOM, FACULTY ROOM
自习室、师资队伍室
GYM
健身房
SHOWERS AND DRESSING ROOMS, BATHROOMS
淋浴及更衣室、浴室
OFFICES AND ADMINISTRATION, STORAGE ROOM
办公行政、储藏室

CENTER FOR NEW TECHNOLOGIES IN ART
艺术新技术中心
LOBBY
大堂
2 LARGE CLASSROOMS
2个大教室
6 MEDIUM-SIZED CLASSROOMS
6个中型教室
EXPERIMENTAL PERFORMANCE SPACE
实验表演空间
DRESSING ROOMS, BATHROOMS
更衣室、浴室
OFFICES AND ADMINISTRATION, STORAGE ROOM
行政办公、储藏室

MUSEUM OF DANCE
舞馆
LOBBY
大堂
CAFETERIA
自助餐厅
AUDIOVISUAL PRESENTATIONS ROOM
视听报告室
PLASTIC ARTS AND DANCE
造型艺术及舞蹈
8 ROOMS FOR PERMANENT EXHIBITION (4 AREAS)
8间长期展览室（4个区域）
4 ROOMS FOR TEMPORARY EXHIBITION
4间临时展览室
STORE
店铺

OTHER COMMON AREAS
其他公共区域
LOBBY / RECEPTION
大堂／接待
CULTURAL AND TOURIST INFORMATION OFFICE
文化和旅游信息办公室
BOOK / MUSIC STORE / SOUVENIRS
书籍／音乐商店／纪念品
CAFETERIA / RESTAURANT
自助餐厅／餐厅
BATHROOMS
浴室
PHONE BOOTHS, AUTOMATIC TELLER MACHINES (ATMS)
电话亭、自动出纳机（ATMS）
ADMINISTRATION, GENERAL STORAGE ROOMS
管理、一般储藏室
UNDERGROUND PARKING (AT LEAST TWO FLOORS)
地下停车场（至少两层）

CULTURAL MANAGEMENT
文化管理
LOBBY / RECEPTION
大堂／接待
ADMINISTRATION
管理
OFFICES
办公室
MEETING ROOMS
会议室
BATHROOMS
浴室
STORAGE ROOM
储藏室

HOSTAL
旅馆
LOBBY
大堂
ADMINISTRATION
管理
RESTAURANT
餐厅
MULTIFUNCTIONAL SPACE
多功能空间
10 SINGLE ROOMS
10间单人房
8 DOUBLE ROOMS
8间双人房
5 MIX DORMITORIES
5间混合宿舍

SOUND STUDIOS
录音室
LOBBY
大堂
ADMINISTRATION / STORAGE ROOM
管理／储藏室
8 RECORDING BOOTHS
8个录制展台
8 REHEARSAL STUDIOS
8间排练室
4 AUDIOVISUAL PRODUCTION ROOMS
4间音像制片房

CREATIVE AREA
创意区
BARS
酒吧
RESTAURANTS
餐厅
CAFETERIAS
自助餐厅
SHOPS (MUSIC, INSTRUMENTS, ART, DANCE)
商店（音乐、乐器、艺术、舞蹈）
ARTIST STUDIOS
艺术家工作室

The differentiated but interconnected spatial sequences of public spaces and, entertainment spaces create a lively, discussion-rich setting for conversations, performances and societal inter-action and at the same time permit the creation of a site that can just as well offer the quiet and concentration that is desirable for the enjoyment. Movement, views and relations in the site manifest themselves clearly as essential parameters to develop the architectural production. The entrance into the building is characterized by continuous spatial concentration and directed visual relations.

According to the terrain and physical environment analysis, the shape of the building is designed as a spatial polyhedron, in order to reduce the impact of wind. The building structure, façade and position make efficient use of natural light and ventilation to reduce building energy consumption. In addition, the design also incorporates passive energy-saving practices, such as exterior insulation, Hollow Low-E glass and automatic shading systems.

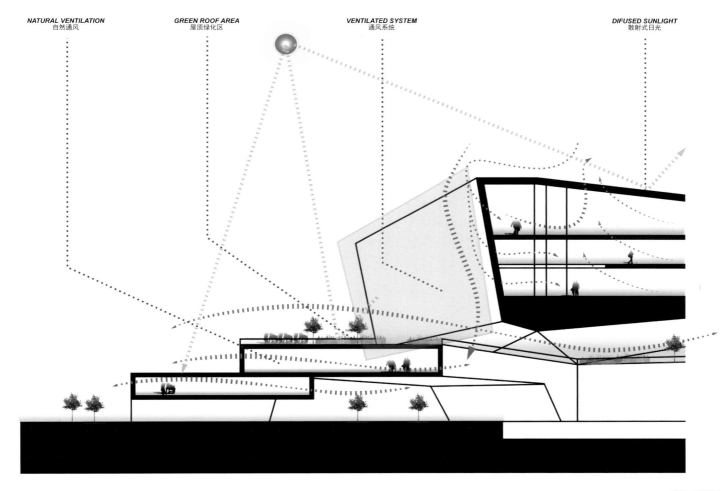

法国圣日耳曼-lès-阿尔帕容新文化中心
St Germain-Lès-Arpajon Cultural Center

设计单位：Ateliers O-S architectes
委托方：圣日耳曼-lès-阿尔帕容市
项目地址：法国圣日耳曼-lès-阿尔帕容市
项目面积：2 113m²
设计团队：Vincent Baur Guillaume Colboc Gaël Le Nouëne

Designed by: Ateliers O-S architectes
Client: City of Saint-Germain-lès-Arpajon
Location: Saint-Germain-lès-Arpajon, France
Area: 2,113m²
Design Team: Vincent Baur, Guillaume Colboc, Gaël Le Nouëne

L 形空间
彩色家具

项目概况

圣日耳曼-lès-阿尔帕容的新的文化中心主要包括多媒体图书馆和音乐舞蹈学校,该项目拥有非常优越的地理位置:文化中心坐落在一个狭长的地带上,可眺望食人魔河的绿色山谷。

项目中的两个建筑由一个倾斜的山谷梯度的倾角相连,形状类似于"U"型夹;它也可看成是项目的两个层面,一个高层次的广场和一个通道交叉建筑物相连接的低层次的广场。这个城市的组成和景观轴线在附近作为吸引人们之处承受着该项目。

建筑设计

"寰球体积和空间"、"功能组织"和"立面"是该建筑设计观念的研究轴,项目设计选择了一个基于生活质量和工作人员工作条件的务实的组织,鼓励交流和宴饮交际。多媒体图书馆是可以从街上直接通达的,并在一个大的俏皮而明亮的开放空间被组织成一个 L 形状:图书馆成为一个活跃的空间,彩色家具作为一个分区,并组织数据空间。

底层是音乐学院和行政办公室,教室被组织在北侧,而两个舞蹈室慨然的向山谷风景的东面打开。

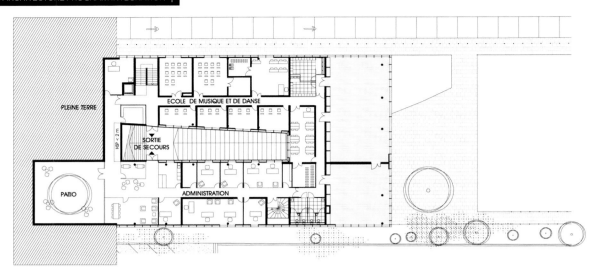
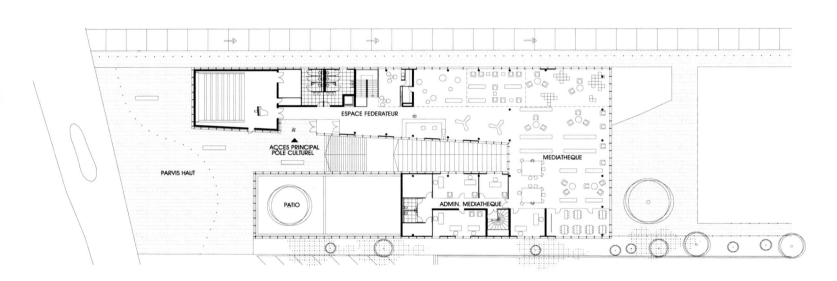

L-shaped Space
Colored Furniture

Project Overview

The new Cultural Center includes Multimedia Library and Music and Dance School. The project location is very advantageous: the cultural center is settling into a long and narrow site and looking at the green valley of the Orge River. The two building volumes in the project are linked by a gradient slope with the shape of "U"; it also can be seen as the two levels, a high-level square and a low level square connected by a passage crossing the building. This urban composition and this tight line in the landscape stand the project as a pole of attraction in the neighborhood.

Building Design

The "volume and space", "functional organization" and the "façades" are the research axis of the building design concept. The project design is chosen a practical organization which is based on the quality of life and staff working conditions to encourage communication and conviviality. The Multimedia Library is directly accessible from the street, and is organized in an L shaped on a large playful and bright open space: the library becomes an active space. The colored furniture acts as a partition and organizes the space of readings.

On the ground floor is the Music School and administrative office. The classrooms are organized on the north side while the two dance studios open generously to the east on the landscape of the valley.

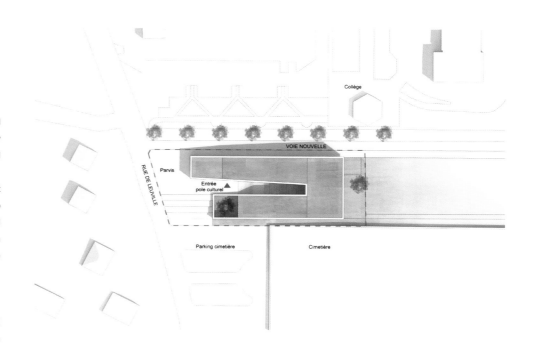

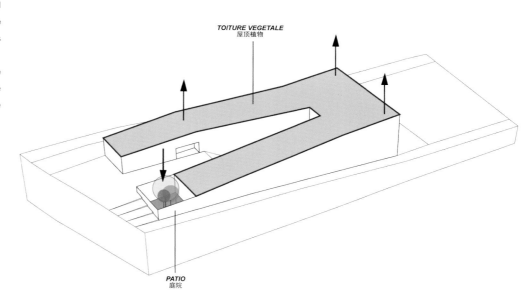

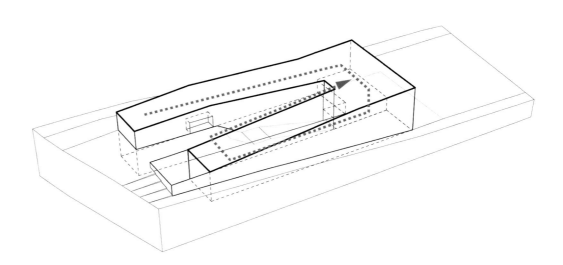

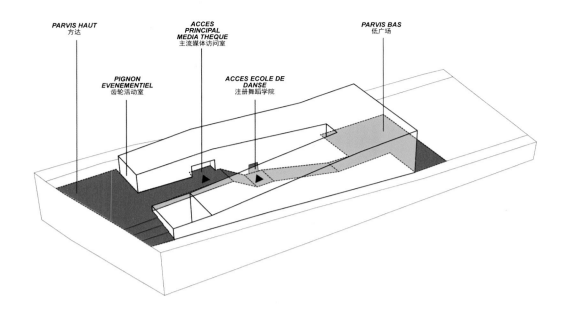
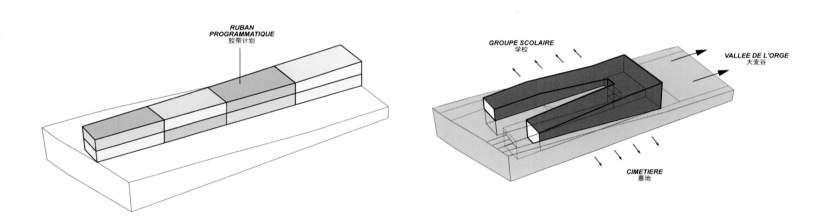
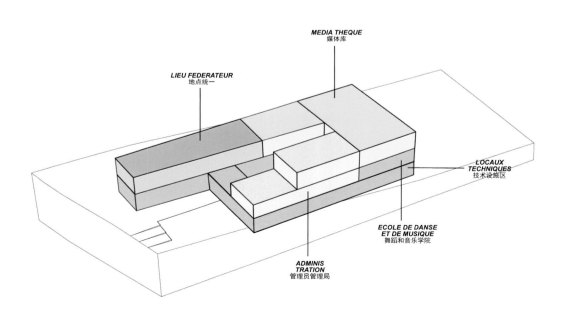

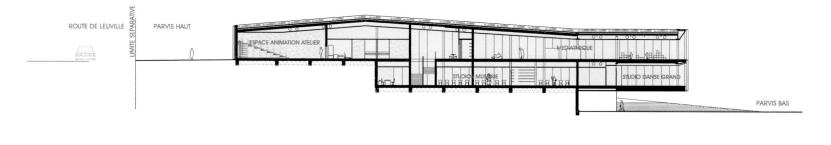
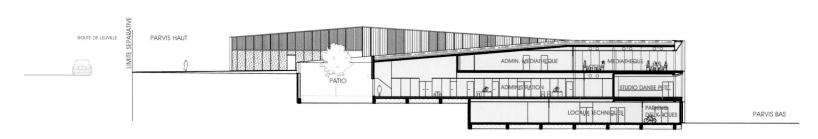
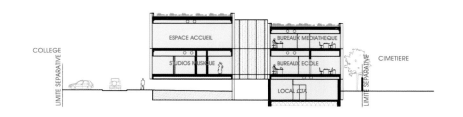

比利时亨克城市中心
GENK

设计单位：Maxwan architects + urbanists
委托方：亨克市
项目地址：比利时亨克市
建筑面积：12 600m²

Designed by: Maxwan architects + urbanists
Client: City of Genk
Location: Genk, Belgium
Building Area: 12,600m²

建筑突破空间的链接

建筑设计

该项目地址沿城市中心的南部高原而立，构成城市核心，并直接导致了该地区不同的自然景观。在城市构造中，它呈现了一种独特的突破，扩大开放了连接城市文化与绿色区域网的可能性。同时，该城市已开始增加了一系列关于城市外围的文化空间。这个地址，躺在城市的主要东西轴上，对在城市核心内提供急需高质量的公共场所和文化空间是至关重要的。这将增加城市的文化活动，同时把现有的传统空间与最近的商业发展连接。

该场址是由质量文物建筑和普通结构的混合，形成目前该场址半公共庭院的一部分。通过与增加的一系列新建筑物的结合去除一些关键的结构，这个空间质量被保持，增强和扩展。现有庭院的孔隙被保持并连接到不同私人和公共氛围的一系列新形成的院落中，从而确立不同的空间关系顺序，在坡度的底部和顶部之间从梯田、一个凸起的甲板、台阶、屋顶花园、滚山花园到一个倾斜的山坡上的变动。

现有庭院通过形成一个穿透两层的院落而丰富下面的文化空间和屋顶上的绿色空间。因此，通过堆叠的空间真正将

自然和文化混合并通过一系列的空隙连接它们。这种丰富的新的文化或自然大楼通过改造现有结构沿主要街道以建立一个强大的现代建筑表面被呈现给城市，该建筑表面设计了一个新的壮观的入口至层状的庭院空间。

Building Breakthrough Space Link

Building Design

The site lies along the southern plateau of the city center that both frames the urban core and leads directly to the diverse natural landscapes of the region. It presents a unique break in the city fabric, opening up the possibility to link the city culture with the green regional networks. Simultaneously, the city has begun to add a series of cultural spaces on the periphery of the city. This site, lying along the major east west axis of the city, is crucial to providing much needed quality public and cultural space within the urban core. It will increase urban cultural activities, while linking to existing heritage spaces to the recent commercial developments.

The site has a mix of quality heritage buildings and mediocre structures, which currently form a semi public courtyard on part of the site. This spatial quality is kept, enhanced, and extended via the removal of a few key structures in combination with the addition of a series of new buildings. The porosity of the existing courtyard is maintained and linked to a series of newly formed courtyards with varying degrees of private and public atmospheres; thus establishing a sequence of different spatial relationships between the bottom and top of the slope ranging from terraces, a raised deck, steps, roof gardens, rolling hill garden, to a sloped hillside.

The existing courtyard is enriched by forming a punctured two layered court—with cultural spaces below and green spaces on the roof; thereby, literally mixing nature and culture by stacking their spaces and linking them by a series of voids. This rich new culture / nature complex is showcased to the city by renovating an existing structure to create a strong contemporary building face along the main street, which frames a new grand entrance into the layered court space.

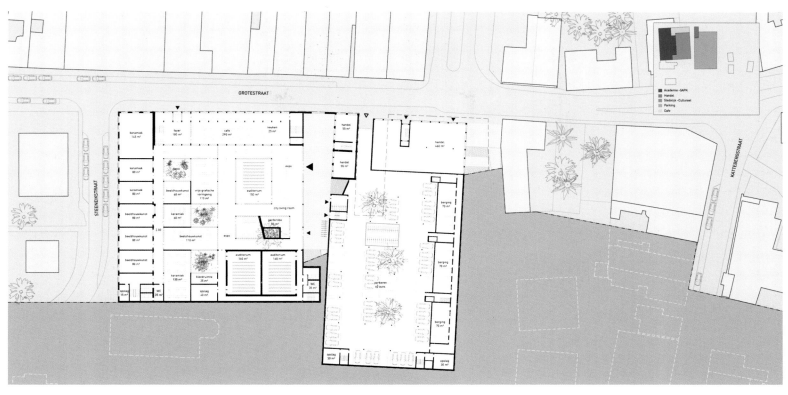

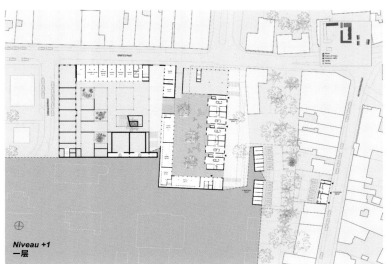

Niveau +1
一层

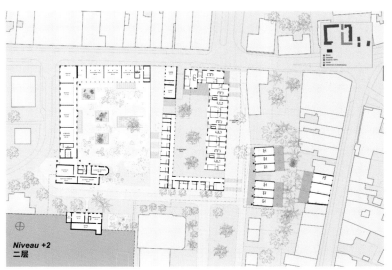

Niveau +2
二层

| 文体建筑方案集成 | CULTURAL AND SPORTS ARCHITECTURE PROGRAM INTEGRATION |

City Meets Landscape
城市满足景观

- CORE OF THE CITY 城市中心
- HOUSING NEIGHBORHOOD 住宅周边
- INDUSTRIAL AREA 工业区
- OPEN SPACE 开放空间
- CANAL 运河
- KATTEBERG – REGINA MUNDI KATTEBERG 的世界之后

Niveau +3
三层

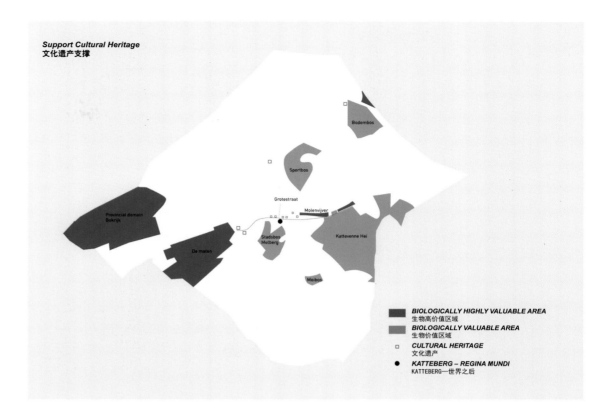

Niveau +4
四层

Niveau +5
五层

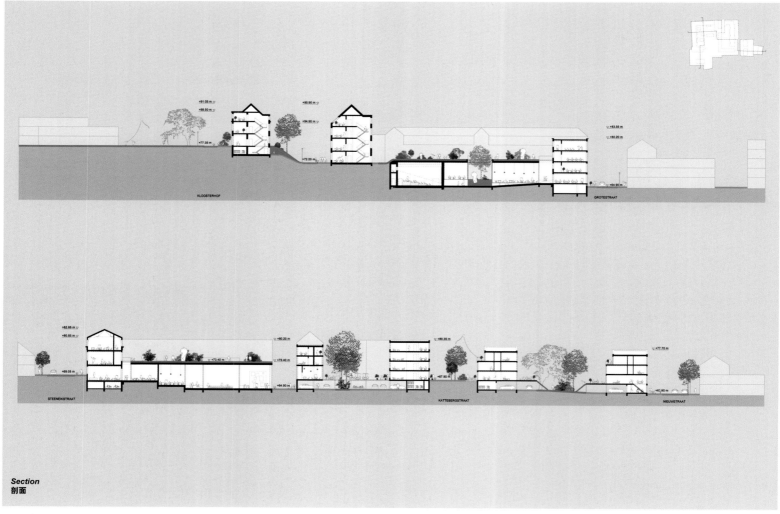

Section
剖面

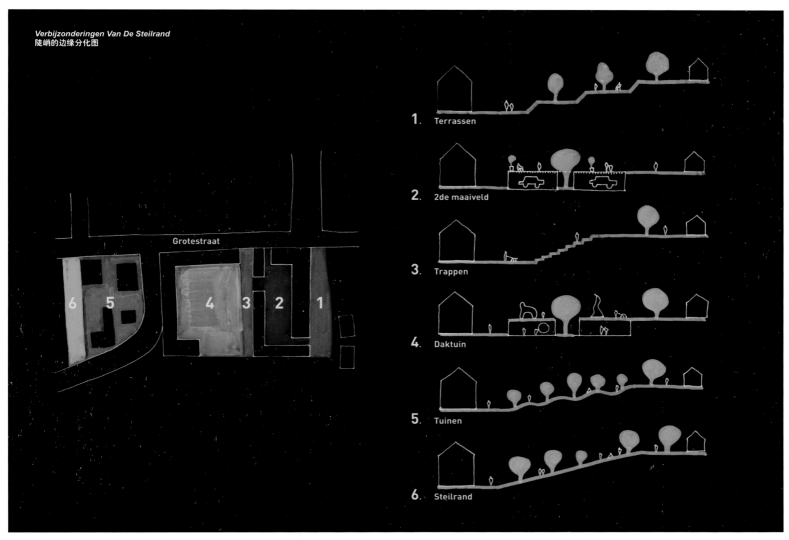

荷兰合作银行 Westelijke Mijnstreek 咨询中心
Rabobank Westelijke Mijnstreek Advice Centre —Sittard

设计单位：麦肯诺建筑师事务所	*Designed by: Mecanoo Architecten*
委托方：Rabobank Westelijke Mijnstreek, Sittard	*Client: Rabobank Westelijke Mijnstreek, Sittard*
项目地址：荷兰锡塔德市	*Location: Sittard, Netherlands*
表面积：6 800m²	*Surface Area: 6,800m²*

灵活的工作空间
可持续的室内气候

项目概况

锡塔德是荷兰林堡省内的一座城市。风光绮丽的城市遍布着铜绿色的橡树和浅黄色泥灰。荷兰合作银行希望在锡塔德车站附近的城市中央新建大型咨询中心作为银行的办公区。

建筑设计

设计意图是在小山上建造一座完全面向社会的开放的合作银行。进入雕塑式天窗装点的曲线天花板下方的入口，便可一眼看见整个大厦的构造。巨大的平台如一条白色绸带在整个大厦内蜿蜒盘旋，邀你摒弃电梯而踏上弯曲的阶梯。一排排阶梯"看台"是中央大厅引人注目的焦点。该区域阶梯式的工作台，打造了灵活的工作空间。经过刷面处理的橡树板包裹着内部的墙壁，与工作台的材料和工作空间的装饰形成统一的整体。一片透明帘子悬挂在围绕阶梯式工作空间布置的 60m 长的轨架上，该布帘采用深浅不一的银色和蓝色拼接而成，通过改变帘子的位置可带来多样的变幻空间和氛围。

包括餐厅在内的整个地面层对外开放。职员和顾客都可使用阶梯式工作空间，感觉很像现代的公共图书馆。楼上的办公区只允许银行职员进入，一层设有进入上层的通道系统。为了提供高质量的音效环境，设计师为一楼的呼叫中心设计了精致的独立空间，并采用玻璃隔断确保其与其他职员的视觉联系。

大厦采用节能装置，并具备高隔热性能和三倍光照条件。天花板可控制室内通风，并通过阳光而非气流控制冷热变化。此外，建筑南面还有一个平板式的空调系统。在炎热的气候条件下，透明屏幕自动覆盖在玻璃立面前，同时机械设备将立面和屏幕之间的热量抽出，以此保障阶梯工作台的宜人气温。

Site Plan
总平面图

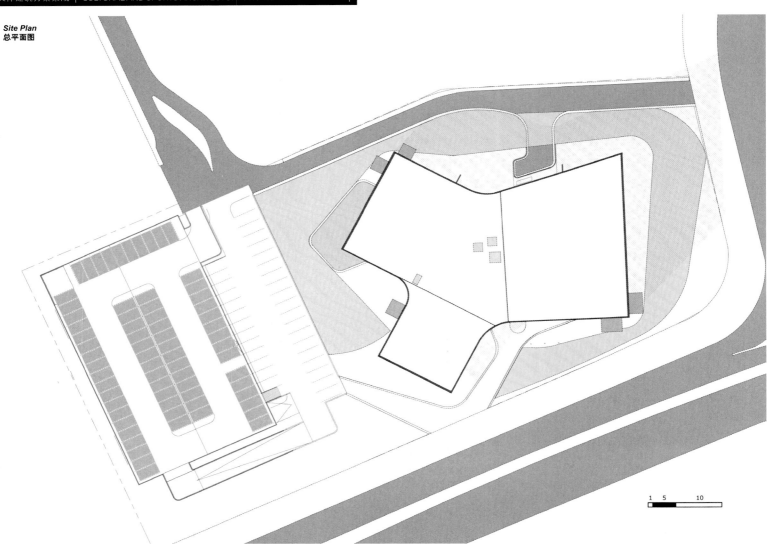

grond

1 erste verdieping

2

3 Derde verdieping

Flexible Workspace Sustainable Indoor Climate

Project Overview

Sittard is located in the rolling landscape of Limburg, the province of the bronze-green oak and the soft yellow marl. The cooperative Rabobank wanted to combine smaller branches within a large centrally-located advice centre near Sittard station.

Building Design

The intent underlying the design is the development of a cooperative bank extending its arms invitingly to society, located on a small hill. Upon entry under the curved ceiling with sculptural skylights, you can survey, in one glance, the whole building and your curiosity is further ignited for activities on the upper floors. These are designed as staggering terraces that meander around the loft as a white ribbon. They invite you to take the zigzagging stairs and not the elevator. The focal point of the central hall is the auditorium which, remarkably, was not in the original brief. In this area, Mecanoo designed a staircase tribune that accommodates flexible workspaces. Brushed-oak paneled walls that wrap the

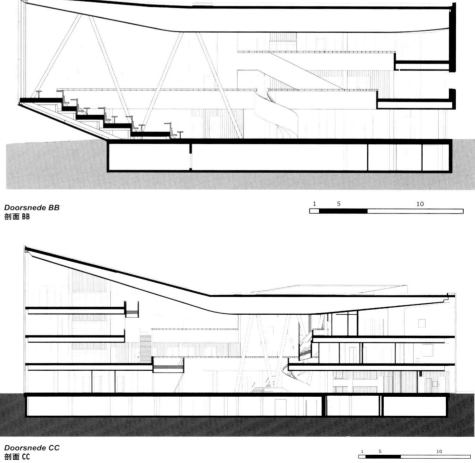

Doorsnede BB
剖面 BB

Doorsnede CC
剖面 CC

interior of the building form a unity with the material of this tribune and its workspace furnishings. Here, a transparent curtain with different shades of silver and blue designed by Petra Blaisse of Inside Outside runs on a 60m long track that winds above the tribune. Diverse spaces and atmospheres can be created by altering the position of the curtain parts.

The entire ground floor including the restaurant is open to the public. Both employees and customers can use the workplaces in the auditorium, just as in a modern public library. The upper office floors are only accessible to staff via a pass system on the first floor. To provide a high quality acoustic environment, the call centre has its own space on the first floor with a beautiful view. A glass separation ensures that it is visually connected to the rest of the staff.

The building has energy-saving installations, high insulation and triple glazing. Climate ceilings control ventilation, heating and cooling through radiation without moving air. In addition, the building has a flat panel air conditioning system facing south. In warm weather a transparent screen automatically drops in front of the glass façade. The screen reflects the solar energy, while the heat of the sun between the façade and the screen is mechanically extracted so that the auditorium remains pleasantly cool.

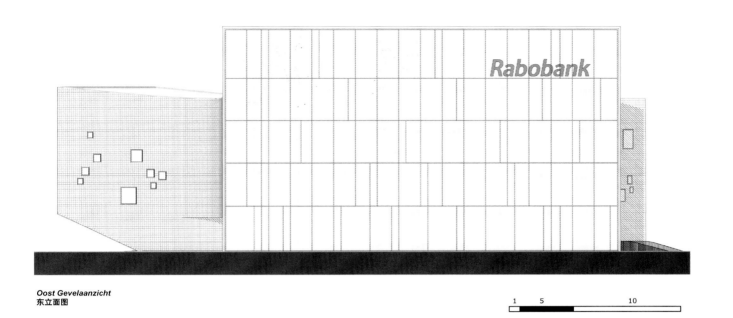

Oost Gevelaanzicht
东立面图

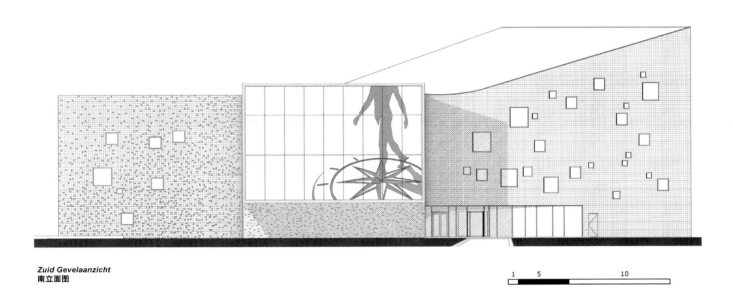

Zuid Gevelaanzicht
南立面图

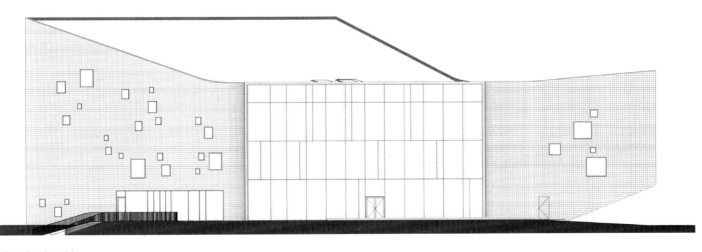

West Gevelaanzicht
西立面图

湖南长沙中联重科新总部展览中心
Zoomlion Headquarters Exhibition Center

设计单位：amphibianArc	Designed by: amphibianArc
委托方：中联重科	Client: Zoomlion
项目地址：湖南省长沙市	Location: Changsha, Hunan
项目面积：10 074m²	Area: 10,074m²
项目获奖：2012年世界建筑节/未来－文化类项目/提名	Award: 2012 World Architecture Festival/Future-Culture Projects/Shortlisted

形态转换
双层幕墙
参数建模立面

项目概况

中联重科新总部展览中心坐落在湖南省长沙市，共四层楼，建筑总高度达到26m。中联重科是中国机械设备制造行业的著名企业之一，已跻身于世界机械设备制造行业前10位。建筑设计理念就是要吻合和发扬中联重科前瞻性、独特性和创新性的企业形象和价值。

建筑设计

项目设计从字面上看最独特的方面是建筑改变或转换形状的能力。整个建筑的双层幕墙系统使这个"转换建筑"成为可能。内层幕墙注重围合和建筑系统。外层幕墙则含有可操作的部分，这部分可以被打开或关闭以模仿不同动物的形态。从一个普通的矩形框作为初始状态，北立面改造成一只老鹰和一只蝴蝶，南立面折叠成一个游泳的青蛙。这些动物的形态反映了公司了解自然和人为发明之间的微妙平衡，以及他们对环保的人类发展的接纳。另外，作为设计策略，我们通过表意形式为领导（老鹰）、短命和脆弱的代表（蝴蝶）和繁荣的代表（蟾蜍）传达中国传统的文化象征意义。

建筑外立面复杂的模式最初是受到蝴蝶或蜻蜓的翅膀图案的启发。基于这些昆虫的翅膀，为了实现这些模式的系统性和有机性，我们采用参数化建模工具来生成和设计立面。用于外墙的材料是钢和玻璃。该模式提供了一种轻盈但坚固的结构。它掩盖并吸收了移动和固定到该处的液压。其结果是完美地奠定了复杂的图案，让阳光穿透展馆，并且在夜间让光线散发到企业园区。

项目为保证其功能性和耐用性，设计将动态和静态结构分开进行工程服务。

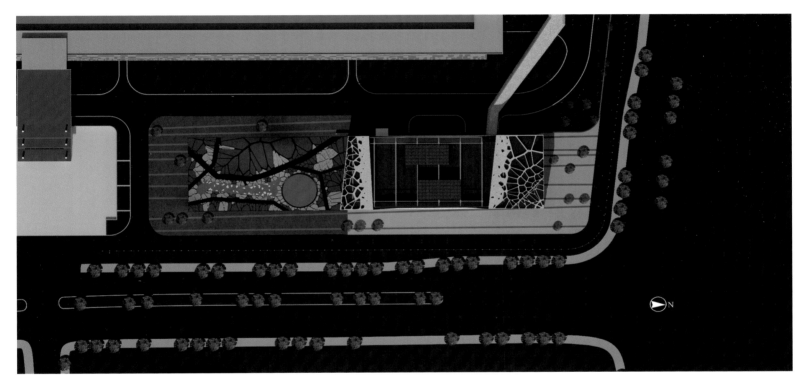

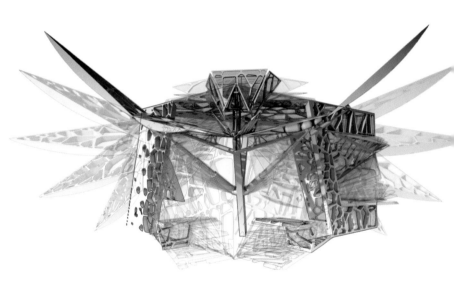

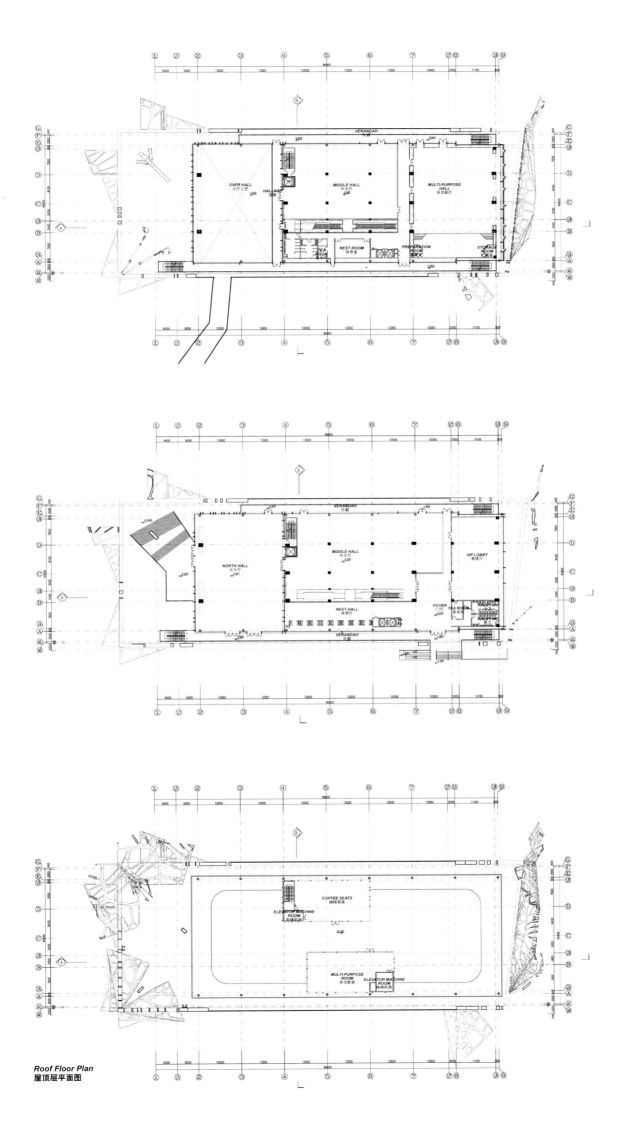

Roof Floor Plan
屋顶层平面图

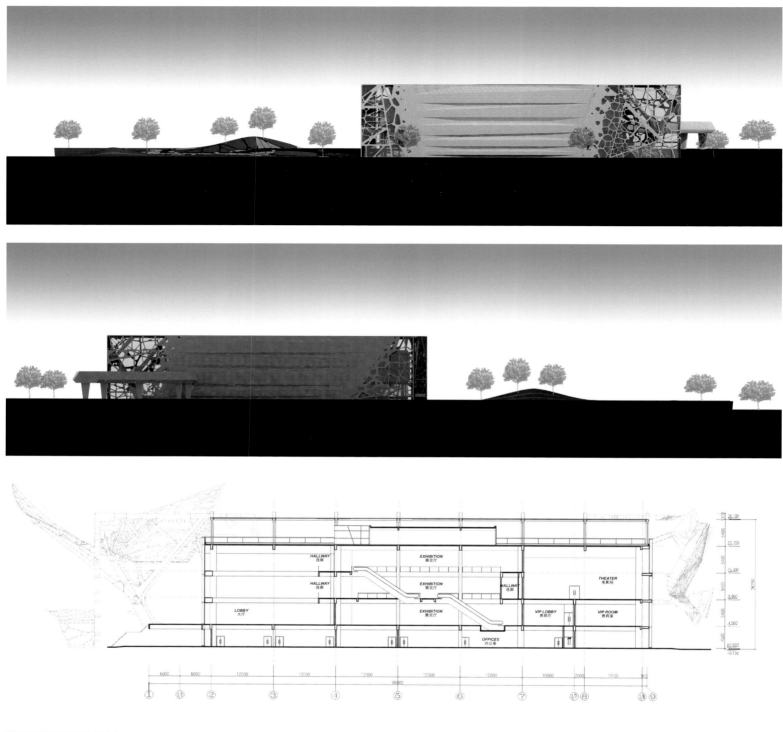

Form Conversion
Double-skin Façades
Parametric Modeling Façade

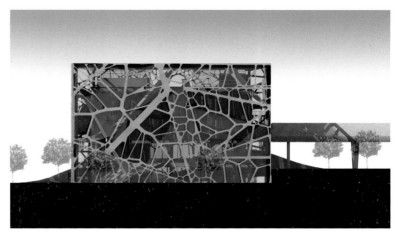

Project Overview

he Zoomlion Headquarters Exhibition Center is located in the city of Changsha, Hunan Province of China. The project has a total of four floors and a total building height of 26 meters. Zoomlion is one of China's leading manufacturers of heavy machinery equipment and ranked top 10 globally in the heavy machinery industry. The concept of the design for its headquarters exhibition center is to match its forward thinking, unique, and mechanistically imaginative corporate image and values.

Building Design

The most unique aspect of our project design is the building's ability to change shape, or transform, literally. The double skin system runs throughout the building to make this "transformer building" possible. The inner skin takes care of the enclosure and building systems. The outer skin contains operable portions which can be opened or closed to mimic different animal forms. From a plain rectangular box as the initial state, the north façade transforms into an eagle and a butterfly, the south folds into a swimming frog. These animal forms reflect the company understands of the delicate balance between nature and artificial invention, and their embrace for environmentally sound human development. Also, as a design strategy, we adopted ideographic forms to convey traditional Chinese cultural symbolism for leadership (eagle), ephemerality and fragility (butterfly) and prosperity (toad).

The intricate pattern on the façade is originally inspired by the wing patterns on butterflies or dragonflies. To achieve the systematic and organic nature of the patterns found on the wings of these insects, we used parametric modeling tools to generate and design the façade. The material for the skin is steel and glass. The pattern provides a light but sturdy structure. It conceals and incorporates the hydraulics which move and hold it in place. The result is a beautifully laid out intricate pattern, which allows daylight to penetrate into the exhibition hall and light emanate out into the corporate campus at night.

To bring this design into actuality and assure its functionality and durability, Kinetic and static structures are separated as engineering services.

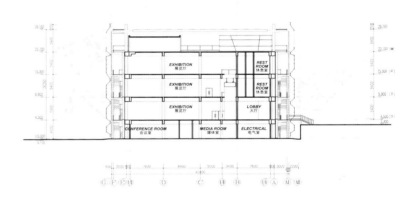

上海国际设计一场
Shanghai International Design No. 1

设计单位：AM 项目，约瑟夫·迪·帕斯夸莱建筑师事务所
同济大学建筑设计研究院（合作）
委托方：杨浦区人民政府
同济大学
项目地址：中国上海市
项目面积：70 500m²

*Designed by: AM project, Joseph di Pasquale Architects
Architectural Design and Research Institute of
Tongji University (Collaborator)*
*Client: Yangpu District People's Government
Tongji University*
Location: Shanghai, China
Area: 70,500m²

立面
知识经济圈
文化可持续

项目概况

上海国际设计一场由设计博物馆及联合国实训基地、同济大学设计研究院、设计创意学院、"设计之心"大楼、"意大利创意设计城"等功能板块构成，目标是打造成为杨浦建设知识创新区的国际化高端平台，以及引领环同济知识经济圈新一轮发展的龙头项目。

| 文体建筑方案集成 | CULTURAL AND SPORTS ARCHITECTURE PROGRAM INTEGRATION |

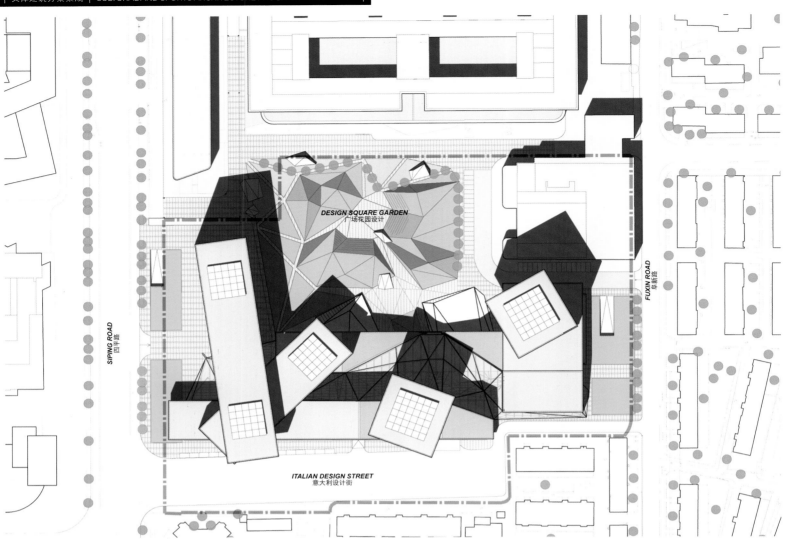

建筑设计

该项目与四平路交接，由于四平路是上海的交通主干道，人流和车流量非常大，项目西侧沿四平路有很长的一个交接面，最能体现国际设计之都的创新精神。西立面作为"上海设计一场"的官方名片，设计称其为"具代表性和创造性价值"。

项目与南侧居住区和东侧保留建筑交接。南侧正对多层住宅北立面和几栋高层塔楼，设计时必须和住宅区留有足够的道路，以致不对其造成遮挡。东侧毗邻阜新路，被保留的建筑与新建筑的距离非常近，而道路系统保持不变。项目东南立面的设计也是"尺度宜人"。项目西侧由一条横跨四平路的人行地下通道实现了与大学校园的无缝对接，被称为"亲密的精神空间"。

建筑的内庭院是意大利文艺复兴建筑的经典。它充分表现出乌尔比诺城大宫殿的亲密和深沉。面对四平路的"马可波罗之门"，设计师运用了"具代表性、创造性、文化可持续性价值"的设计手法，通过城市肌理分析，设计把重点放到了这个主立面（准确体现业主"将杨浦区和意大利联系起来"）的构想上。

设计采用"文化可持续"的设计符号——"门"的概念，这扇向西侧开敞的大门，展现出项目本身对大学、城市和西方（意大利）开放的姿态。

在景观的设计上，不是单纯地补白空地，而是在新建筑和现有建筑之间有意安排了一个景观区，融合了意大利几何型园林和中国传统山水园林特点。

| 文体建筑方案集成 | CULTURAL AND SPORTS ARCHITECTURE PROGRAM INTEGRATION |

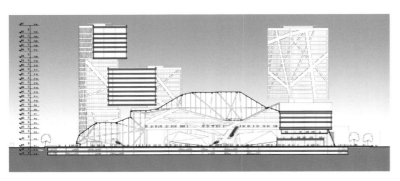
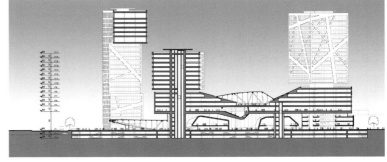
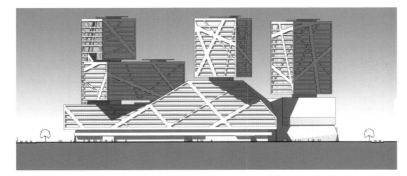

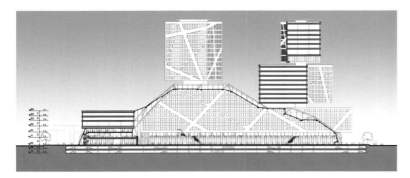
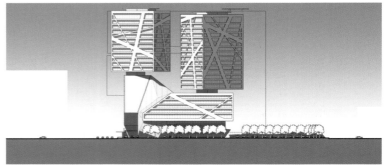

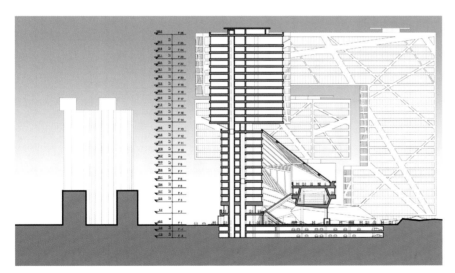
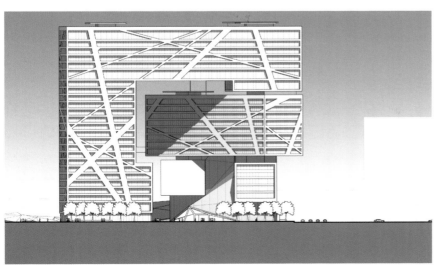

Façade
Knowledge Economy Circle
Sustainable Culture

Project Overview

The project consists of the Design Museum and the United Nations Training Base, Design and Research Institute of Tongji University, of Design and Innovation College, "Heart of Design" building, "the Italian City of Creative Design" and other functional blocks, and the goal of the project is to build an international high-end platform in Yangpu Knowledge Innovation Zone, and lead a new round of development projects in the knowledge and economic circle in Tongji.

Building Design

This side is facing west is the most important street bordering the site in terms of urban ranking. Siping road is the connection with the main city road network and is the most representative and symbolic stage for the new complex. This side is the official face of the building to the district and the entire Shanghai city. The design called this approach "the symbolic creative value".

The relation with the residential texture connects the south and the east side of the site. On the south the site faces the north elevation of high residential towers. This means that on this side it will have a wide road and anyway the new buildings will not keep away the sun from the residences. The east side facing Fuxin road is very important to establish a correct and proportioned relation with the existing buildings because the width of the road can not be changed and some existing building will be very close to the new complex. The design in the southeast is called "the human scale value". The west side of the project connects with the inner life of the campus through the pedestrian tunnel that passing below Siping road and that connects the Tongji campus with the site, which is called this approach "the intimate spirit space".

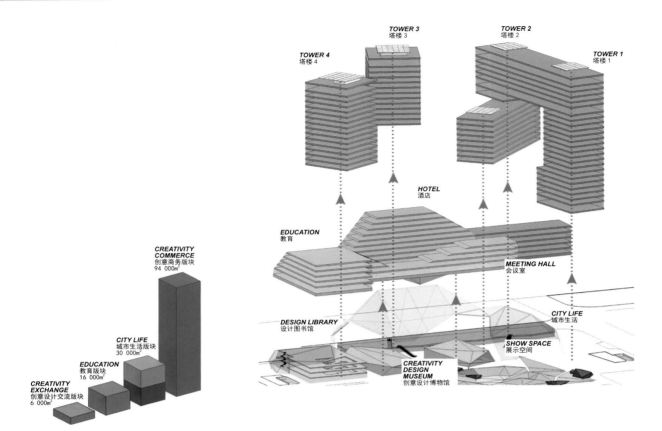

The design has been taken inspiration from the best of classical Italian architectural tradition: the Palazzo Ducale Di Urbino designed by Luciano Laurana in late XV century, which definitely is the most important Renaissance Palace in Italian history of architecture. Facing Siping Road "the Marco Polo Gate", the symbolic value and cultural sustainability approach are used. After the context analysis, the main problem was to find a symbol for monumental Siping road front suitable for this project especially to match the expectation of the client for this complex to be a link between the design district in Shanghai and Italy.

The design is used the design symbol of "sustainable culture"—the concept of "door". This door opened towards to the west side shows the project itself is opened for the University, the City and the West (Italy).

For the landscape concept we started from the idea that the gardening should not be just a filling of the forgotten empty spaces between the buildings. It should be the key link element between the existing buildings and the new buildings. The landscape design philosophy is a merge between the Italian garden style and the Chinese garden style.

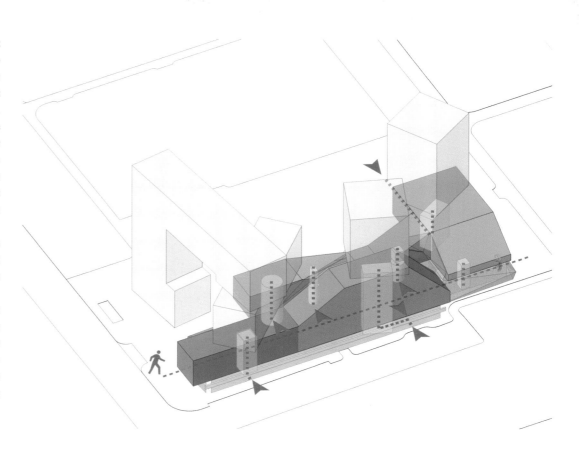

塞浦路斯尼科西亚 RAUF RAIF DENKTAS 纪念馆和博物馆
Rauf Raif Denktas Memorial and Museum

设计单位：Tuncer Cakmakli Architects
委托方：Turkish Republic of Northern Cyprus
项目地址：塞浦路斯尼科西亚
用地面积：94 000m²
建筑面积：1 500m²

Designed by: Tuncer Cakmakli Architects
Client: Turkish Republic of Northern Cyprus
Location: Lefkosa, Cyprus
Site Area: 94,000m²
Building Area: 1,500m²

历史之旅
植物水池
综合楼

项目概况

Rauf Raif Denktas 纪念馆和博物馆是土耳其当时的领导人在他在世时的一个项目。在本设计中有关土耳其参考成为 Denktaş 的纪念馆和纪念博物馆的设计立足点，项目建造的目的是为了一起纪念 Türk Mukavemet Teşkilatı、Cumhuriyet 种植园、纪念公园和自然博物馆，它象征着土耳其地理位置的抽象纪念碑。

建筑设计

土耳其创始人 R.R. Denktaş 的墓是一个巨大的塔和一个坚实的体块，坐落在一个方形底座上。此外，它象征着土耳其地理位置的抽象纪念碑。简单而纯粹的纪念碑包含与光影的和谐的纪念休息室。

在纪念公园，纪念墙为纪念馆、纪念碑塔和纪念广场创建一个分界线，它是由有植物的水池所包围的。该博物馆的设计引入广场，提供游客在塞浦路斯土耳其社会及集成的纪念仪式广场和纪念公园的历史之旅。室内和室外空间和园林的存在服务于教育和研究设施，对游客开放提供享受的同时也是一个大的综合楼。

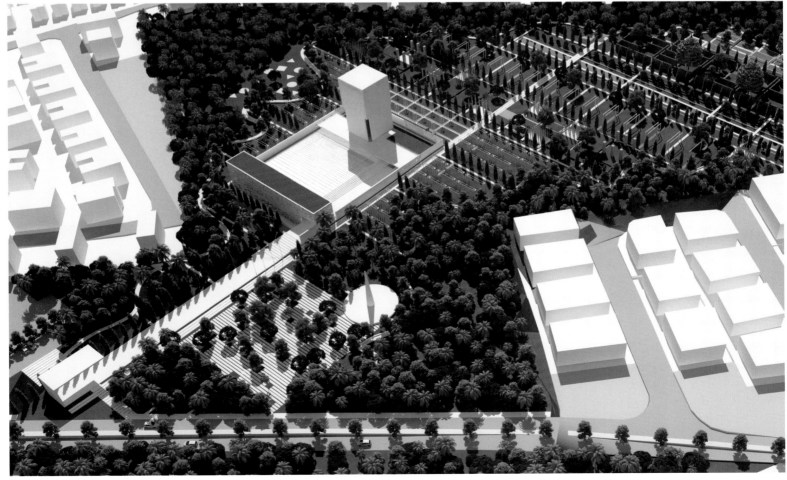

Site Plan
总平面图

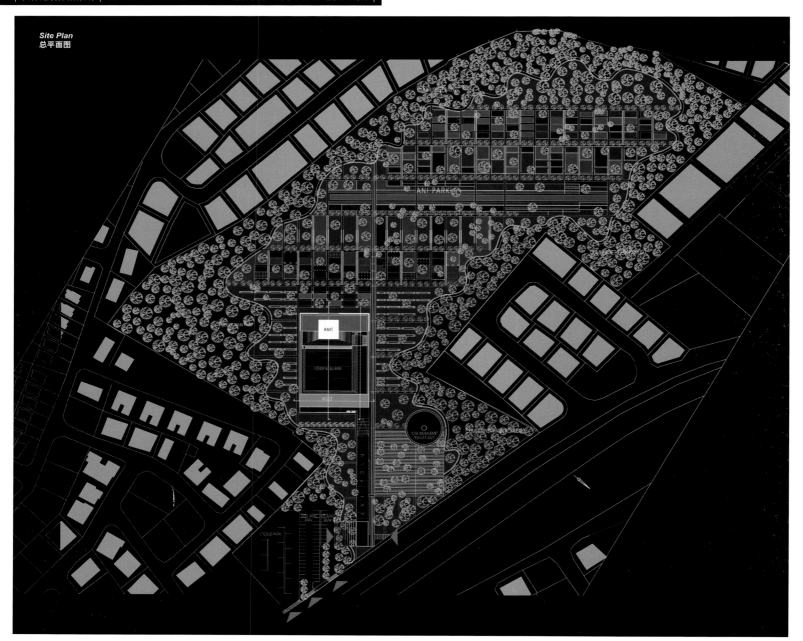

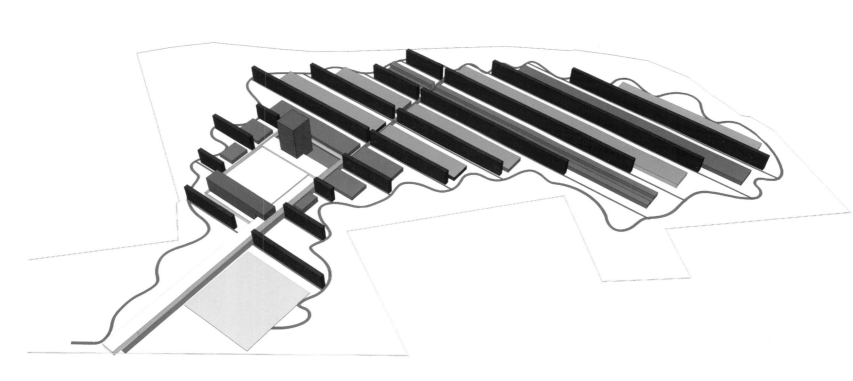

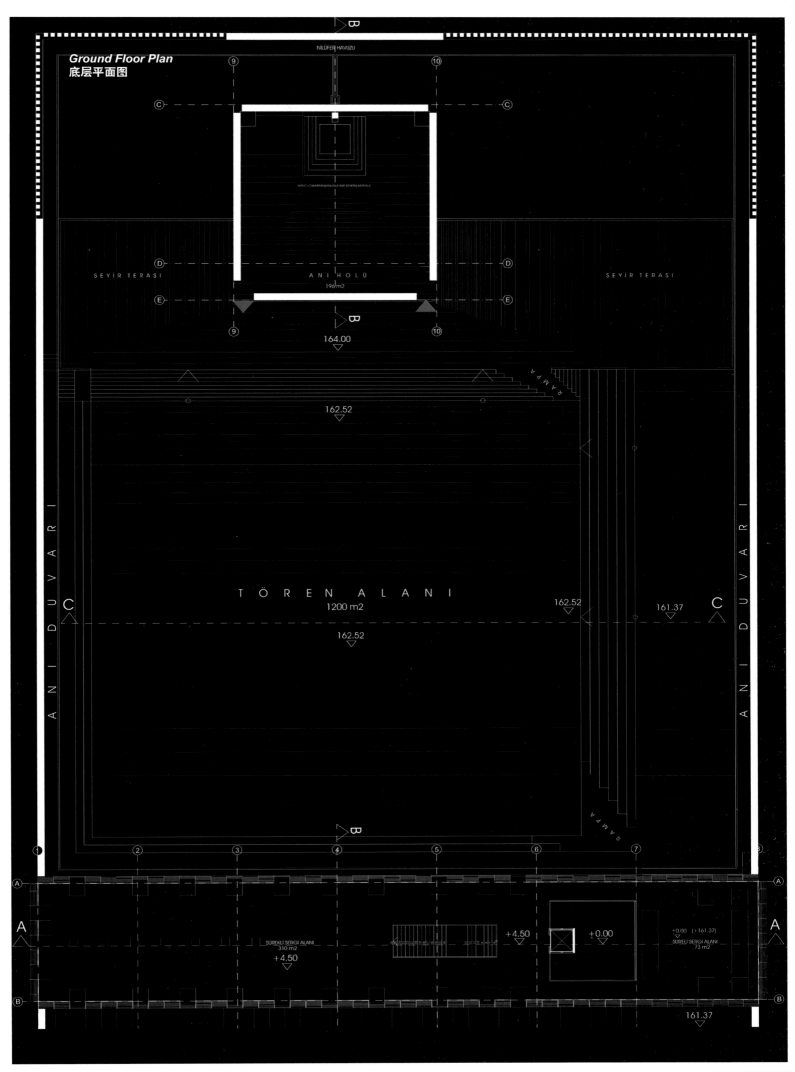

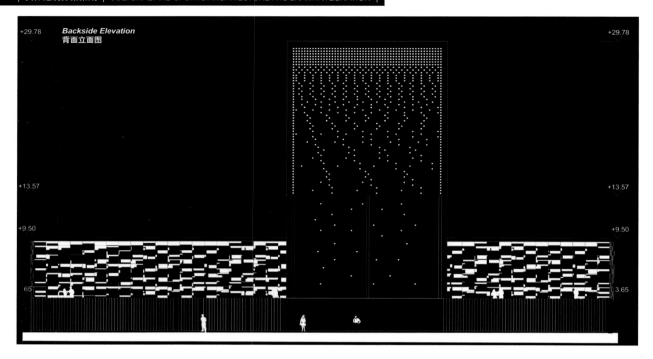
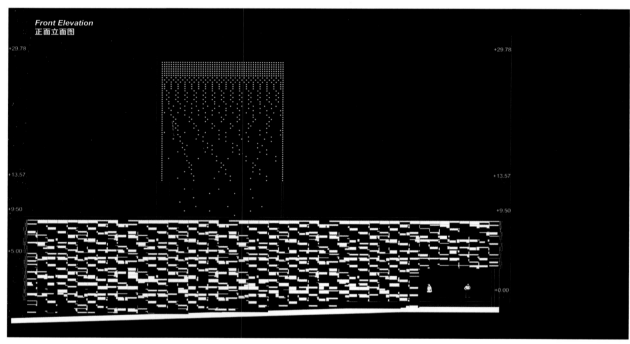
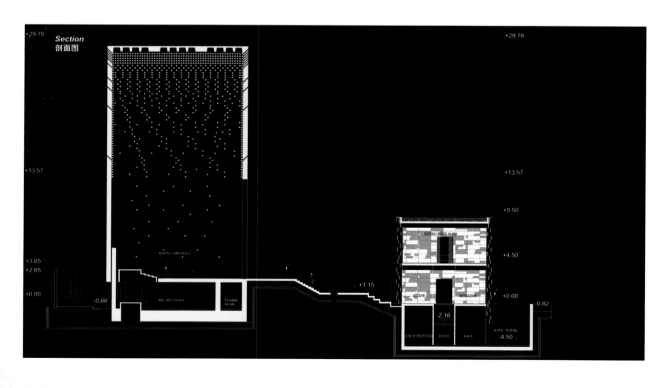

History Tour
Plant Pool
Complex Building

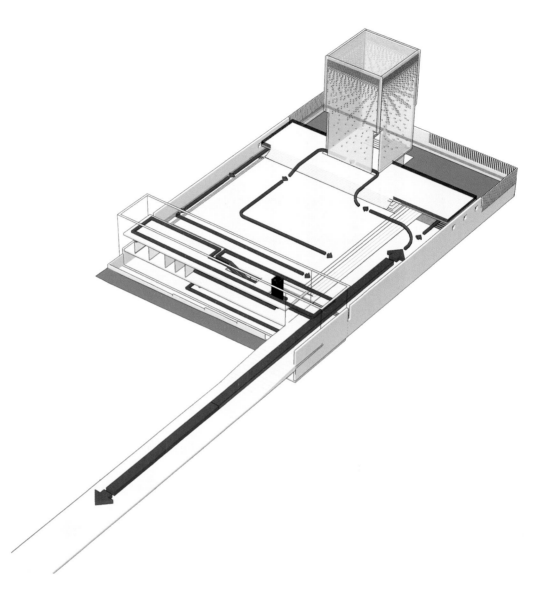

Project Overview

R.R. Denktaş was one of the leaders of the Turkish world during his lifelong. In this design regarding to the reference to the Turkish world, it is aimed to form Denktaş's memorial and memorial museum with the memorial of Türk Mukavemet Teşkilatı, Cumhuriyet Plantation Park, memorial park and nature museum, together. Also, it symbolizes the abstract monuments of the Turkish geography.

Building Design

The founder president's—R.R. Denktaş—tomb is a monumental tower and a solid mass that sits on a square base. Also, it symbolizes the abstract monuments of the Turkish geography. The simple and pure monument contains the memorial lounge with the light and shadow harmony.

In the memorial park, a memorial wall creates a boundary for memorial museum, monumental tower and memorial square and it is surrounded by a pool with plants. The museum design introducing the square provides the visitors with a trip in the history of Cyprus Turkish Society and integrating memorial-ceremony square and memorial park and there exist outdoor and indoor spaces and gardens which serve education and research facilities, is a big complex for visitors to enjoy at the same time.

美国路易斯安那州博物馆与名人堂
Louisiana State Museum and Sports Hall of Fame

设计单位：Trahan Architects
委托方：路易斯安那州规划和控制基金办公室
项目地址：美国路易斯安那州纳契托什
建筑面积：2 601.29m²
主创人员：Victor F. "Trey" Trahan III, FAIA
　　　　　Brad McWhirter, AIA　　Ed Gaskin, AIA
　　　　　Mark Hash　　　　　　Michael McCune, AIA

Designed by: Trahan Architects
Client: State of Louisiana, Office of Facility, Planning & Control
Location: Natchitoches, Louisiana, USA
Building Area: 2,601.29m²
Key Personnel: Victor F. "Trey" Trahan III, FAIA;
*　　　　　　　Brad McWhirter, AIA; Ed Gaskin, AIA;*
*　　　　　　　Mark Hash; Michael McCune, AIA.*

建筑合并 棱角分明 流动的曲线

项目概况

路易斯安那州博物馆与名人堂位于路易斯安那州纳契托什合,它是在合并两个历史悠久的、对比鲜明的藏馆之基础上建造的,将两种建筑进行综合,是为了更好地提升游客的体验价值。项目从设计上协调了体育与历史、过去与未来、容器与内容物之间的对话。

建筑设计

该建筑的前身分别位于一所大学体育馆里面和一个19世纪的法院中,因此,该建筑在设计过程中首先需要考虑的是如何解读体育文化历史。

建筑内部组织是现有城市蜿蜒流转的延伸,该建筑设计调整了历史性商业中心以及毗邻居民区的规模和性质,尽可能地融合两个藏馆,使其成为一个整体,而不是两个分散的独立主题。建筑外观以百叶窗和隔板打磨,简洁、大方、棱角分明,由百叶窗控制的光线、风景和通风,除了使门面更有生气外,更加有利于增强建筑装饰表面衔接的清晰度。不同于建筑外观的"有棱有角",建筑内部以流动的曲线为主题,引诱游客遗忘步行游览,并进入令人回味的内部展览空间。动态门厅被雕刻出1 100块铸石面板,无缝地集成所有的系统,用来自上面的自然光线冲洗。流面伸入画廊,作为电影和显示的"屏幕"。在建筑上层,用铜百叶遮蔽,有路径可俯瞰城市广场以及连接到内部公共空间的阳台。

Building Merger Angular Flow Curves

Project Overview

The project is located in Natchitoches, Louisiana, and it is built on the basis of merging two historic and contrasting collections in order to elevate the visitors' experience value. The project on the design has balanced the dialogues between sports and history, the past and the future, the containers and the contents.

Building Design

The building is formerly housed in a university coliseum and a nineteenth century courthouse, so the first consideration in the design process of the building is that how to interpret athletics culture and history.

The internal organization is an extension of the existing meandering urban circulation, while the design mediates the scale and character of the historic commercial core and adjacent residential neighborhood to merge the two collections as much as possible so that it becomes an integral whole rather than two separate and independent themes. The exterior is decorated with the shutters and clapboards which are simple, generous and angular. The louvered controls light, views and ventilation, animates the façade, and it is good for enhancing surface articulation achieved by architectural ornamentation. The interior building mainly uses the flow curves as the theme other than the "angular" building façade, enticing visitors to leave the walking tour and into the evocative exhibit spaces within. The dynamic foyer is sculpted out of 1,100 cast stone panels, seamlessly integrating all systems and washed with natural light from above. The flowing surfaces reach into the galleries, serving as "screens" for film and display. At the upper level of the building, the path arrives at a veranda overlooking the city square, sheltered by copper louvers, further connecting the interior to the public realm.

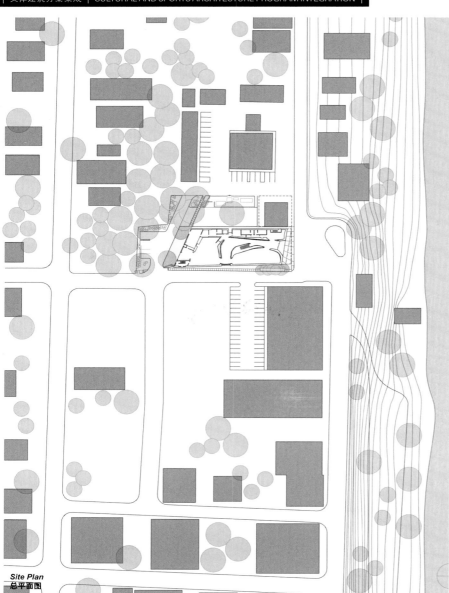

Site Plan
总平面图

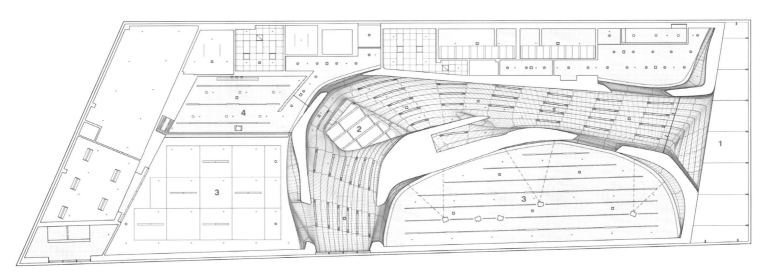

First Floor Rcp
一层平面图

N 0 10 20

1. PORCH
1. 走廊
2. FOYER
2. 门厅
3. GALLERY
3. 画廊
4. CLASSROOM
4. 教室
5. VERANDA
5. 阳台
6. ADMINISTRATION
6. 行政区

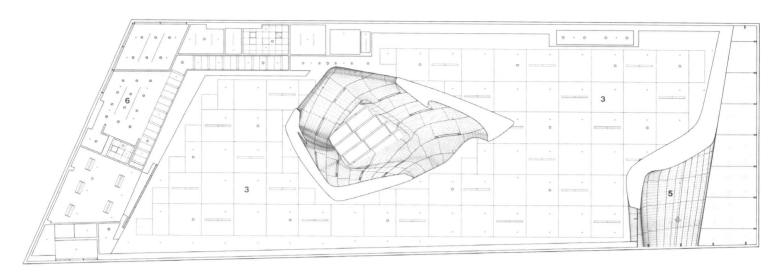

Second Floor Rcp
二层平面图

| 文体建筑方案集成 | CULTURAL AND SPORTS ARCHITECTURE PROGRAM INTEGRATION |

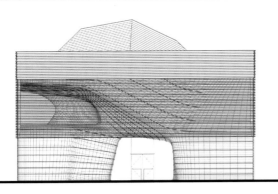
East Elevation
东立面

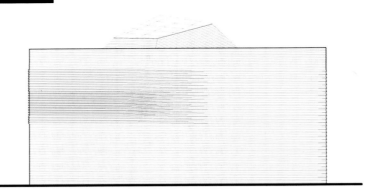
West Elevation
西立面

North Elevation
北立面

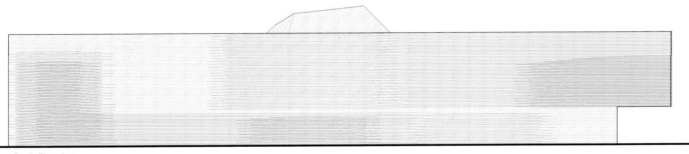
South Elevation
南立面

| 文体建筑方案集成 | CULTURAL AND SPORTS ARCHITECTURE PROGRAM INTEGRATION |

Section
剖面

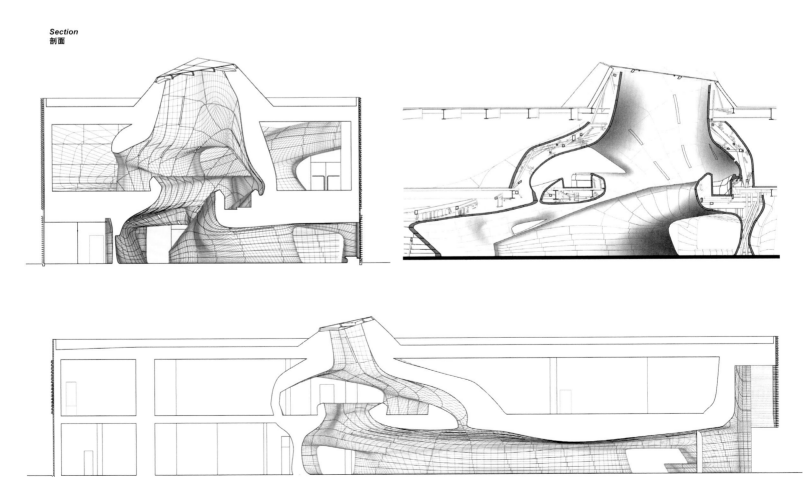

德国汉诺威史普格尔博物馆扩建
Sprengel Museum Extension

设计单位：LAN Architecture	Designed by: LAN Architecture
委托方：汉诺威市	Client: City of Hannover
项目地址：德国汉诺威市	Location: Hannover, Germany
项目面积：4 200m²	Area: 4,200m²

网状幕墙
展览花园

项目概况

每个建筑都有自己的历史，并提供了一个当代视角辐射整个城市的景观。史普格尔博物馆的扩建工程具有巨大的变化潜力，博物馆作为其中的一部分，旨在加强与周围环境、建筑和地形之间的相互作用。设计师们通过塑造建筑的垂直度来满足这一宗旨：创造灵活多样的展览空间、提升该机构的能见度、更新原建筑的形象以及与现有架构之间的对话等。

建筑设计

项目设计的主张是它只是作为一个博物馆的整体而存在，它不是一个重新组合、复合架构的概念，立体主义美学的分散景观特征的多样性在该建筑体系的逻辑结构中得到广泛应用。

该建筑采用垂直配置，体量之间的堆叠组成空间结构，为其模块化和灵活性提供了一种巨大的可能性；建筑外层布满了双重点的网状物，造型新颖奇特，为外部观察者创造出一种不断变化的景观运动游戏，使他们能够在视觉上与新大楼进行互动。建筑内部，在展览室中开辟了休息空间，自然光线透过玻璃幕墙，漫射其中。俯瞰城市景观的阳台充当休息的避风港，并通过博物馆成为游客行程中不可或缺的一部分。带旋转木马的展览花园使这一领域更加完整化，它为游客提供了一个可碰触的自然环境，并营造了一种温馨的感官体验。

扩建的新馆在功能上是多方面的，并且对游客绝对开放，它的存在更加丰富了这座建筑的文化感和使命感。

Mesh Curtain Wall Exhibition Garden

Project Overview

Each building has its own history, and provides a contemporary perspective to radiate out the landscape over the city.

The Sprengel Museum extension has enormous potential for change. The Museum as for a part among them, aims to strengthen the interaction with its immediate surroundings, buildings and topography; the designers met this aim by shaping the verticality of the building: the creation of diverse and flexible exhibition spaces, the improvement of the institution's visibility, the renewal of its image, a dialogue with the existing architecture.

Building Design

The potential of the design is its conception as a museum not as a recomposed and composite architecture. The multiple and fragmented viewpoints characteristic of the Cubist aesthetic are extensively applied here within an architectural logic.

The building adopts the vertical configuration and the volumes stack to be structural space which provides an enormous possibility for modularity and flexibility. The envelope is studded with a double network of points, with the innovative shape creating a kind of kinetic game; constantly changing view for outside observers, enabling them to visually interact with the new building. Inside the building, they open up intermediate resting spaces between the exhibition rooms where the natural light can go through the glass curtain walls and diffuse light among them. Balconies looking out over the cityscape act as havens of rest, and become an integral part of the visitor's itinerary through the museum. An exhibition garden with a carrousel completes it, providing a tactile natural environment with a warm sensory experience.

The new Museum varies its function and resolves to be open to visitors, whose presence gives the building a richer quality for the building's culture and mission.

德国拜罗伊特理查德·瓦格纳博物馆扩建
Richard Wagner Museum Extension

设计单位：OFIS Arhitekti	Designed by: OFIS Arhitekti
委托方：理查德·瓦格纳博物馆	Client: Richard Wagner Museum
项目地址：德国拜罗伊特	Location: Bayreuth, Germany

斜坡
神秘入口
天窗

Site Plan
总平面图

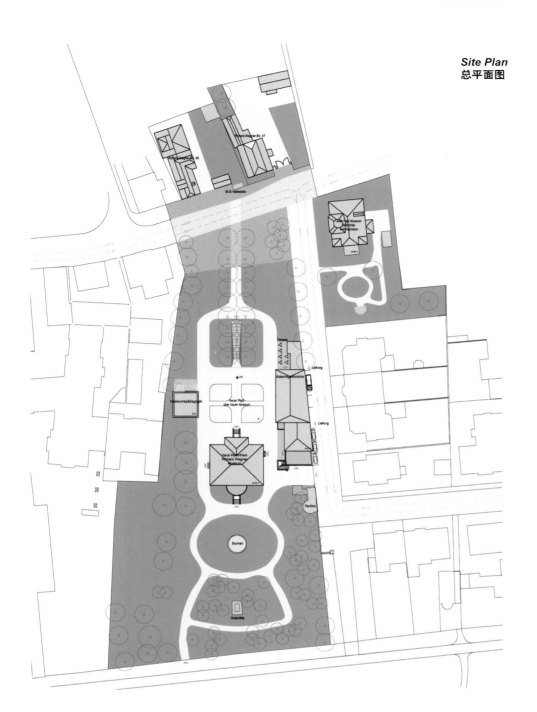

项目概况

理查德·瓦格纳博物馆坐落在拜罗伊特，德国东南部城市，它紧邻连接柏林与慕尼黑的 9 号高速公路和 70 号高速公路；铁路四通八达，与欧洲高速轨道网相接，其交通位置非常优越。

而这个历史古迹代表面临着与自己独特身份的重排机会。设计规划在主要广场下面安排一个入口和展览厅。

建筑设计

建筑的主要材料是混凝土和玻璃。博物馆的展览厅被安置在地面广场下面，入口是一个斜坡引导的，它隐藏在原有的林木之间，这样设计既保留了原有的绿化率，同时也给博物馆提供了一个幽静的环境，增添了一份神秘感。展览厅的光线来源于开向地面的四个大天窗，人们在广场上也可以看到展览厅，因此，形成了地上和地下互动的效果。

地面上设计了呈四个方向展开的天窗，斜坡上辅助设计了多个座位，视觉上它们与地面的综合楼相呼应，在这个空间里，游客可以休息，同时也能进行小型的户外活动和音乐会，进一步丰富了博物馆的功能。

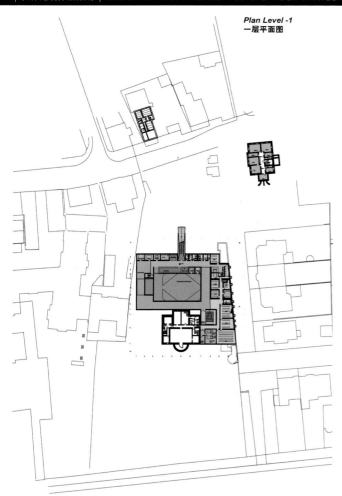

Plan Level -1
一层平面图

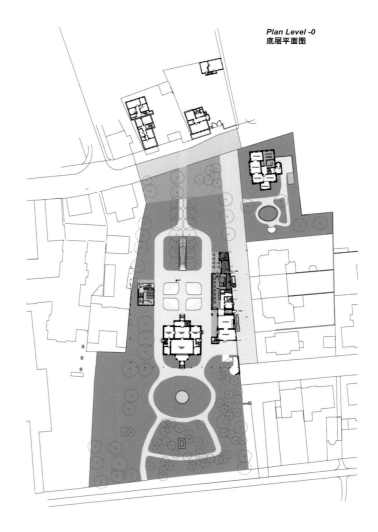

Plan Level -0
底层平面图

Ramp
Mysterious Entrance
Dormer

Project Overview

The project of Richard Wagner Museum is located in Bayreuth, a southeast city, Germany. It is close to connecting Berlin and the No. 9 highway and No. 70 highway of Munich; the convenient railway is connected with the European high-speed rail network, therefore, its transport position is very superior.

However, the historical monument is facing the rearrangement opportunities with itself subtle unique identity. The design and planning arranges for one entrance and exhibition hall under the main square.

Building Design

The main materials used for the new building are concrete and glass. The exhibition is housed in the museum plaza below ground, and the entrance is a ramp guide which is hidden among the existing trees, so this design not only retains the original green rate, but also offers a quiet environment for the museum, adding a mystery. The light in the hall comes from the four large skylights opening toward the ground. The people in the square can also see the hall; therefore, it forms interactive effects between the ground and underground.

There are skylights in four directions on the ground, and the slopes aid to design several seats. They visually echo with the complex on the ground floor. In this space, visitors can rest, and also can carry out small-scale outdoor events and concerts, which further enrich the museum's functions.

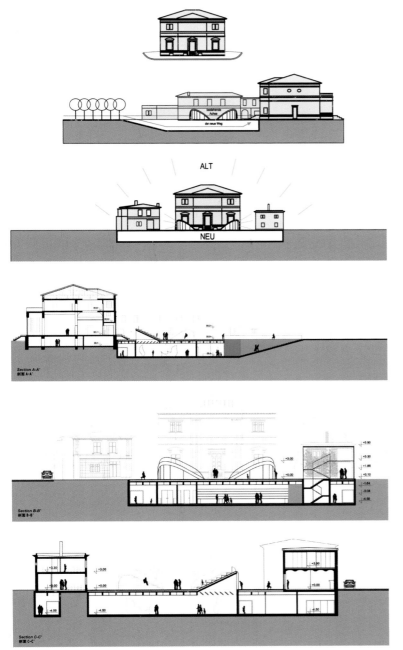
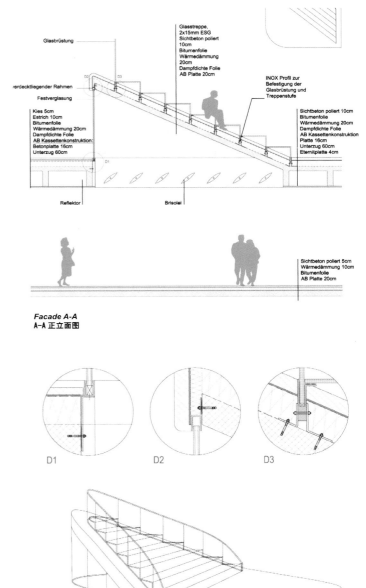

奥地利韦尔斯 Welios
Wels Welios

设计单位：Archinauten	*Designed by: Archinauten*
委托方：OÖ Science-Center Wels Errichtungs-GmbH.	*Client: OÖ Science-Center Wels Errichtungs-GmbH.*
项目地址：奥地利韦尔斯市	*Location: Wels, Austria*
建筑面积：2 389m²	*Building Area: 2,389m²*

项目概况

　　本案坐落在奥地利上奥地利州的韦尔斯市。welios 是一个科学博物馆，专注于可再生能源的节能策略。welios 被设计成一个信息节点，采用开放的、动态的坡道，倾斜墙和流畅线条的空间概念。

**信息节点
零能耗建筑
水泥体结构**

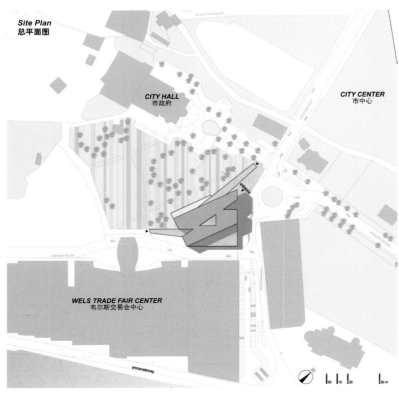

Site Plan
总平面图

CITY HALL
市政府

CITY CENTER
市中心

WELS TRADE FAIR CENTER
韦尔斯交易会中心

建筑设计

Welios，设计将它使人们和媒体进行互动。连接城市和游乐场的公众人行路沿中心轴穿插在建筑物中，形成一个长长的入口，它既非内部结构也非外部构建，游客和行人可以在此得到满足。welios 是一个零能耗建筑，也是一个嵌入了新的实验系统的试点项目。

展览室的 X 形配置允许专题空间的灵活性安排——有时是看不到外面的景观的。然后开放的中庭与另一个中庭连接客房。其他房间可以看到外面，阳光也可以照射到房间内。这种方式的空间序列可以被布置成符合展览概念的不同尺寸和特征。

该建筑是一个坚固的水泥体结构。内饰——天花板、墙壁——干性结构。紧凑式结构形式和高度绝缘外墙施工有着高程度的能效。一个闪闪发光的白色扩展金属幕墙作为一个同质外壳覆盖整个建筑体，被限制在"开放"的凹部。外墙的窗口在金属幕后欲隐欲现，同时也是作为遮光元素。线条图案是用钢带制成立面的"力线"。背面——具有节能 LED 灯条钢带的照明使立面设计不断变化，达到从轻微的微光到动态的效果。通过这种方式，科学中心的存在在黑暗中也变得必要，并具有一个显眼的外观。

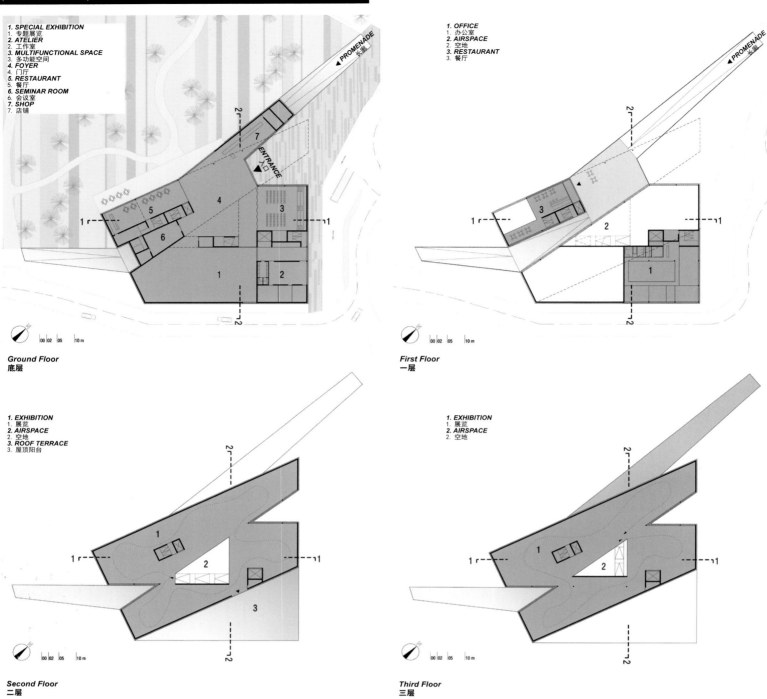

Ground Floor
底层

1. SPECIAL EXHIBITION
   1. 专题展览
2. ATELIER
   2. 工作室
3. MULTIFUNCTIONAL SPACE
   3. 多功能空间
4. FOYER
   4. 门厅
5. RESTAURANT
   5. 餐厅
6. SEMINAR ROOM
   6. 会议室
7. SHOP
   7. 店铺

First Floor
一层

1. OFFICE
   1. 办公室
2. AIRSPACE
   2. 空地
3. RESTAURANT
   3. 餐厅

Second Floor
二层

1. EXHIBITION
   1. 展览
2. AIRSPACE
   2. 空地
3. ROOF TERRACE
   3. 屋顶阳台

Third Floor
三层

1. EXHIBITION
   1. 展览
2. AIRSPACE
   2. 空地

Information Node
Zero Energy Building
Cement Structure

Project Overview

The project is located in the city of Wels, Austria. Welios is a science museum focusing on renewable energy-saving strategies. It is designed as an informational node, an open and dynamic spatial conception of ramps, sloping walls and flowing lines.

Building Design

It permits people and media to interact a public footpath which connects city center and fairgrounds perforates the building along one of the central axes. It forms a long threshold: neither interior nor exterior, where visitors and pedestrians almost meet. Welios is a zero energy building, a pilot project which embeds new experimental systems.

The X-shaped configuration of the exhibition rooms allows for a flexible arrangement of the thematic spaces—sometimes the view to the outside is closed, then the open atrium connects the rooms with one another. Other rooms allow a view to the outside, sunlight falls into the room. In this way sequences of spaces can be arranged with various sizes and characteristics in keeping with the exhibition concepts.

The building is a solid construction as a cement body. The interior—ceilings, walls—is made in dry construction. The compact construction form and the highly insulated outside wall construction promise a high degree of energy efficiency. A shimmering white, expanded metal façade covers the entire body of the building as a homogeneous casing. Glass walls are limited to the "opened up" recesses. Window openings in the external façade are blurred behind the metal curtain, which serves as a shading element at the same time. The line pattern is made with steel bands—the "force lines" on the façade. The backlighting of these bands with energy-saving led light strips enables a variable design of the façade, ranging from a slight shimmer to dynamic effects. In this way, the science center is also imbued with the necessary presence in the dark and with an unmistakable appearance.

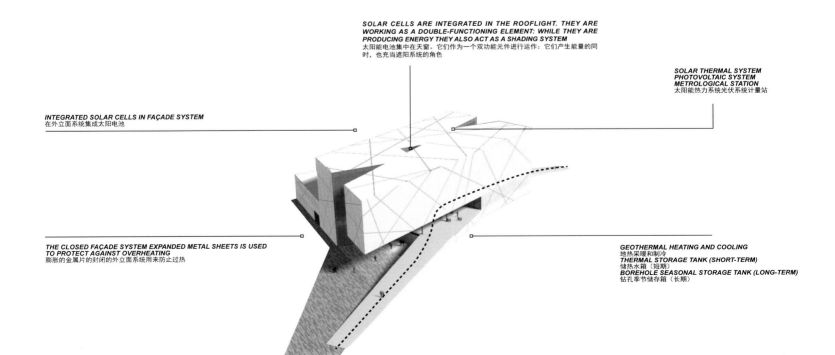

| 文体建筑方案集成 | CULTURAL AND SPORTS ARCHITECTURE PROGRAM INTEGRATION |

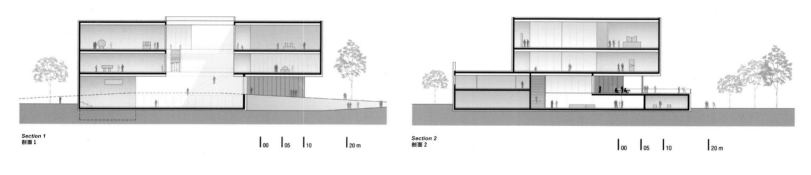

Section 1
剖面 1

Section 2
剖面 2

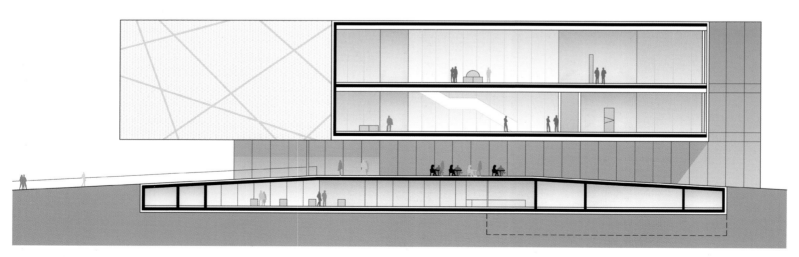

Section Promenade
长廊剖面

斯洛文尼亚马里博尔美术馆
Slovenia Maribor Art Gallery

设计单位：Architekt Lukas Göbl Office for Explicit Architecture
项目地址：斯洛文尼亚马里博尔
项目面积：17 000m²

Designed by: Architekt Lukas Göbl Office for Explicit Architecture
Location: Maribor, Slovenia
Area: 17,000m²

跨学科枢纽
轴状建筑结构
动态循环

项目概况

马里博尔美术馆的设计初衷旨在为各种文化机构进行跨学科的对话创造空间。该美术馆独特的形式结合了功能和效率、技术和架构、教育和知识、以及创造性及显示屏。其目的是要成功整合这些不同的兴趣领域，实现使用选项互惠互利的捆绑。

建筑设计

该建筑混凝土和钢结构的外皮是由白瓷砖制成的。透过整个外立面，黑色的一个关节网络创建一个网格图案，几乎使建筑成为一个图，设计使眼前的空间环境融合成建筑的一部分。公共走道纵横交错于整个建设工地，甚至进入到一楼部分。由于该建筑可从两侧进入其中，而且城市河流以及公共广场就位于基地北侧，因此使马里博尔美术馆融入全市现有交通和人行道系统得以保障。此外，该架构的形式动态地反映地形因子与一系列露头和凹槽。紧凑的形式有针对性地在特定的位置开辟，使马里博尔美术馆的跨学科性质立即变得清楚。

美术馆内部的空间分布遵循了城市设计概念。公共和半公共空间如门厅、咖啡厅、餐厅、图书馆、阅读室位于底楼，各自设有独立入口设施。售票区和欧洲文化城2012年的信息中心都在正厅。在底楼的该建筑中心和在一楼的儿童博物馆共享一个小的、单独的休息室。这个入口是专为灵活性最大和在必要时可连接的大门厅而设计的。该建筑西部设有多个办事处和公寓。轴状建筑结构的中部连接不同的展览空间。纵轴是进入画廊展览室的主要通道。总体而言，展览区域涵盖五个楼层，这是由斜坡的系统相连，意味着不断循环的动态——知识、教育和游客的流动。

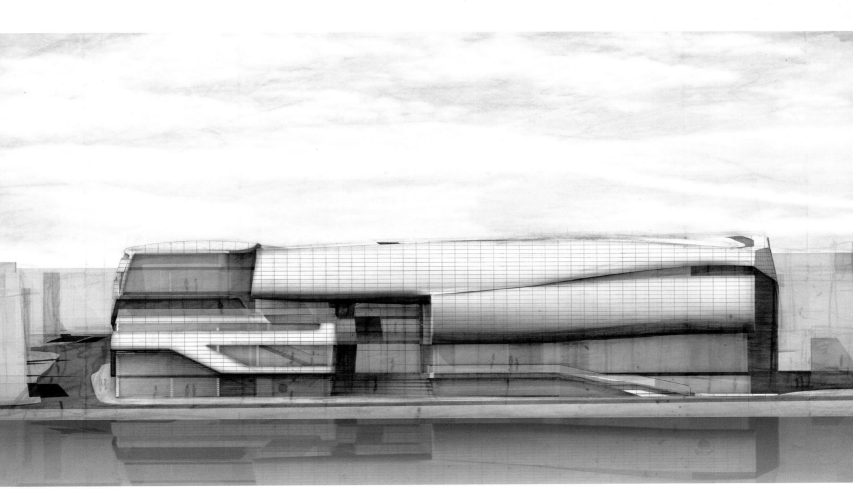
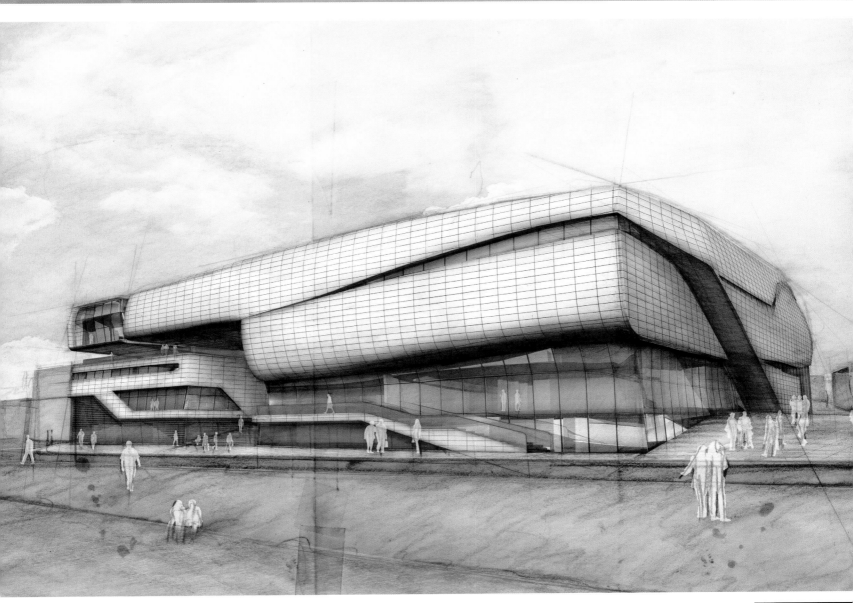

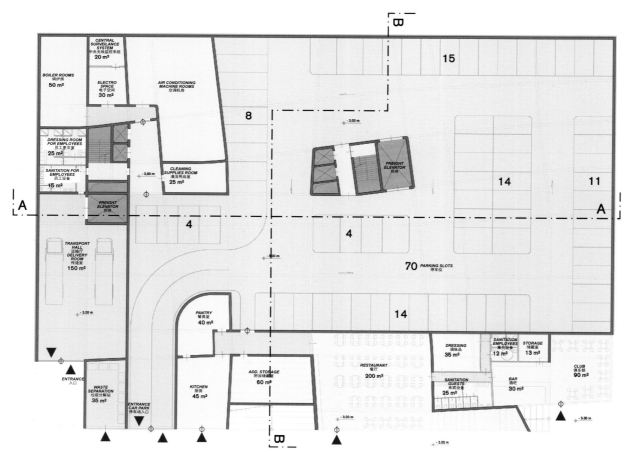

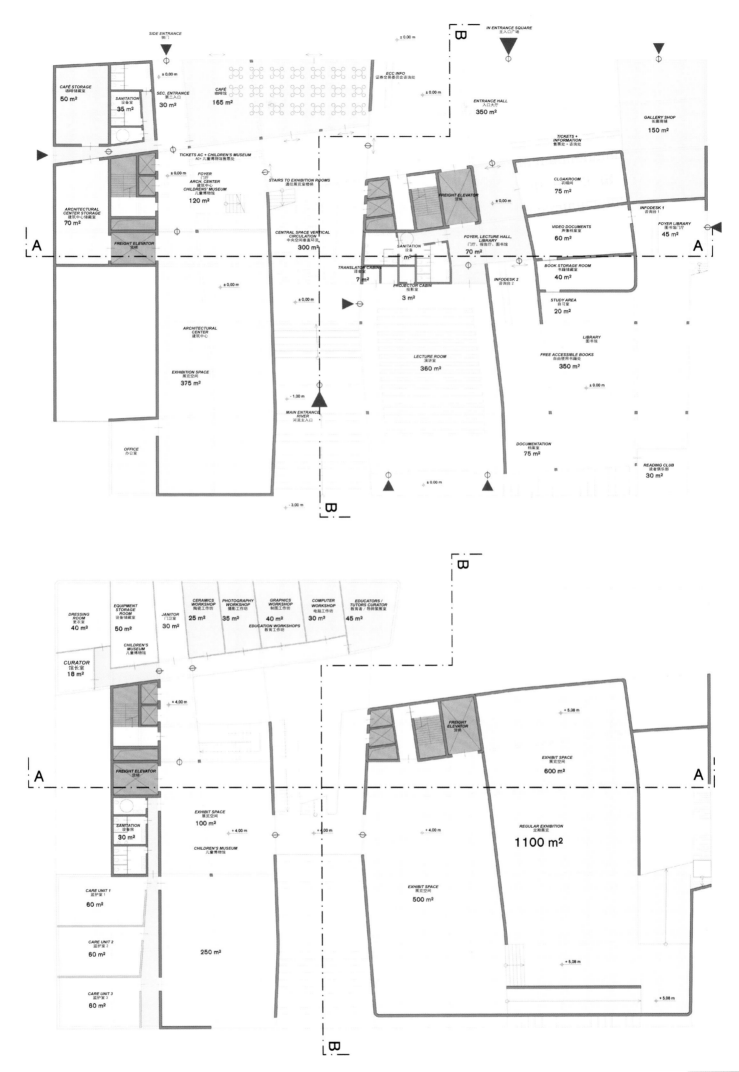

| 文体建筑方案集成 | CULTURAL AND SPORTS ARCHITECTURE PROGRAM INTEGRATION |

Cross-disciplinary Hub
Shaft-like Structures
Dynamic Cycle

Project Overview

The original intention of the Maribor Art Gallery design aims to create space for the interdisciplinary dialogue of various cultural institutions. The distinctive form of the Maribor Art Gallery combines functionality and efficiency, technology and architecture, education and knowledge, and creativity and display. The aim was to successfully integrate these different fields of interest to attain a mutually beneficial bundling of use options.

Building Design

The outer skin of the concrete and steel structure is made of white ceramic tiles. Across the entire façade, a network of black joints creates a grid pattern that almost makes the building become a diagram. The immediate spatial context is literally part of the building. The public walkways crisscross the entire building site and even enter parts of the ground floor. Since the building is accessible from both sides—the river as well as the public square on the north side—the integration of the Maribor Art Gallery into the city's existing traffic and sidewalk system is guaranteed. In addition, the architecture's form reacts dynamically to topographical factors with a series of outcroppings and recesses. The compact form is pointedly opened up at specific locations, making the interdisciplinary nature of the Maribor Art Gallery immediately clear.

The spatial distribution of the Art Gallery's interior follows the over urban design concept. Public and semi-public spaces such as the entrance hall, cafe, restaurant, library, and reading room are located on the ground floor, each with separate entrance facilities. The ticketing area and Information Centre of European Culture City 2012 are in the main hall. The Architecture Centre on the ground floor and the children's museum on the first floor share a small, separate foyer. This entryway is designed for maximum flexibility and can be joined with the large entrance hall if necessary. The western part of the building houses several offices and apartments. A central, shaft-like building structure connects the different exhibition spaces. This vertical axis is the main means of access to the gallery exhibition rooms. In total, the exhibition area covers five levels, which are connected by a system of ramps, implying the dynamics of constant circulation—flows of knowledge, education, and visitors.

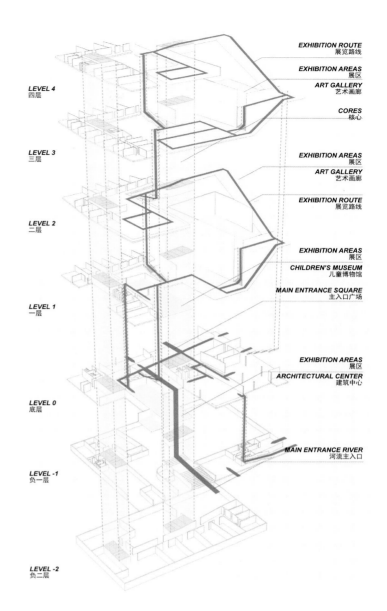

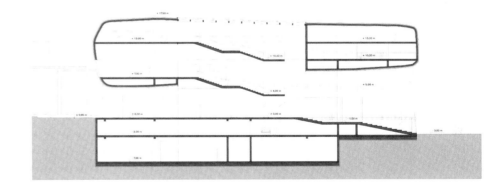

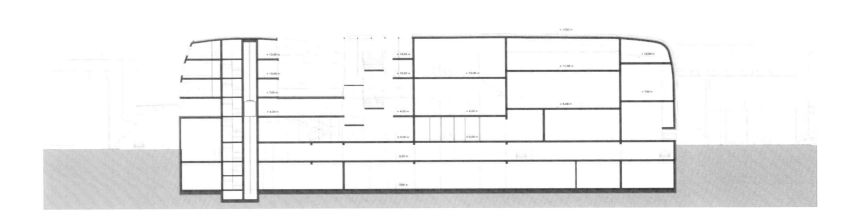

四川成都安仁桥馆
Museum-Bridge in Anren

设计单位：非常建筑
　　　　　深圳市鑫中建建筑设计顾问有限公司
委托方：四川安仁建川文化产业开发有限公司
项目地址：四川省成都市
项目面积：2403.2m²

Designed by: Atelier FCJZ
　　　　　　Shenzhen Xinzhongjian Architectural Design Consulting Co., Ltd.
Client: Sichuan Anren Jianchuan Cultural Industry Development Co., Ltd.
Location: Chengdu, Sichuan
Area: 2,403.2m²

项目概况

桥馆的正式名称为十年大事记馆（1966~1976），是用于展示樊建川先生所藏文化大革命时期文物的陈列馆。馆址位于四川成都大邑县安仁古镇，建川博物馆聚落中。因基地横跨河流，该建筑同时是一座步行桥，从而得名桥馆。

建筑设计

桥馆是人们反思20世纪中期一个时代历史的场所，设计以该时代的建筑语汇为线索建立起时间的连续性。同时，本项目也成为一项回溯西方古典主义在中国的发展过程以及尝试在当代设计中传承古典精神的探索。因而在此考量下形成的博物馆形态是一个简洁的带有深远出檐的单一体量。在设计过程中不断反思古典主义的原则。博物馆盒子没有典型古典主义建筑所具备的基座，而是支撑在倾斜的柱子上。更为明显的是，混凝土表面的粗糙肌理颠覆了古典主义的优雅感，整个混凝土结构粗砺有质感。设计希望桥馆"粗野古典主义"与生俱来的、富有历史沧桑感的外观能引导参观者们追问历史的真义。

设计师将步行桥看作街道的延续，试图弱化基础设施、建筑与城市之间的界限。博物馆盒子的悬挑出檐覆盖了步行桥，创造了供参观者和过路者止步的室外建筑空间，同时也用作博物馆的入口。博物馆室内通过四个横跨屋顶的天窗来进行采光，展览陈列由樊建川先生设计。桥下的开放空间将用作展览、茶座和演出场地，屋顶平台将来可用作雕塑展示。

现实主义
粗犷古典之美
城市基础设施建筑

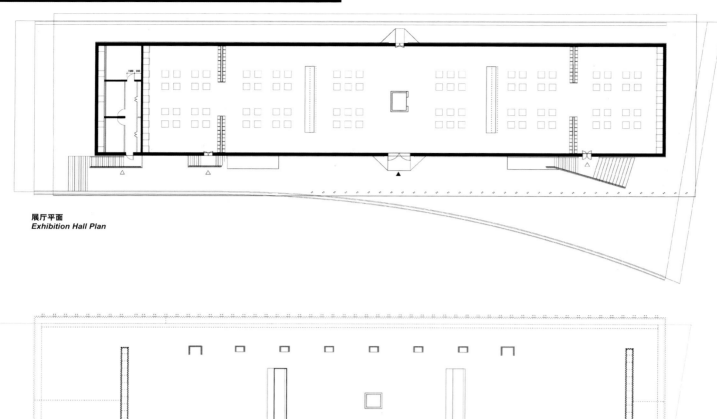

展厅平面
Exhibition Hall Plan

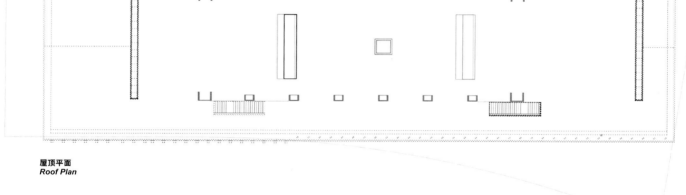

屋顶平面
Roof Plan

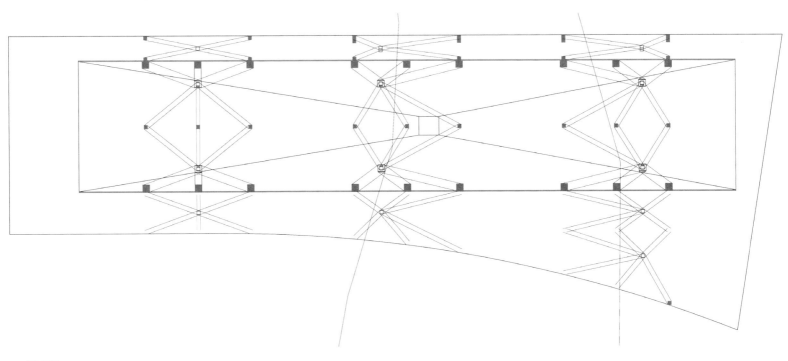

桥下平面
Underbridge Plan

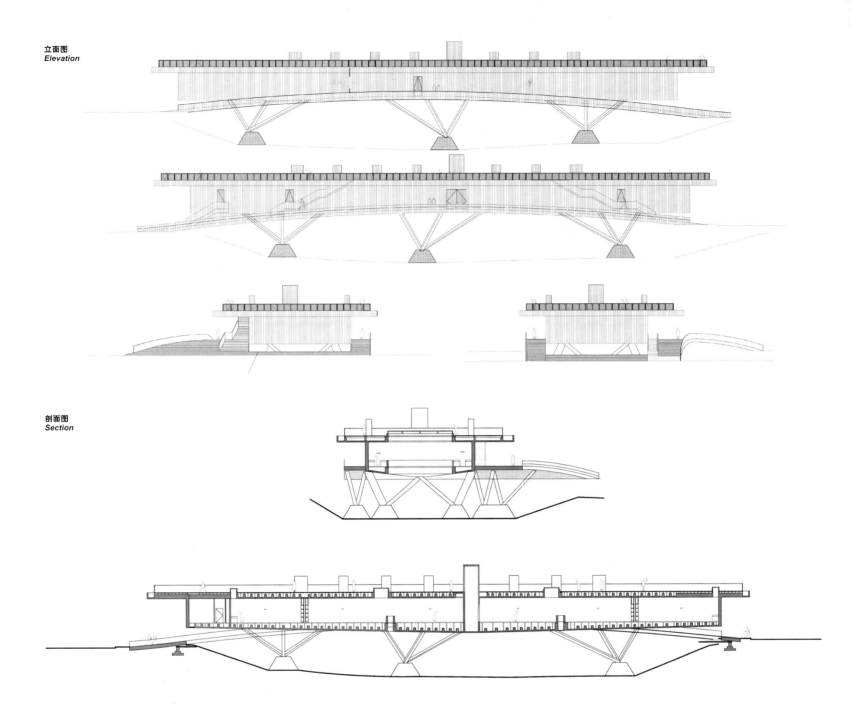

立面图
Elevation

剖面图
Section

Realism
Rugged Classic Beauty
Urban Infrastructure Construction

Project Overview

Officially named 1966-1976 Major Events Pavilion, this is a museum that houses the artifacts from the Great Cultural Revolution period from Mr. Fan Jianchuan's collection. Located in the Jianchuan Museum Town in Anren, Sichuan, over a river, the museum building is simultaneously a pedestrian bridge. Therefore, it is nicknamed Museum-Bridge.

Building Design

The Museum-Bridge will be where one goes to reflect one of these decades in the early half of the twentieth century, the design could be constructed the continuity of time by offering clues of the architectural manifestations of that historical period. Furthermore, the project evolves into a mission to trace back the development of Western Classicism in China and to see the possibility of carrying on the classical spirit in contemporary design. The observation materializes in the museum per se in the form of a simple volume with a large cantilevered roof.

Yet, the classical principles are questioned through out the design process, for instance, the museum box has no base or foundation as a typical classical building would do but supported on slanted columns. More visibly, the elegance of the Classicism is subverted by the roughness of the concrete surface. As the result of the entire structure is built with concrete by the unskilled local labor force, becoming a celebration of roughness. The design wish the "Brutalist Classicist" architecture of the Museum-Bridge may gain an instantly aged appearance that could aid the visitors in search for the meaning of the history.

The design makes the bridge as the continuation of the street to further blur the boundary between infrastructure, architecture, and city. The bridge is covered by the cantilevered eaves of the museum box to create outdoor architectural space and a stopping point for visitors and passersby, meanwhile the eaves serves as the portal of the museum. The interior of the museum is lit by four sky lights that cut across the ceiling and the exhibition installation is being designed by Mr. Fan. The under-bridge open space will be used for exhibition, tea, and performance and the roof terrace for sculpture display in the future.

俄罗斯彼尔姆芭蕾舞馆
Opera Ballet Theater, Perm, Russia

设计单位: Neutelings Riedijk Architects
委托方: 彼尔姆市
项目地址: 俄罗斯彼尔姆市
建筑面积: 33 000m²

Designed by: Neutelings Riedijk Architects
Client: City of Perm
Location: Perm, Russia
Building Area: 33,000m²

景观的延续
机器地板

项目概况

彼尔姆位于卡马河畔、乌拉尔山西麓，是俄罗斯彼尔姆边疆区的首府，1723年建城，1781年设镇，自19世纪以来便是俄罗斯重要的工业城市。该城市是俄罗斯第13大城市，是彼尔姆边疆区重要的政治、工业、科学和文化中心。

彼尔姆市柴可夫斯基芭蕾舞剧院是俄罗斯最好、最古老的剧院之一，不论在历史上还是现今都是俄罗斯最值得游览的剧院，它的芭蕾舞剧团是俄罗斯最受欢迎的剧团。

建筑设计

19世纪彼尔姆市柴可夫斯基歌剧院和芭蕾舞剧院的扩展被认为是轻轻地行走在讲台上，形成了公园景观的延续。

两个物体是漂浮在讲台以上，如公园里的雕塑，毗邻现有的剧院。一边在垂直元素设有排练，另一边在一个水平元素设有新剧院。通过向公园开放排练厅、大厅和剧院室，使在这个单一综合体内为观众提供众多戏剧成为可能。创新的机器地板和天花板使得一些影院设置成为可能，还能让新剧院举办各类当代戏剧作品。

146–147

| 文体建筑方案集成 | CULTURAL AND SPORTS ARCHITECTURE PROGRAM INTEGRATION |

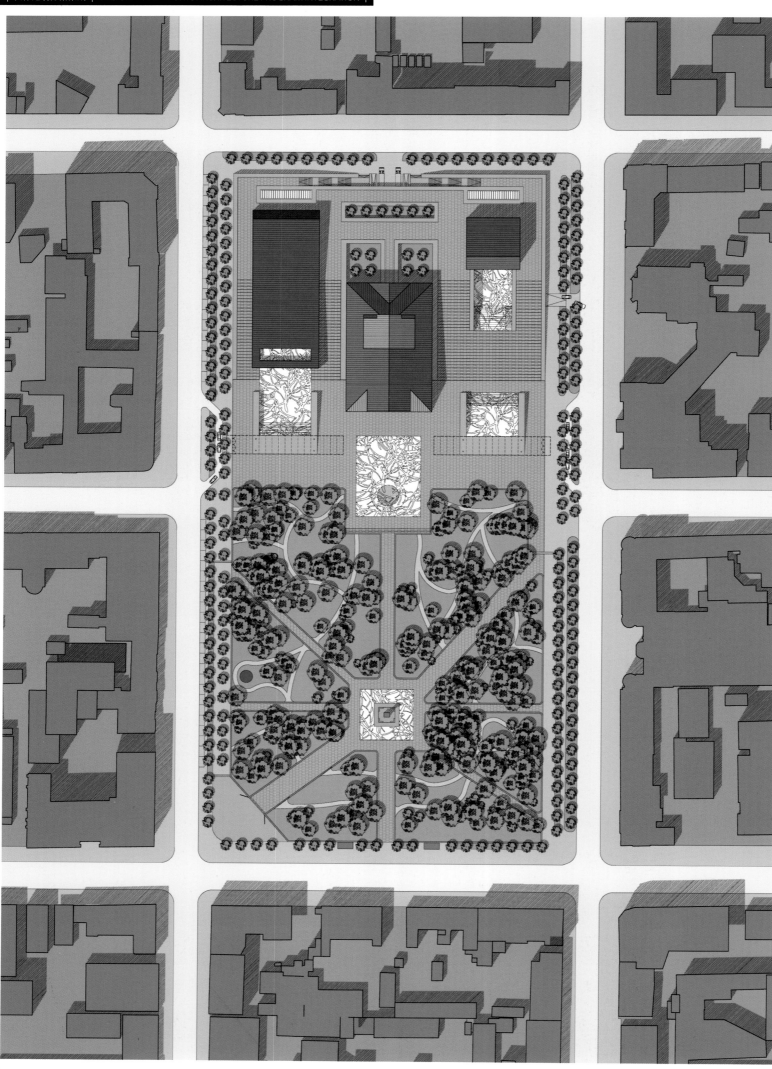

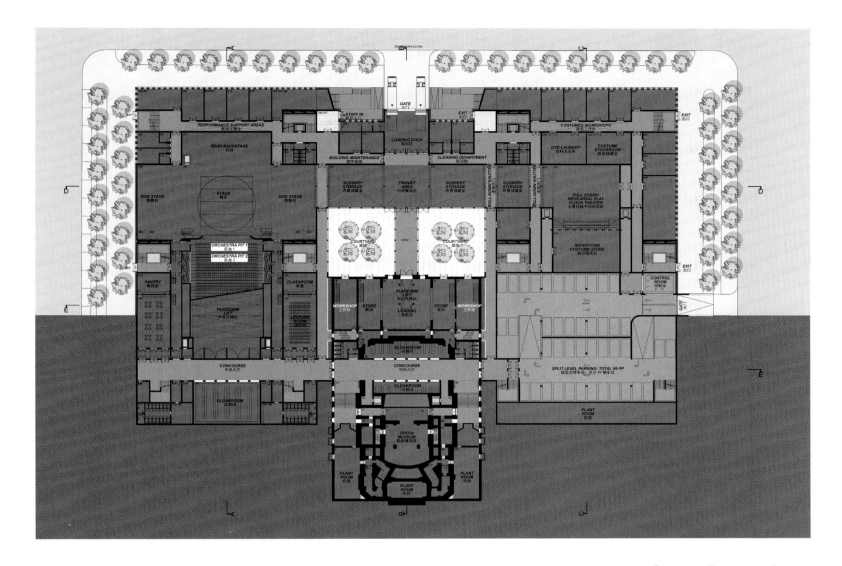

Landscape Continuation Machine Floor

Project overview

Perm is located in the riverside of Kama River and the west side of Ural Mountain, and is the capital of Perm Krai, Russia. The city was built in 1723 and the town was established in 1781, so it has been an important industrial city in Russia since the 19th century. Perm is the 13th large city in Russia and is an important political, industrial, scientific and cultural center. Perm Tchaikovsky Ballet Theatre is one of Russia's oldest and best theaters. Whether now or in the history, it is the theater where is the most worth visiting in Russia. The ballet troupe is Russia's most popular.

Building Design

The extension of the nineteenth century Perm State Tchaikovsky Opera and Ballet Theater is conceived as a gently stepped podium forming a continuation of the park landscape. Two objects are floating above the podium, as sculptures in the Park, next to the existing theater. One side houses the Rehearsal Studios in a vertical element, the other side houses the new theater room in a horizontal element. By opening the rehearsal rooms, the foyer and the theater room to the park, a multitude of theatrical possibilities within one single complex are offered. A revolutionary machine floor and ceiling makes several theater settings possible, and allows the new Theater to host all kinds of contemporary theatre productions.

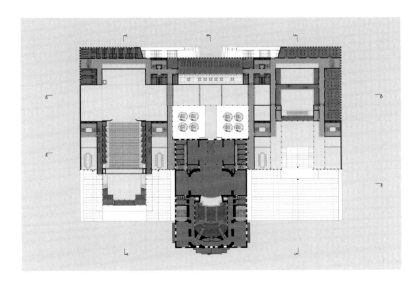
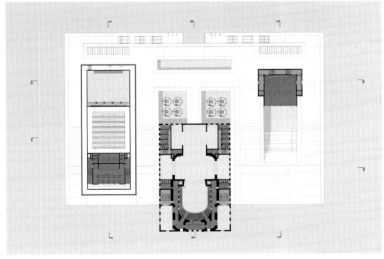

| 文体建筑方案集成 | CULTURAL AND SPORTS ARCHITECTURE PROGRAM INTEGRATION |

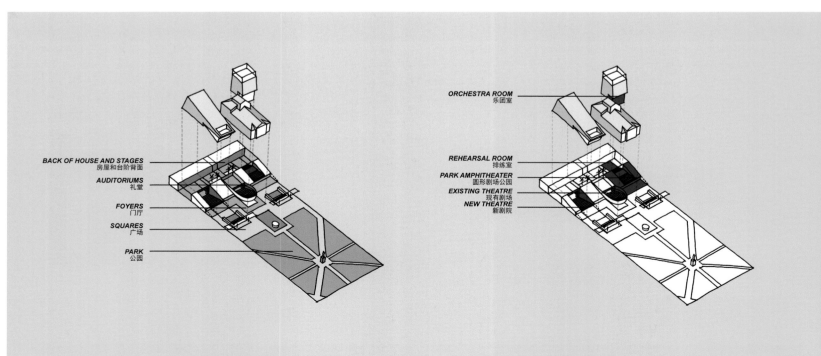

Organizational Concept
组织概念

Auditoriums
礼堂

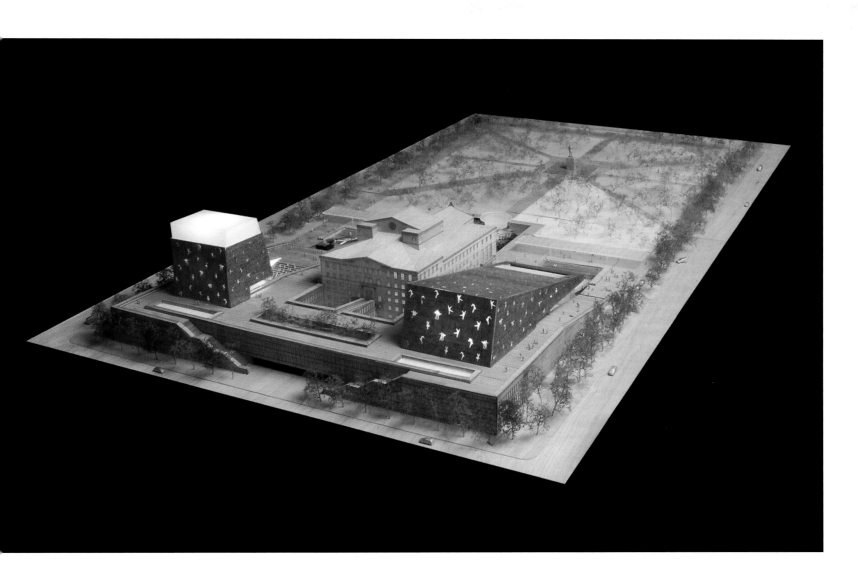

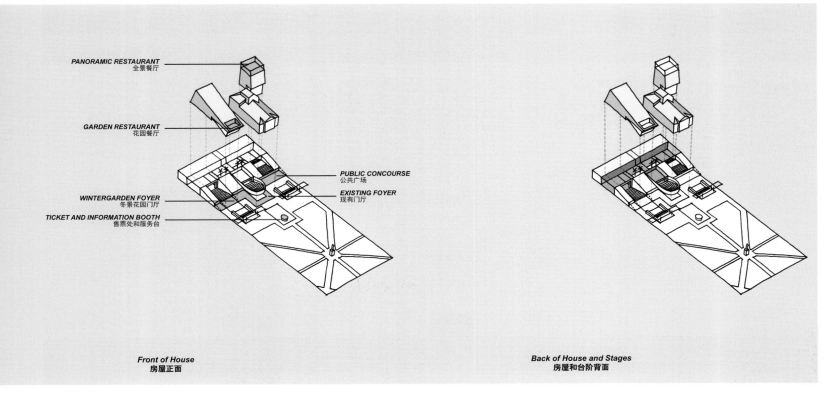

Front of House
房屋正面

Back of House and Stages
房屋和台阶背面

PANORAMIC RESTAURANT 全景餐厅
GARDEN RESTAURANT 花园餐厅
PUBLIC CONCOURSE 公共广场
WINTERGARDEN FOYER 冬景花园门厅
EXISTING FOYER 现有门厅
TICKET AND INFORMATION BOOTH 售票处和服务台

| 文体建筑方案集成 | CULTURAL AND SPORTS ARCHITECTURE PROGRAM INTEGRATION |

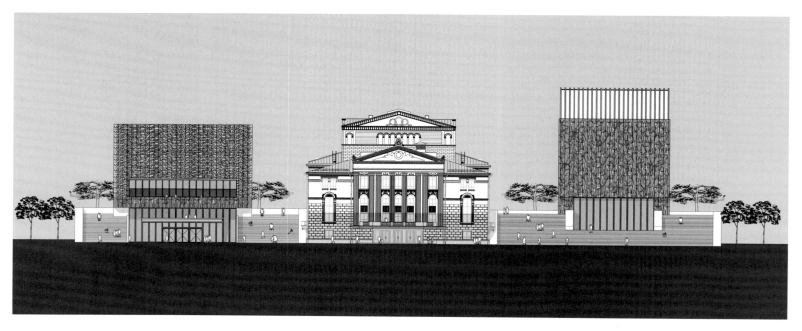

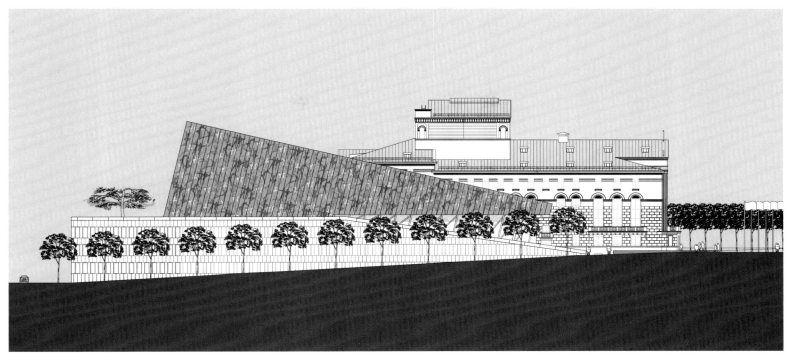
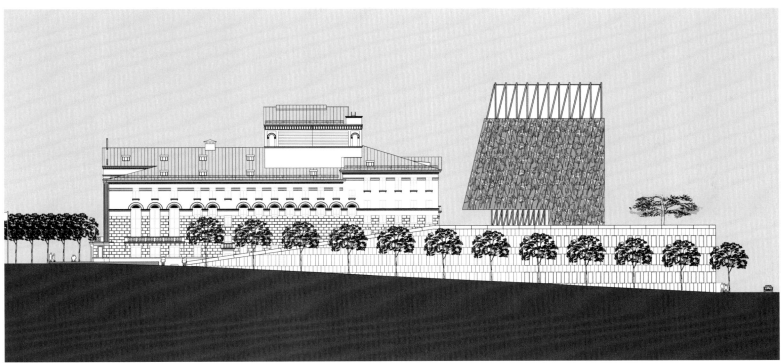

| 文体建筑方案集成 | CULTURAL AND SPORTS ARCHITECTURE PROGRAM INTEGRATION |

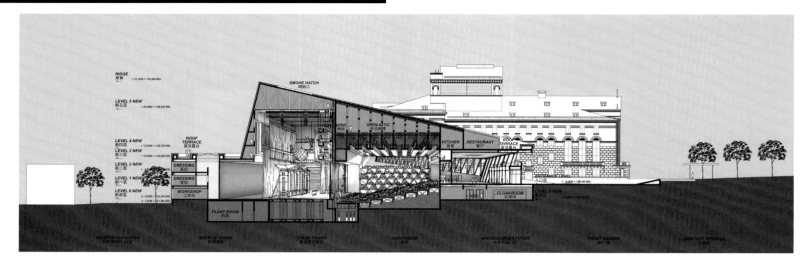

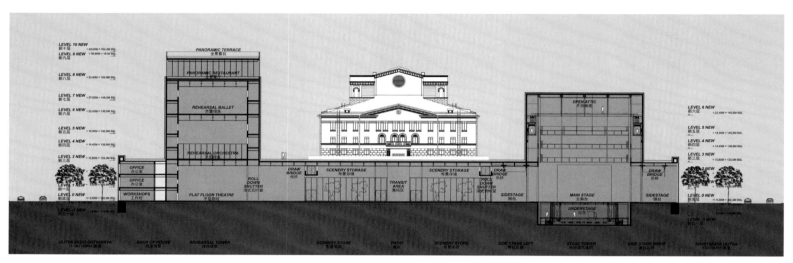

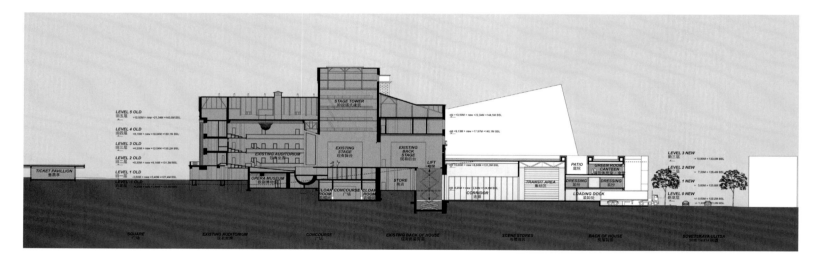

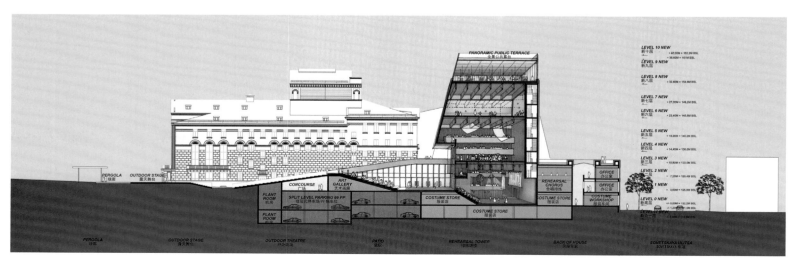

挪威 Våler 新教堂
New Church of Våler

设计单位：CEBRA	*Designed by: CEBRA*
委托方：Vaaler Parish Council	*Client: Vaaler Parish Council*
项目地址：挪威 Våler	*Location: Våler, Norway*
建筑表面积：1 099m²	*Surface Area: 1,099m²*

十字架
技术照明
木芯

项目概况

在挪威东南部的Våler村，2009年的一场大火，摧毁了当地的老教堂。由于教堂是当地社区一种非常重要的建筑——无论是作为一个社交聚会点还是作为具有特征性的景观元素。因此，Vaaler Parish Council决定在当地重建一座新教堂，以满足当地人民的生活和信仰需要。

建筑设计

新教堂在设计过程中将当地的文化历史背景与现代建筑的表现元素相结合，从而创建了一个使用光线和木材为室内主要设计元素的倾斜十字形状的标志性建筑。

十字架，这是一个强烈的视觉符号，它凭借其简单的线条将水平线与垂直线完美地结合在一起，在宗教信仰里这意味着世俗与上天。十字架也代表教会的基本功能。当十字形状倾斜和教堂的屋顶形成一个倾斜的平面，其图案在视线水平以及从空气中变得可见，建立了与景观点相同的图像。

教会在使用照明的方式是定向的——对文化和教会盛事的表演。该项目在设计过程中为了在配套礼仪和仪式程序中捕捉和利用照明，采用了一系列如棱镜、暗箱和挂镜的技术，从而创造体验的额外维度。

Våler教堂室内设计的一个重要的灵感来源，就是对于木材和木雕的使用。轻木补体和阶段以照明深入木刻的方式来搭建内部空间。照明使材料活灵活现，作为回报，内部通过浮雕和采光井的应用，以各种方式反映和影响照明。中路的木芯作为教会的入口广场，主要空间本身的垂直体验和水平空间清晰开放的环节，使整个教堂在优雅正式的语言中汇集在一起。

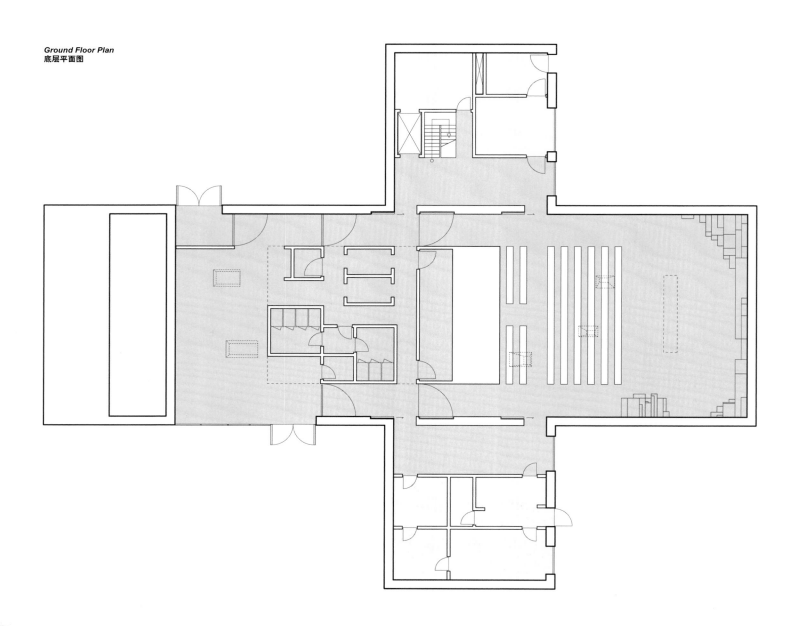

Ground Floor Plan
底层平面图

First Floor Plan
一层平面图

Roof Plan
屋顶平面图

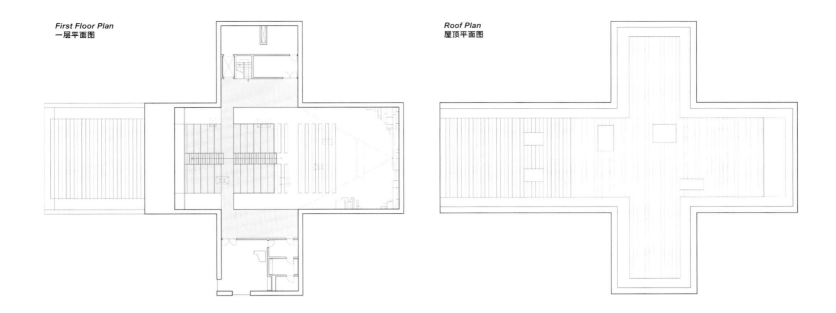

Cross
Intelligent Natural Illumination
Heart of Wood

Project Overview

The village of Våler, in the south eastern part of Norway, burned down the village's old church in 2009. The church is of great importance for the local community—both as a social gathering point and as characterizing landscape element. Therefore, Vaaler Parish Council has decided to rebuild a new church to satisfy the needs of life and belief for the local people.

Building Design

The new church in the design process is combined the local culture-historical context and the modern architectural elements thus to build a symbolic building with a tilted cross shape using light and wood as the main design elements inside the building.

The Cross is a strong visual symbol, which beautifully combines the horizontal with the vertical in its simplicity—and in its meaning the worldly with the heavenly. In the same way, the cross also represents the church's fundamental function. When the cross shape is tilted and the church's roof forms a sloping plane, the motif becomes visible at eye level as well as from the air, creating the same image from both points of view.

The church is orientated in such a way that it uses illumination—for the staging of cultural and church events. The project incorporates a series of techniques such as prisms, camera obscure and peg mirror in the design process in order to capture and utilize the light for supporting the liturgical and ceremonial proceedings, thus creating an additional dimension of experience.

A significant source of inspiration for the design of the interior of the Våler church is the usage of wood and woodcarvings. Light and wood complement and stage one another, modeling the interior spaces in the same way that light gives depth to woodcarvings. The light makes the materials come alive and in return the interior reflects and colors the light in various ways through the application of reliefs and light wells. The centrally placed wooden core acts as a lucid and open link between the horizontal spatiality of the church's entrance square and the vertical experience of the main space itself, thus allowing the whole nave to be brought together in a single elegant formal language.

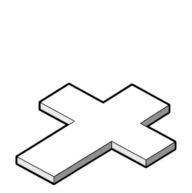 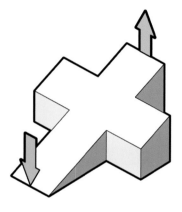 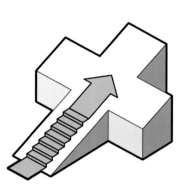 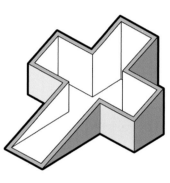

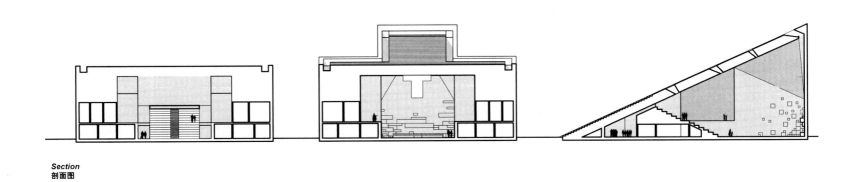

Section
剖面图

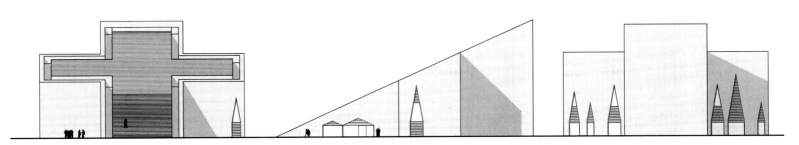

Elevation
立面图

丹麦奈斯特韦兹竞技场
Næstved Arena

设计单位：CEBRA	*Designed by: CEBRA*
委托方：奈斯特韦兹自治市	*Client: Næstved Municipality*
项目地址：丹麦奈斯特韦兹	*Location: Næstved, DK*
项目面积：10 000m²	*Area: 10,000m²*

碗形形状
植物容器
文化聚会

项目概况

奈斯特韦兹竞技场在丹麦东部，位于一个绿色的城市里。一侧被一个足球场和体育馆包围，另一侧被居民区包围，建筑面积为 10 000m²，观众容量为 2 500 到 4 000 人，是一个集体育、音乐和文化活动为一体的新的多用途场地。

建筑设计

竞技场设计的重点是将氛围和社区精神意识以及有效的功能和物流结合起来，以创造新文化聚会场所。建筑体量的圆形几何形状引导入口并进入竞技场的核心区域。竞技场的碗形形状及周边看台的倾斜面使建筑的规模发挥作用——它们有助于在体育赛事和音乐会中创建激烈和起伏的氛围，在那里你会感到自己在同一时间作为个体的渺小和作为社区的庞大。

竞技场中包含的功能，为其综合楼和多样的使用模式形成基地。当用途在活动的不同类型和规模之间变化时，该建筑内部功能之间的关系相应地进行切换并进行调整。竞技场扎实的根基和严格的体量被不同高度起伏凸起的外观所软化，这是为了调整相应的内部功能。植物作为容器，被纳入外墙的穿孔铝板。因此，建筑物相当庞大的外观减少了，建筑逐步过渡为一个公园。

Site Plan
总平面图

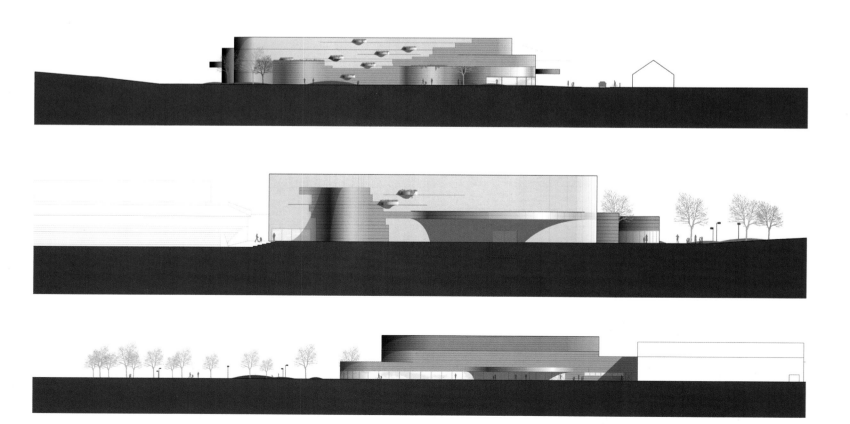

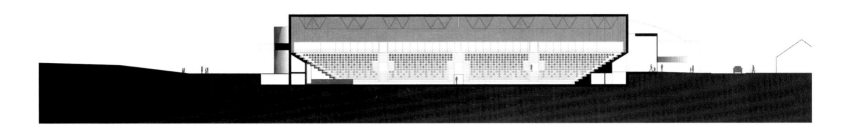

Bowl-like Shape
Plant Container
Cultural Gatherings

Project Overview

Næstved Arena is in eastern Denmark and located in a green urban pocket surrounded by a football stadium and sports hall on one side and a residential area on the other. The 10,000 m² arena has a capacity of 2,500 to 4,000 spectators, and is a new multipurpose venue integrated for sports, music and culture events.

Building Design

The primary focus of Næstved Arena's design is to combine atmosphere and sense of community spirit with effective functionality and logistics in order to create a new cultural meeting place. The building volume's rounded geometries lead towards the entrance and into the heart of the arena. The arena's bowl-like shape and the surrounding stands' sloping planes bring the building's scale into play—they contribute to creating the intense and heaving atmosphere during sports events and concerts, where you feel small as an individual and large as a community at one and the same time.

The arena contains a collection of functions, which form the base for a complex and diverse range of usage patterns. When the usage changes between different types and scales of events, the relations between the building's internal functions switch and adapt accordingly. The arena's basic and stringent volume is softened by a façade of undulating protrusions of varying heights, which are adjusted to the corresponding internal functions, as well as plant containers, which are incorporated into the façade's perforated aluminum panels. Thus, the building's quite massive appearance is reduced and the building forms a gradual transition to the park.

俄罗斯莫斯科卢日尼基奥体中心游泳池
Swimming Pool of Luzhniki Olympic Complex

设计单位：Arch Group
委托方：卢日尼基奥林匹克体育中心
项目地址：俄罗斯莫斯科
项目面积：32 627m²

Designed by: Arch Group
Client: Luzhniki Olympic Complex
Location: Moscow, Russian
Area: 32,627m²

时代感
独特屋顶设计

项目概况

卢日尼基奥林匹克体育中心是俄罗斯和欧洲唯一有地方特色和代表建筑的一个建筑群。本案游泳池的建造和卢日尼基体育场各自形成了一个单体。因此，有必要保留现有建筑的3个立面，以免破坏这个建筑群。

泳池实际上是入口处有着地铁站的奥林匹克综合体内的主要建筑。这意味着，东南立面成为了重建泳池的主要对象，它将成为行业最高标准的现代体育和娱乐设施，以超强的时代感吸引着无数游客。

建筑设计

由于建筑有一个非常富有表现力和可识别连接其他两个廊柱立面的轮廓，所以设计完全保留现有西北立面。

高性能玻璃体块为建筑整体搭建简单的矩形结构，扮演着背景和从属的角色，缓解新旧建筑带来的视觉冲突，起到粘合剂作用。屋顶是建筑最具表现力的元素。它是一个单一的平面，犹如覆盖在整个综合楼上，轮廓平滑。从地面仰视，屋顶几乎虚掩不见，它只是形成了建筑的轮廓，但从立交桥俯视时却清晰可见。

建筑的屋顶从北到南、从豪华运动竞技场的远角到自身的落角倾斜而下，这样能最大限度地接近豪华运动竞技场；屋顶的窗户能最大程度地转向太阳。屋顶在三处主要领域之上，设有大的封闭式活动玻璃开口，以确保其良好的日照。夏天，玻璃开孔开放，让游客尽情享受日光浴。

水上乐园区位于沿建筑物光线较好的西南侧，这里日照充足。区域的两侧是历史悠久的柱廊结构，横跨柱廊可欣赏到卢日尼基的公园全景。综合楼的其他所有功能形成了一个单一的五层体块，沿相反的东北侧延伸，很好地区分了娱乐区和入口处。

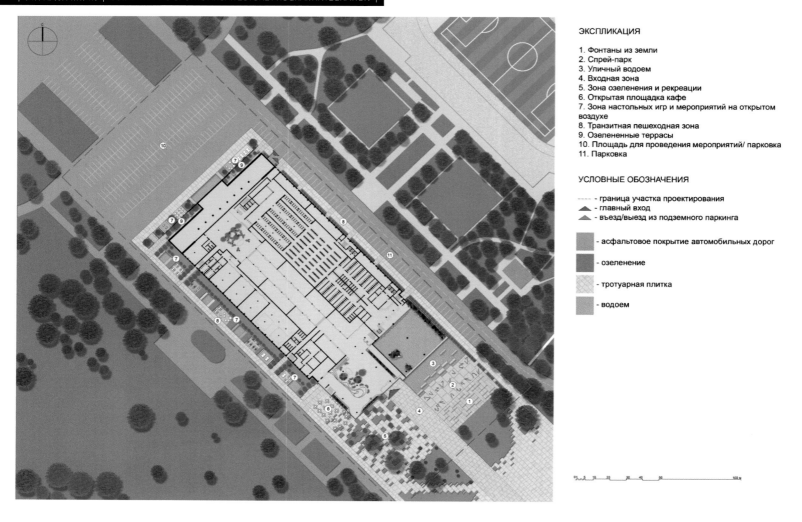

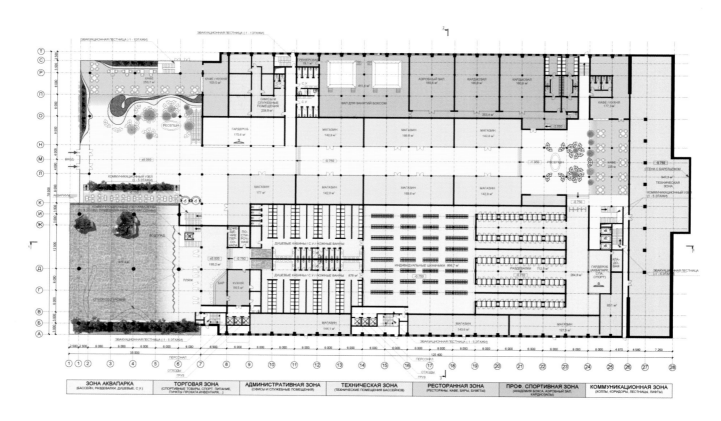

ПЛАН 1 УРОВНЯ. М 1:500.

ПЛАН 2 УРОВНЯ. М 1:500.

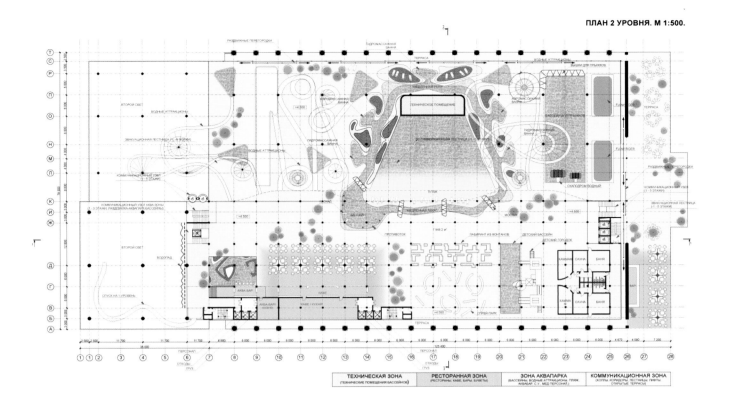

| ТЕХНИЧЕСКАЯ ЗОНА | РЕСТОРАННАЯ ЗОНА | ЗОНА АКВАПАРКА | КОММУНИКАЦИОННАЯ ЗОНА |

ПЛАН 3 УРОВНЯ. М 1:500

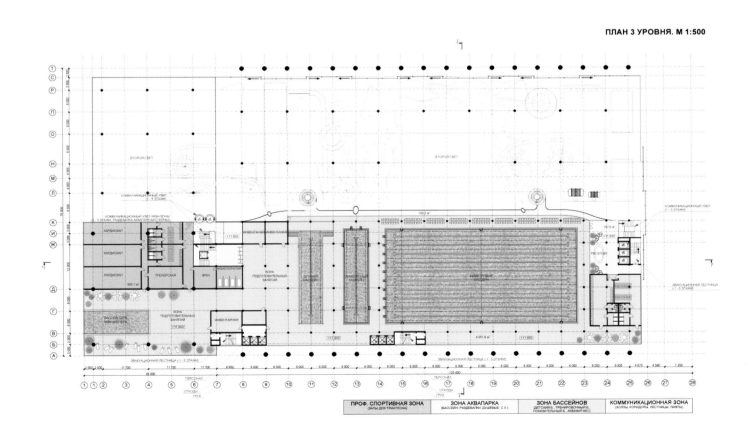

| ПРОФ. СПОРТИВНАЯ ЗОНА | ЗОНА АКВАПАРКА | ЗОНА БАССЕЙНОВ | КОММУНИКАЦИОННАЯ ЗОНА |

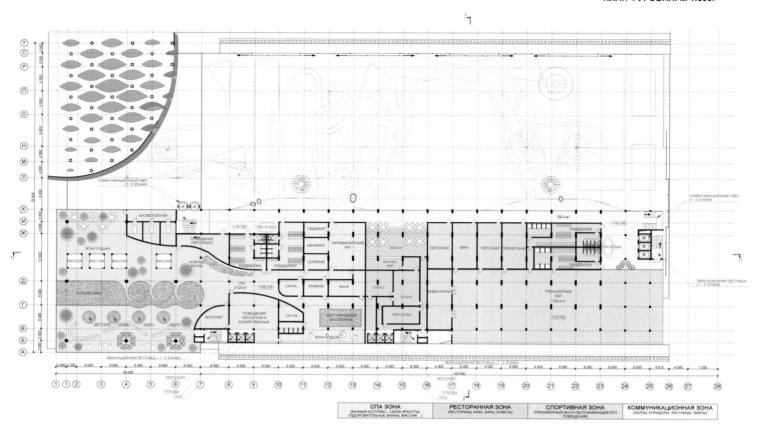

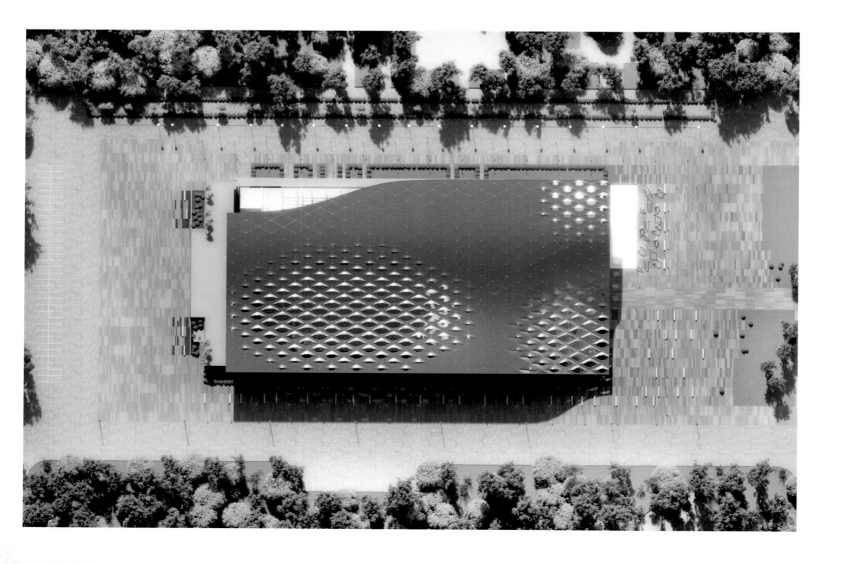

ПЛАН 5 УРОВНЯ. М 1:500.

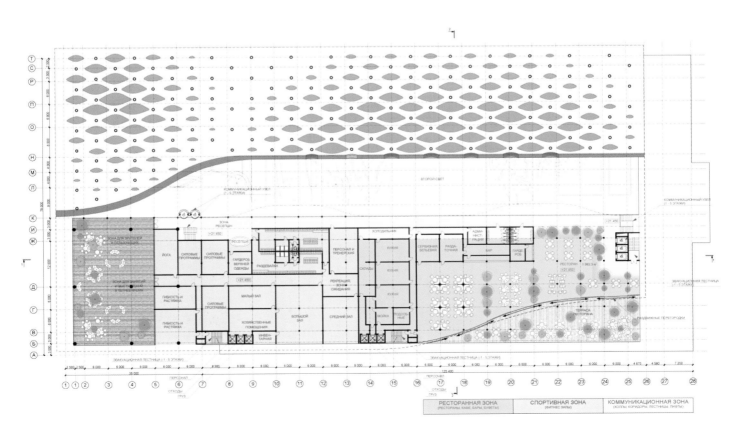

| РЕСТОРАННАЯ ЗОНА | СПОРТИВНАЯ ЗОНА | КОММУНИКАЦИОННАЯ ЗОНА |

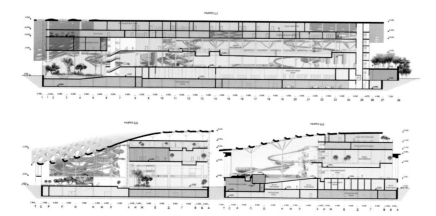

Period Feel
Unique Roof Design

Project Overview

Olympic Complex Luzhniki is unique place for Russia and Europe that represents a single architectural ensemble of buildings. One of the two buildings makes up a single composition with a Grand Sports Arena (Luzhniki Stadium). Therefore, it is necessary to keep three façades of the existing building so as not to disrupt this ensemble.

This pool building is actually the main object at the entrance to the territory of the Luzhniki Olympic Complex from the Vorobyovy Gory metro station. This means that the south-eastern façade becomes the main for the renovated pool building. This façade becomes a modern sports and entertainment facilities that meet the highest industry standards to attract visitors.

Building Design

Existing north-west façade we keep completely because it has a very expressive and recognizable silhouette that connects the other two colonnaded façades.

The new glass volume as a whole has a simple rectangular structure, built in a square of historic façades, and playing background and subordinate role, to create no visual conflict of old and new. It performs the function of a binder between the old architecture and contemporary. Roofing is the most expressive element of modern building. It is a single plane as if lowered down on the entire complex and take it smoothed contours. From the ground, the roof itself is almost not visible, it only forms a silhouette of the building but it's clearly visible from the metro bridge and automobile overpass.

Roof falls from north to south, from the far corner of Grand Sports Arena to

СХЕМА ДОСТУПА МАЛОМОБИЛЬНЫХ ГРУПП НАСЕЛЕНИЯ К ОСНОВНЫМ ЗОНАМ КОМПЛЕКСА.

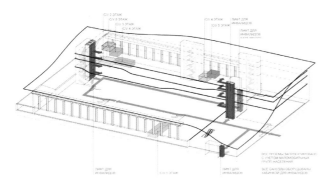

the opposite corner, to minimally close Grand Sports Arena and make the maximum turn of windows in the roof to the sun. In three places of the roof over the main areas there are large openings closed by glass to ensure their good insolation. Glazing apertures can be open to allow visitors to sunbathe in the summer.

Water park zone is located along the well-lit south-west side of the building. On both sides zone is framed by historic colonnades. In this case we get a panoramic view of Luzhniki parkland across the colonnade. All other functions of the complex form a single five-story volume which stretches along the opposite north-east side so as not to obscure the entertainment zone and the entrance.

СХЕМА УСТРОЙСТВА ИЛЛЮМИНАТОРОВ В ЧАШАХ БАССЕЙНОВ НАД ТОРГОВОЙ ГАЛЕРЕЕЙ.

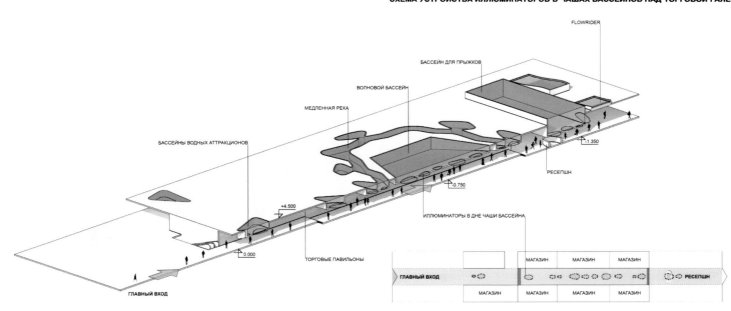

智欲圆而行欲方

Cautious Ideas

阿富汗喀布尔国家博物馆
Afghanistan National Museum

设计单位：THEEAE LTD.
项目地址：阿富汗喀布尔市
项目面积：16 000m²

Designed by: THEEAE LTD. (The Evolved Architectural Eclectic Limited)
Location: Kabul, Afghanistan
Area: 16,000m²

文化中心
希望、启迪
开放空间

项目概况

这所新设计的国家博物馆将成为喀布尔的标志性建筑。它不仅仅是居民教育发展的主力,而且将迅速成为周边国家之间的文化中心。它的建立也将成为该地区的文化催化剂,鼓舞、促进社区内住宅的开发。

文体建筑方案集成 | CULTURAL AND SPORTS ARCHITECTURE PROGRAM INTEGRATION

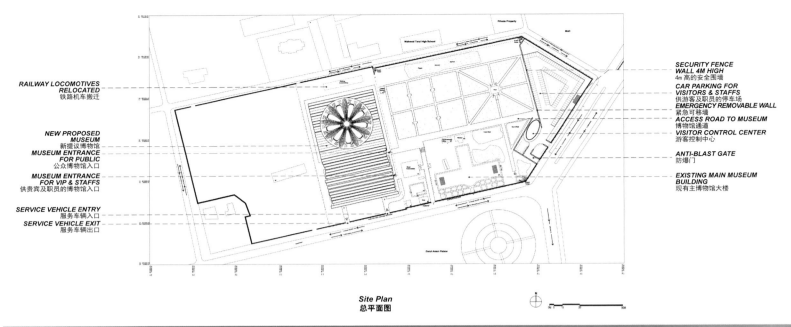

Site Plan
总平面图

- RAILWAY LOCOMOTIVES RELOCATED 铁路机车搬迁
- NEW PROPOSED MUSEUM 新提议博物馆
- MUSEUM ENTRANCE FOR PUBLIC 公众博物馆入口
- MUSEUM ENTRANCE FOR VIP & STAFFS 供贵宾及职员的博物馆入口
- SERVICE VEHICLE ENTRY 服务车辆入口
- SERVICE VEHICLE EXIT 服务车辆出口
- SECURITY FENCE WALL 4M HIGH 4m高的安全围墙
- CAR PARKING FOR VISITORS & STAFFS 供游客及职员的停车场
- EMERGENCY REMOVABLE WALL 紧急可移墙
- ACCESS ROAD TO MUSEUM 博物馆通道
- VISITOR CONTROL CENTER 游客控制中心
- ANTI-BLAST GATE 防爆门
- EXISTING MAIN MUSEUM BUILDING 现有主博物馆大楼

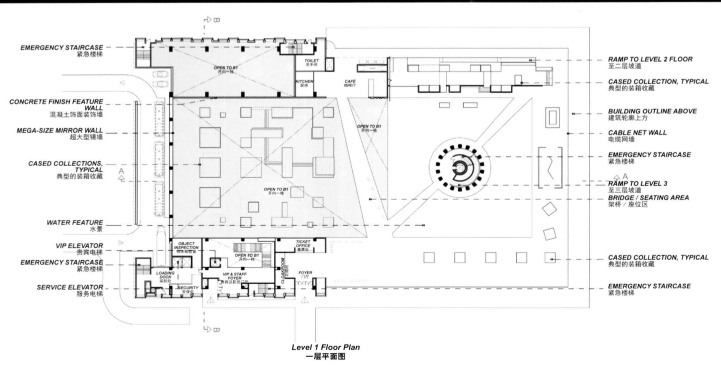

Level 1 Floor Plan
一层平面图

- EMERGENCY STAIRCASE 紧急楼梯
- CONCRETE FINISH FEATURE WALL 混凝土饰面装饰墙
- MEGA-SIZE MIRROR WALL 超大型镜墙
- CASED COLLECTIONS, TYPICAL 典型的装箱收藏
- WATER FEATURE 水景
- VIP ELEVATOR 贵宾电梯
- EMERGENCY STAIRCASE 紧急楼梯
- SERVICE ELEVATOR 服务电梯
- RAMP TO LEVEL 2 FLOOR 至二层坡道
- CASED COLLECTION, TYPICAL 典型的装箱收藏
- BUILDING OUTLINE ABOVE 建筑轮廓上方
- CABLE NET WALL 电缆网墙
- EMERGENCY STAIRCASE 紧急楼梯
- RAMP TO LEVEL 3 至三层坡道
- BRIDGE / SEATING AREA 架桥 / 座位区
- CASED COLLECTION, TYPICAL 典型的装箱收藏
- EMERGENCY STAIRCASE 紧急楼梯

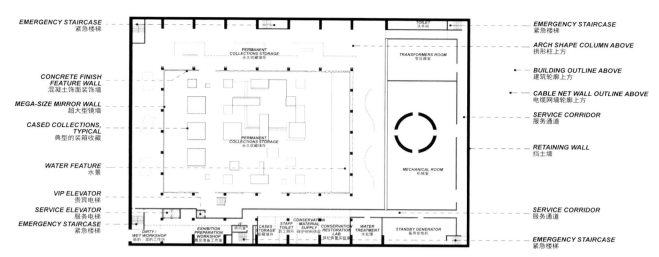

Basement Level Floor Plan
底层平面图

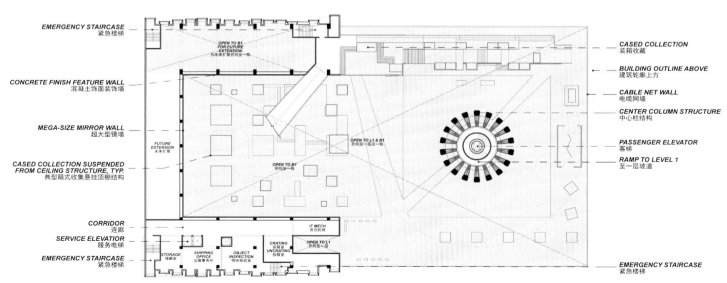

Level 2 Floor Plan
二层平面图

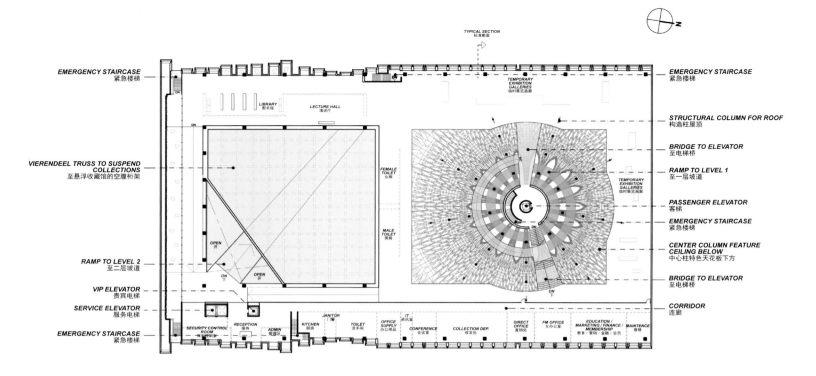

Level 3 Floor Plan
三层平面图

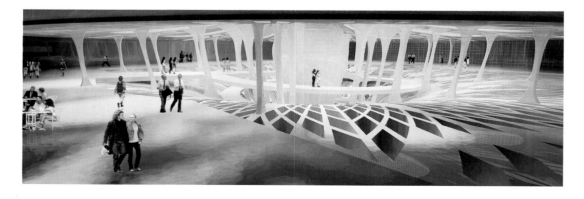

建筑设计

新博物馆的设计旨在希望能找回丢失的财富，而保留其文化资产为该地区人民带来希望与启迪。建筑设计关注的不仅是这些收藏品，还有情感空间，人们需要一个安全的场所来休息并且享受大自然的乐趣。

外立面材料，用特别的露石混凝土箱设计有图案的木框架。该露石混凝土根据现有的环境以及阿富汗的历史元素来进行构图框架。该框架本身是尽可能多的把光线引入室内空间，从而透过阿富汗的历史建筑来发现它。

内部主要的公共空间大多向自然敞开，巨大的支撑柱设计成拱形来表现文化的丰富性。由此形成一个独特的空间为参观者（公众、非公众、VIP会员等）提供休憩。巨型柱子下面的收藏品以合理的距离陈列在玻璃盒中，不仅是出于安全着想，也是为防止内部人造水池带来的潮湿影响。

收藏品和安全储存区从地下室到三层都是敞开的。南侧的内墙设计安装有镜子，使空间看起来更加宽敞，感觉就像置身于阿富汗的山谷之中。博物馆东西面盒状墙体的设计起到弗伦第尔桁架系统的作用，以支撑加长的悬臂。悬臂由这个桁架和北区中心的巨型柱子支撑。更有趣味的空间是三层的公共收藏区域，自然光由楼顶的天窗照入，并延伸到主收藏区。

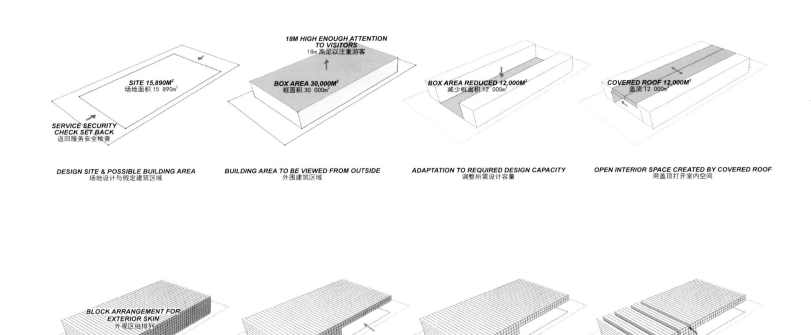

Design Process
设计过程
Physical Model Study Development
物理模型研究发展

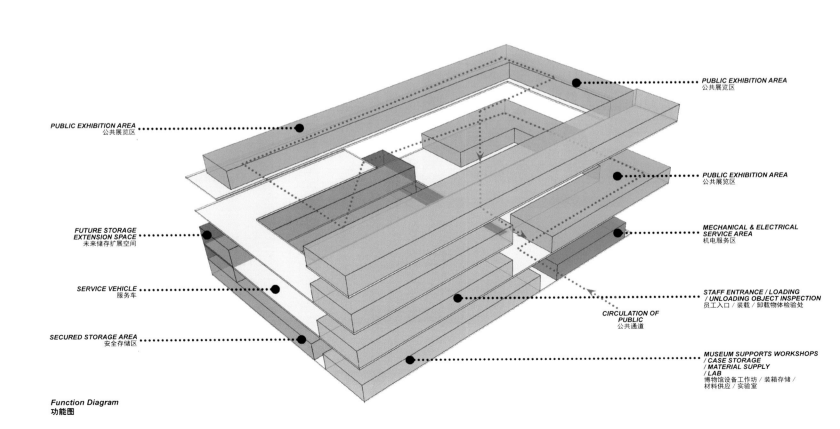

Function Diagram
功能图

| 文体建筑方案集成 | CULTURAL AND SPORTS ARCHITECTURE PROGRAM INTEGRATION |

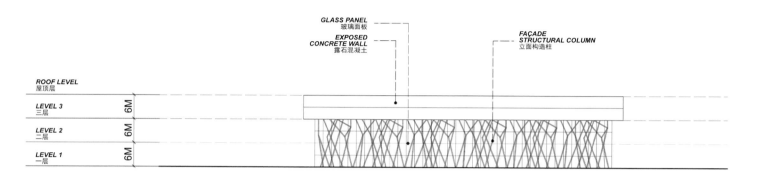

South Elevation
南立面图

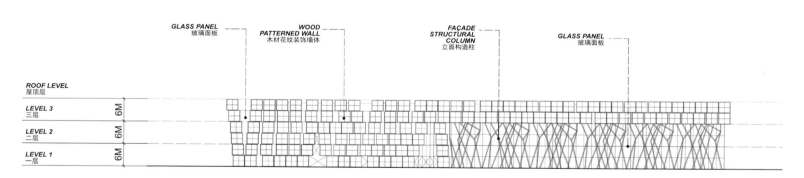

North Elevation
北立面图

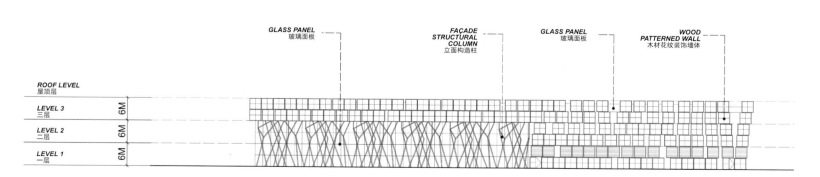

East Elevation
东立面图

West Elevation
西立面图

Cultural Center Hope and Inspiration Open Space

Project Overview

The new design of the National Museum of Afghanistan will be a landmark for Kabul. The new National Museum is not only the anchor of residents' educational development, but will be a catalyst for the cultural quarter in which it sits, inspiring and promoting residential development in the community.

Building Design

Its new museum to catch up the lost richness and its cultural asset is necessitated to give hope and inspire people in that region. The major concern for the architecture was not only about the collections but also emotional realm of space that requires a place to give a rest and the joy of the nature in its heritage safe and secured.

From the outside façade materials, it is simply designed with each patterned wood frame in individual exposed concrete box unit. The exposed concrete was to relate the existing environment, as well as the patterned frame mimics the historical elements of Afghanistan. The frame itself is to bring the lights to the interior space as much as its space is to be found from Afghanistan's historical building.

The major public space inside is much opened into the nature. The supporting mega-column is the shape of arch that transformed to express the richness of culture. Thus, it provides unique space to give a feeling of rest to public and non-public, VIP members, who visits the museum. The collections under the mega-column are well arranged in a proper distance and they are cased in glass boxes not only for the security, but also for the protection of the moisture from the artificial pond inside.

The area of the collections and storage space that is secured is opened from basement area to Level 3 ceiling space which is around 22m height. Plus, the south interior wall is designed to install mirror entirely to give the space even looking more opened than what it is so that the visitors will experience as if they are in the valley of mountain in Afghanistan.

The east and west boxed walls are designed to work as Vierendeel truss system to support the extensively prolonged cantilever to the main entrance collection area. This cantilever is supported by this truss and the mega-column in the center of the north area. The space that creates more interesting is the level 3 public collection area. The natural lights penetrate through the skylight above. It still continues to pass down to the main collection space.

Façade Design Idea
立面设计理念
Typical Section at Façade
典型的立面剖面图

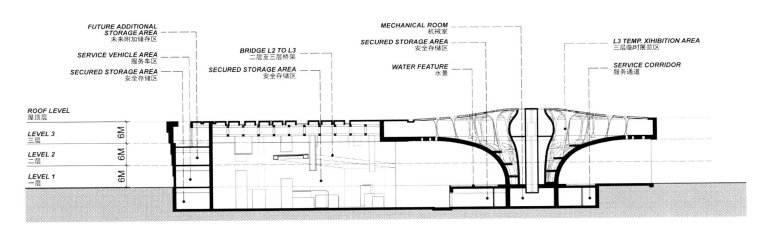

Longitudinal Section (A)
纵剖面（A）

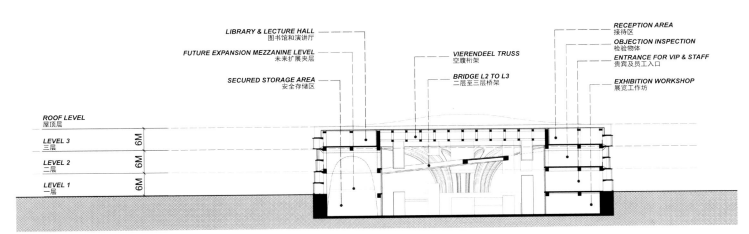

Transversal Section (B)
横剖面（B）

委内瑞拉库马纳会展中心
Cumana Convention Center

设计单位：Saraiva + Associados	***Designed by**: Saraiva + Associados*
委托方：Martifer	***Client**: Martifer*
项目地址：委内瑞拉库马纳	***Location**: Cumana, Venezuela*
项目面积：20 208m²	***Area**: 20,208m²*

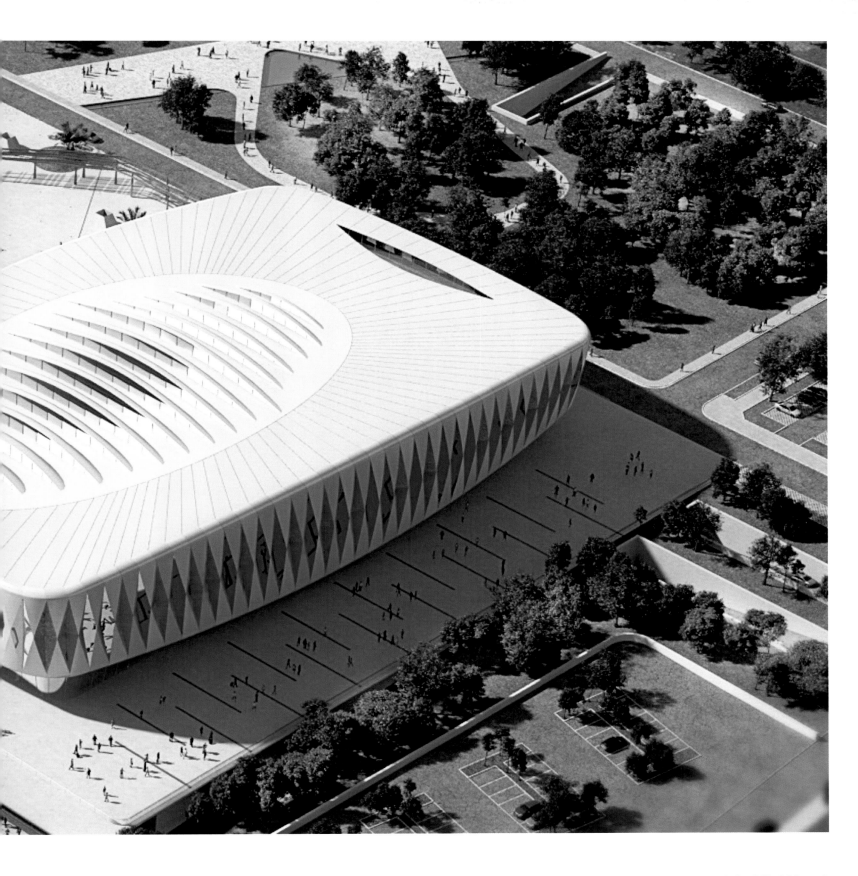

柔和流线模型
多功能建筑
"大眼鲷鱼"雕塑

项目概况

未来会展中心位于委内瑞拉苏克雷州首府库马纳,如曼萨纳雷斯河苏克雷州海岸线一样,设计将在敏感的城市区域完成一座现代的、可持续的文化建筑和模型参考。该项目包括一个多用途主场和一个由本地品种树木和灌木组成的大型绿地,可以提供给当地居民和游客使用。

建筑设计

项目的核心建筑元素是一个用于大型活动的多功能区域，如会议、展厅、交易会、音乐会和体育赛事。建筑的设计促进了与外界空间即圣路易斯海滩的视觉关系。一个可容纳1 000人的大礼堂位于主场的后面，从外部就可进入。这种基础设施可以独立使用，同时也可以在主要的会议中心等活动中使用。会议中心的内部两侧高达22m，最多可以容纳10 000人，可举行各种活动，并符合空气处理、亮化和美化的技术要求。有多功能会议室的商务中心也使得所需的程序更加完整化。

广场从地面升高，并通过一个可容纳4 000人的礼堂形状的大型下行透水区，位于东象限的侧面。广场被树木所环绕，构成海岸线附近一个新的景观和阴影区。"大眼鲷鱼"（本地物种）的雕塑装饰了公共空间，这也是从当地动物群选取利用的一个设计元素。通过一个连接到一楼大空隙地区的平台，由此进入里面有停车场和基础服务设施。另外，项目综合楼底层设有大约500个车位的停车场。

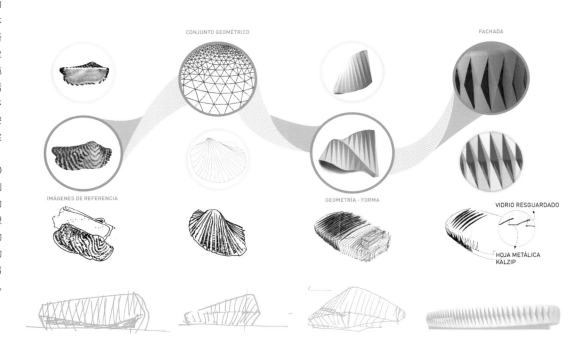

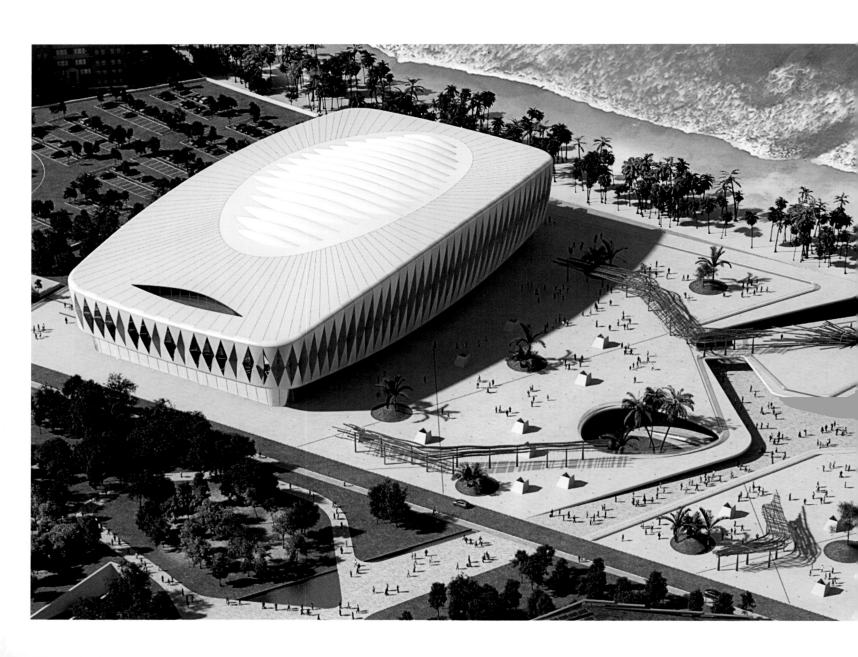

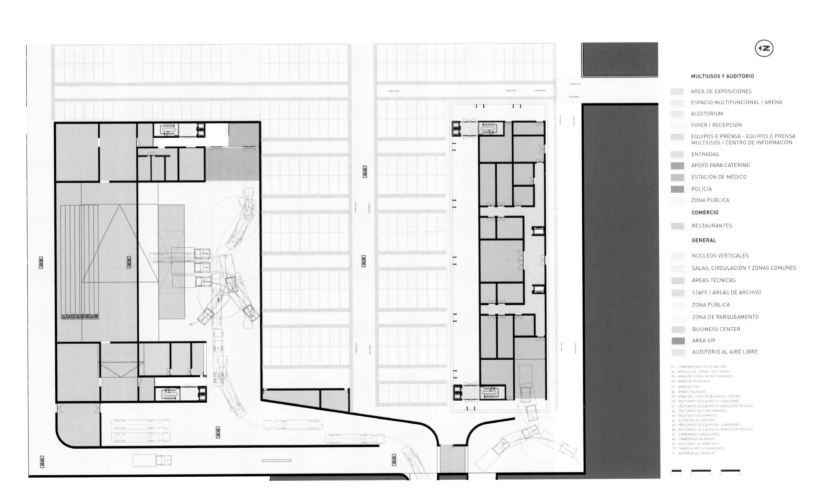

MULTIUSOS Y AUDITORIO

AREA DE EXPOSICIONES
ESPACIO MULTIFUNCIONAL / ARENA
AUDITORIUM
FOYER | RECEPCIÓN
EQUIPOS E PRENSA - EQUIPOS E PRENSA MULTIUSOS / CENTRO DE INFORMACIÓN
ENTRADAS
APOYO PARA CATERING
ESTACIÓN DE MÉDICO
POLÍCIA
ZONA PÚBLICA

COMERCIO

RESTAURANTES

GENERAL

NÚCLEOS VERTICALES
SALAS, CIRCULACIÓN Y ZONAS COMUNES
ÁREAS TÉCNICAS
STAFF | ÁREAS DE ARCHIVO
ZONA PÚBLICA
ZONA DE PARQUEAMENTO
BUSINESS CENTER
AREA VIP
AUDITÓRIO AL AIRE LIBRE

| 文体建筑方案集成 | CULTURAL AND SPORTS ARCHITECTURE PROGRAM INTEGRATION |

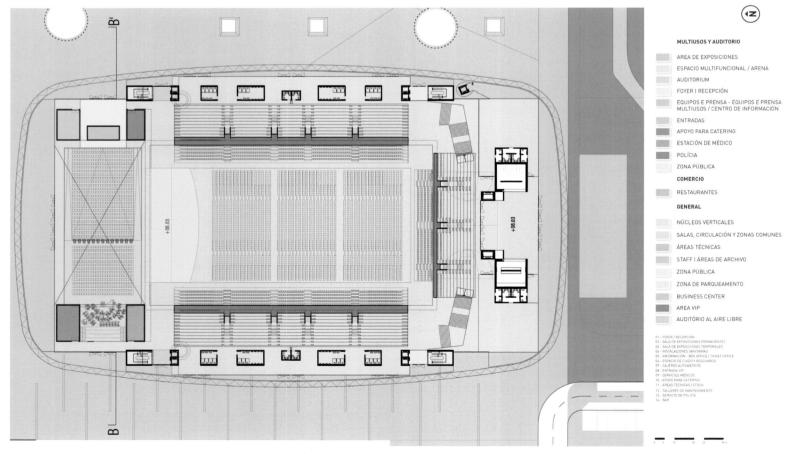

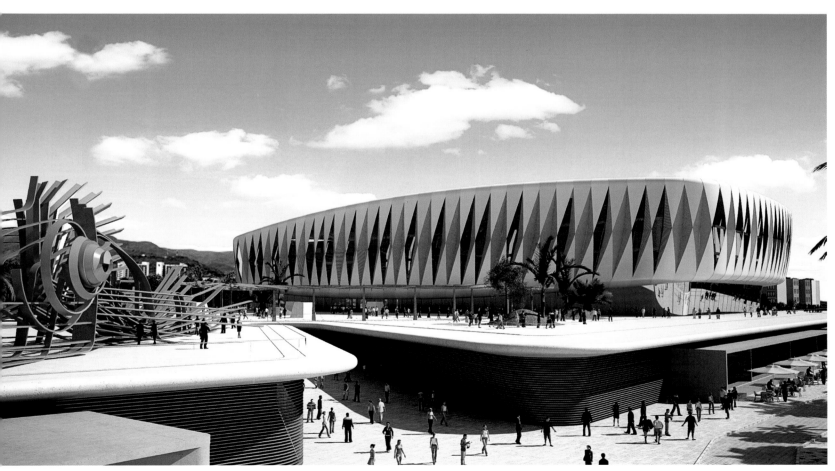

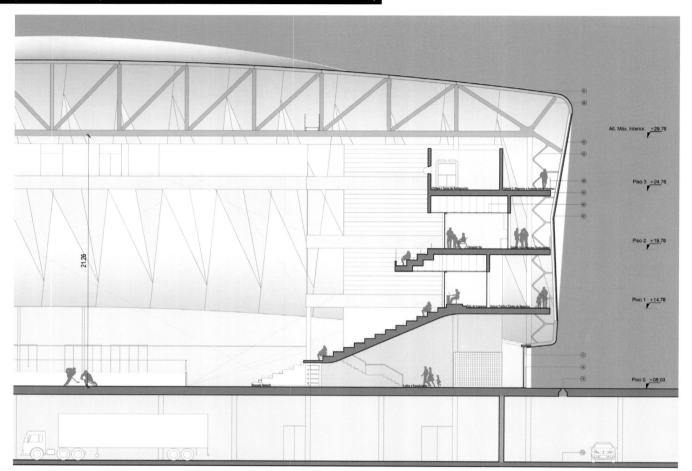

MULTIUSOS
ARENA BOX

Capacidad entre 500 e 3600 personas, Modular

Distribución Central Sentada, de Pie, con o sin Bancada

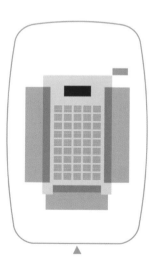

MULTIUSOS (DISPOSICIÓN 2)
CONGRESOS

Capacidad Máxima 7366 Plazas
Distribución Central Sentada
2500 Plazas - 1.6 Personas / m2
Tablas AB & C
4118 Plazas + 748 Plazas en Primer Balcón
e 14 Cabinas VIP

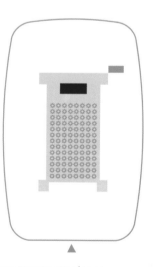

MULTIUSOS (DISPOSICIÓN 2)
GRAN BUFFET

Capacidad 832 Plazas
Distribución Central Sentada

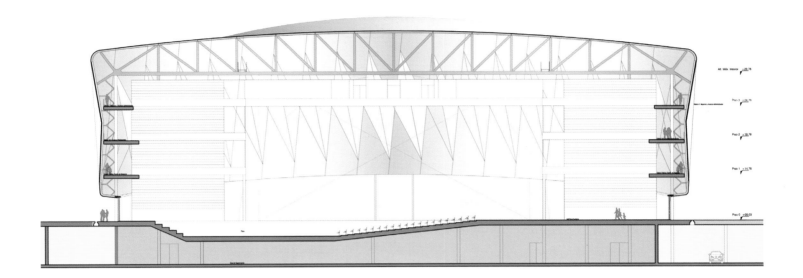

Gentle Flow Line Model
Multi-purpose Building
"Priacanthus Arenatus" Fish Sculpture

Project Overview

The future Convention Center of Cumana is located in Cumana, the capital of Venezuela, and is planned to become a contemporary, sustainable and model reference in a sensitive area such as the coastline of the State of Sucre, where the city of Cumana is located. The project includes a multi-purpose arena and a large green space composed of trees and shrubs of local species, which can be enjoyed by local population and visitors.

Building Design

The central element of the project is a multifunctional area for large events such as conferences, showrooms, trade fairs, concerts and sport events. The design of the building promotes a visual relationship with the outside space namely St. Louis's beach. A large auditorium for 1,000 people is located behind the main stage, accessible from the outside. This infrastructure can be used autonomously and at the same time with other events taking place in the main congress center. The wing of the convention center is 22m height in the interior, allowing all sorts of events and complying with the technical requirements of air handling, lighting and landscaping. It can receive a maximum 10,000 people. A Business Center with multi-purpose meeting rooms completes the required program.

The square is elevated from the ground level and is flanked in the East quadrant by a large descending permeable area with the shape of an auditorium for about 4,000 people, surrounded by trees. It creates a new landscaped and shaded zone near the coastline. The sculpture of a fish the "Priacanthus arenatus" (a local species) ornaments this public space taking hand of an element from the local fauna. This platform is accessible by a large void that connects it to the ground floor, where the parking and services infrastructures are located. In addition, the parking area underneath the complex has approximately 500 places.

■ AREA DE PISO ■ AREA DE JUEGO

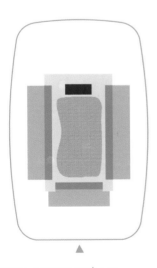

MULTIUSOS (DISPOSICIÓN 1)
GRANDES ACTUACIONES MUSICALES

Capacidad Máxima 10866 Plazas
Distribución Central de Pie
6000 Plazas - 3,4 Personas / m2
Tablas AB & C
4118 Plazas + 748 Plazas en Primer Balcón
e 14 Cabinas VIP

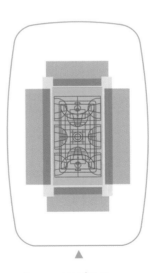

MULTIUSOS (DISPOSICIÓN 3)
COMPETICION DESPORTIVA

Capacidad Máxima 5620 Plazas
Distribución Central Sentada
Tablas AB & C + Tabla Temporária
4872 Plazas + 748 Plazas en Primer Balcón
e 14 Cabinas VIP

北京国家美术馆
National Art Museum of China

设计单位：盖里建筑师事务	**Designed by: Gehry Architects**
委托方：中国国家美术馆	**Client: National Art Museum of China**
项目地址：中国北京市	**Location: Beijing, China**
项目面积：30 000m²	**Area: 30,000m²**

微缩的城市
五角星造型
"中国红"

项目概况

在过去的 20 年里，博物馆建筑发生了日新月异的变化。它们的规模将不仅仅是一座大楼，更是一座微缩的城市。而中国国家美术馆就是世界上第一座这种新典范的博物馆，是第一座被构思为小型城市的博物馆。

项目以"城市"这一概念为核心进行规划，混合着"官方"或基层的分区。它的中心和边缘包含可规划为中国和国际区域、现代和历史区域、商业和政府区域等，就像一座城市一样，区域之间可以重新定义、修复甚至替换。

将美术馆规划为一座城市并不是说它不能提供博物馆的本质功能，和所有城市一样，其单独的部分可能很小，但是能提供博物馆所特有的多种服务。

建筑设计

主基座建筑是一些传统的正交博物馆空间，还有更现代化的，形态更自由的建筑，仿佛正在讲述或探索着中国的艺术故事。建筑基于一个五角星造型，主要流通线路从外围的多个入口点到达中心区，五角星连接到"灯笼"，这也是该美术馆的立体象征。红色的立面表皮，完美地展现了"中国红"这一传统的色彩元素。

单一的切线将"灯笼"与鸟巢联系起来，与城市的错综复杂性相比，"灯笼"结构的六层主楼提供了宽敞、开放的空间，以致于建筑并未与展览会的布局相冲突。建筑的内部组织也很合理，其灵活的表皮从金属框架中伸展出来，使它看起来就像一个谜。

Miniature City Pentagram Shape "China Red"

Project Overview

In the past two decades, the museum building has been occurred rapid changes. Their size will be more than just a building, it is a miniature city. The National Art Museum of China is the world's first museum with this new paradigm and is a museum conceived as a small city.

Project is with the concept of "city" as the core for planning and mixing with the "official" or grassroots partition that its center and the edge contain the planning zones for the Chinese and international areas, modern and historic areas, commercial and government areas, etc., so it likes a city which can be redefined, repaired or replaced between the regions.

The Museum planned for a city is not to say it can not provide an essential function of the museum, but it is same as all the cities that its separate parts may be small, but it can provide a variety of unique services to the museum.

Building Design

The main base buildings are some of the traditional orthogonal museum spaces, as well as more modern and freer form constructions, as if it is being told or exploring the Chinese art stories. The architecture is based on a five-pointed star shape, and when the main circulation lines from the multiple periphery entry points reaches to the central area, five-pointed star is connected to the "lantern", which is the three-dimensional symbol of the museum. Red façade epidermis perfectly demonstrates the traditional color elements of this "China Red".

Single tangent will make the "Lantern" link with the nest, and compared with the intricacies of the city, the six-story main building of the "lantern" structure provides a spacious and open space, so that the building does not conflict with the exhibition layout. The internal organization of the building is also very reasonable, and its flexible skin stretched out from the metal frame, making it look like a mystery.

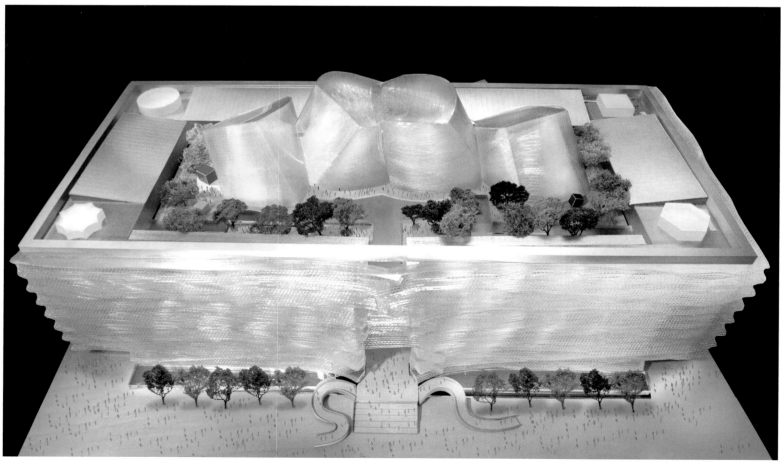

广东广州三馆一场
Guangzhou Three Museums One Square

设计单位：UNStudio	Designed by: UNStudio
委托方：广州市文化广电新闻出版局 广州市科技和信息化局 广州市城市建设投资集团	Client: Administration of Culture, Press, Publication, Radio and Television Guangzhou Municipality Science and Information Technology of Guangzhou Municipality Bureau Guangzhou City Construction Investment Group
项目地址：广东省广州市	Location: Guangzhou, Guangdong
项目面积：240 000m²	Area: 240,000m²

艺术文化集群
城市绿洲

项目概况

三馆一场项目位于广州新中轴线南段的起点，与已建成的花城广场以珠江相隔，与广州城市新地标广州塔将成为重要的旅游观光区。基地范围北起阅江西路，南至新港中路，西起艺景路，东至艺苑东路与赤岗北路。项目由岭南广场、广州博物馆、广州美术馆、广州科学馆组成，设计将打造成广州文化的核心地区，建成后将大大提升广州市文化软实力和国际竞争力。

建筑设计

三馆一场项目的设计理念是在公园般的优美环境中创建一个艺术文化集群，以展现广州的悠久文化、历史背景和锐意创新。

岭南广场设计中,建筑与景观高度融合成为一体,用一种新的尺度与现代语言重新阐释岭南园林与建筑。岭南广场作为广州新中轴线的延续,建筑与景观紧密相连,融为一体,它以新的尺度阐释这种亲密关系。广场亦如仙境般的世外桃源、山、水、林、田构筑成一片城市中的自然绿洲。

三馆都嵌入在景观中作为"一个托盘对象"。这3个建筑都被定为在博物馆广场的三个"角落",通过自己的建筑语言确定各自的身份。

科学馆,专为教育和娱乐设计,是代表知识进化的建筑体块的扩大。在设计中,功能已被转移切换形成一个紧凑的平面排列,形成一个与公共空间交织的立体包层。博物馆的设计是受传统岭南园林石头的启发。由于地面五层、地下两层,博物馆被设计成一个紧凑的体块,以切口和接缝的形式局部衔接。这些接合处进一步解决了展览和公共职能之间的交流问题。美术馆设计通过循环流通结合了三个不同大小的建筑体块形成建筑的组织元素。博物馆的三个"豆荚"是通过游客体验的公共路线联系在一起。沿着该路线,不同的光线条件是根据展出的艺术作品的要求来定位的,正如该路径作为讲故事的工具,是用于中国绘画的。

| 文体建筑方案集成 | CULTURAL AND SPORTS ARCHITECTURE PROGRAM INTEGRATION |

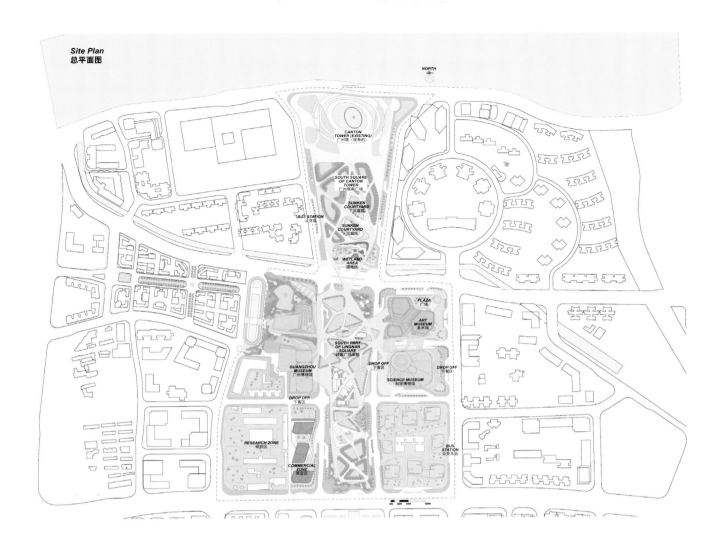
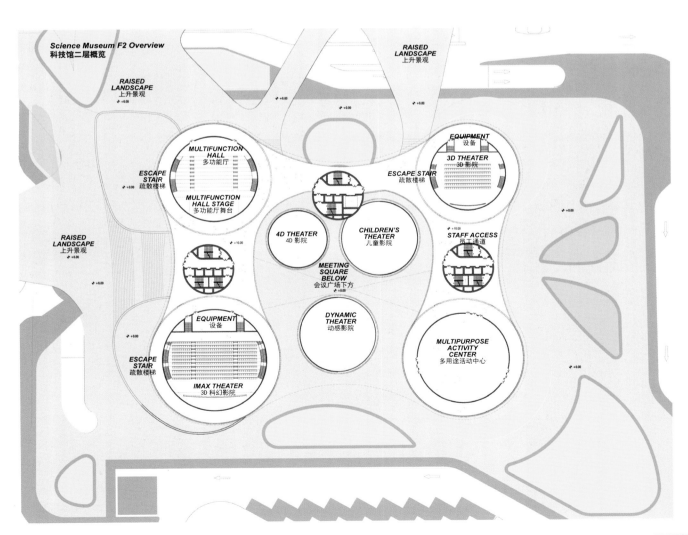

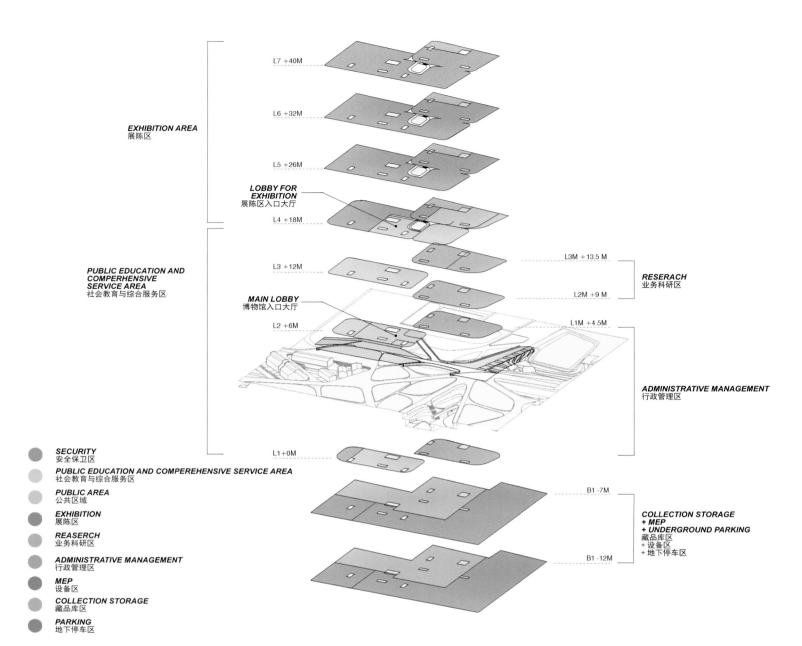

Arts and Culture Cluster
Urban Oasis

Project Overview
Situated at the South end of the Guangzhou New Urban Axis, the project is going to become an important tourist and sightseeing zone together with Canton Tower, the recent city landmark in Guangzhou. The site area is bounded by Yue Jian West Road to the North, Xin Gang Middle Road to the South, Yi Jing Road to the West, and Yi Yuan East Road and Chi Gang North Road to the East. The plot is composed of Lingnan Square (Park), Guangzhou Museum, Guangzhou Art Museum and Guangzhou Science Museum, which will be developed into a core cultural district in Guangzhou. After completion of the project, cultural soft power and international competitiveness of Guangzhou city will be dramatically enhanced.

Building Design
The design concept of Three Museums One Square is to create a state-of-the art cultural cluster in a park-like setting so as to offer an opportunity to express the city of Guangzhou's culture, background and innovation.

Lingnan Square is a landscape park separated from the surrounding city by a massive wall of trees. The open field of this green oasis links the whole area with Canton Tower, one of the most iconic buildings of Guangzhou. A meandering walkway and a river connect the landmark buildings on both

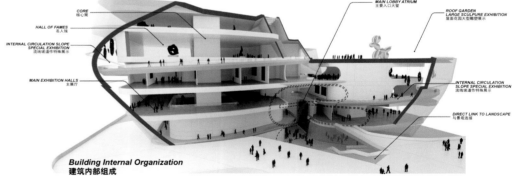

Building Internal Organization
建筑内部组成

sides of this open field. The new continuation of Guangzhou main axis is closely related to Lingnan culture where architecture and landscape become inseparable. This close relationship between architecture and nature is carried to the new scale of Guangzhou. The buildings with their organic and sculptural appearance resemble mountains in an open rice field as it is shown in so many Chinese paintings.

Three museums are embedded on the landscape as "objects on a tray". These three buildings are positioned on each corner of the museum square, pronouncing its identity through their architectural language.

The science museum, designed for education and entertainment, is a proliferation of building mass representing the evolution of knowledge. In design, the functions have been shift sliced to form a compact planar arrangement, creating a three-dimensional envelope that intertwines functions with public space. The design for the Guangzhou Museum is inspired by the stone laying in traditional Lingnan gardens. With five stories above ground two below, the museum is designed as a compact volume with localized articulation in the form of cut-outs and seams. These areas of articulation further address the exchange between exhibition and public functions. The design for the art museum combines three differently sized building blocks through a looping circulation that forms the organizing element of the building. The three "pods" of the museum are linked through the public routing which curates the visitor experience. Along this route different lighting conditions are positioned according to the requirements of the art work on display—just as the path is used in Chinese painting as a tool for storytelling.

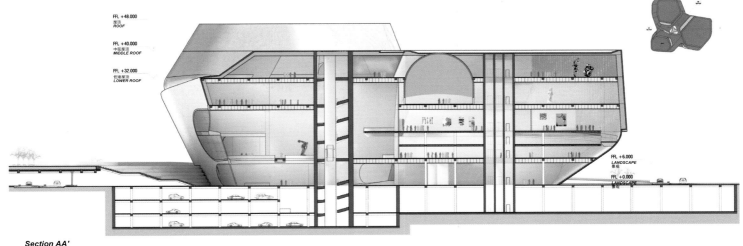

Section AA'
剖面 AA'

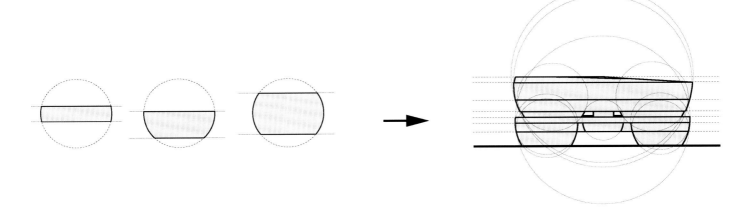

FOUR BOWL SHAPE VOLUME ON THE GROUND, PROVIDING SEPARATED ACCESSES TO THEATRES AND MULTI-FUNCTIONAL HALL
地面上的四个碗形体量容纳剧院, 多功能厅, 并提供各自独立的入口

USING A PUBLIC BELT CONNECT MORE THEATRES AND EDUCATIONAL HALL
利用公共空间联系各个剧院和大厅

MORE THEATRES HANGING UNDER EDUCATIONAL LEVEL, CREATING DYNAMIC SPACE
更多的小剧院悬挂在公共教育设施层下方, 创造出更多样的空间

EXHIBITION HALLS ARE LOCATED ON THE UPPER LEVELS, CONNECTED BY A MAIN EXHIBITION ENTRANCE HALL
展览空间位于建筑顶部几层, 利用一个主要入口大厅进行联系

Internal Pagoda
室内的塔

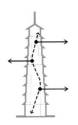
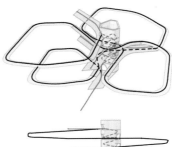
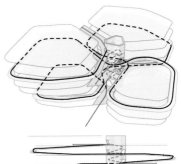

PAGODA PROVIDING MULTIPLE VIEWS & DIRECTIONALITY
赤岗塔提供访客多方面的视野与角度

CENTRAL CIRCULATION CORE AS THE INTERNAL PAGODA
中央流线核心如同室内的塔

LINK THE CIRCULATION TO A COMPLETE LOOP
联结流线成为一个回路

A UNROLLED STORY LINE ALONG THE VISITING PATH
空间流线如故事卷轴

江苏镇江科技馆
Zhenjiang Science and Technology Museum

设计单位：上海同为建筑设计有限公司	**Designed by:** Shanghai Twuad Architectural Design Co., Ltd.
委托方：镇江市科学技术协会	**Client:** Zhenjiang Municipal Science and Technology Association
项目地址：江苏省镇江市	**Location:** Zhenjiang, Jiangsu
建筑面积：8 200m²	**Building Area:** 8,200m²

功能与流线
空间与表皮

项目概况

该项目位于具有千年历史的现代城市——镇江，这是个蕴含深厚历史积淀，却又充满勃勃生机的城市。该设计游走于它的过去和现在，试图抽取出一些类型学元素，加以适当变形、转换、场所化，以期为其完成一种最佳的现代诠释。

除此之外，设计师们力求通过该建筑创造一个对话场景：即对古代的追忆与现代的感知以及未来的畅想紧密交织；人们在它们不断的对话中，感受着过去、现在和未来。

建筑设计

平面功能与流线。整个建筑平面呈方形，地上四层，地下一层，地上总建筑面积约14 500m²。平面一层主要为架空空间，与地下一层共同构成为下沉式科普广场。场馆的主入口设在南面的二层，由一层通过大台阶及自动扶梯上至二层主入口。大台阶东西两侧分别为售票处和纪念品商店；大台阶地下一层可做设备房。二层平面功能主要为短期展厅，科普活动室以及若干常设展厅；三层平面功能主要为常设展厅以及科普报告/电影厅；四层平面则为管理办公用房、业务研究用房，以及屋顶花园。参观流线主要由两种流线构成：南面二层的主入口进入和下沉科普广场的楼梯、电梯上至二楼及以上；后勤流线主要由建筑北面东西两侧的垂直交通上至四层。

建筑空间与表皮。建筑竖向空间设计为：地下一层、一层以及三、四层层高均为5m，二层高6m，建筑高度21m。6个筒形垂直交通体共同串联各层空间，另有5个竖向的喇叭形中空玻璃体穿插在各层空间，使各层内部空间丰富多变，并为各层空间带来流动的光照和通风。此外，局部空间形成上空的内庭，使得整个建筑形成极具流动性和多样性的展示空间。四层中部设有屋顶花园，为其后勤办公人员提供优质的空间感受。建筑表皮的设计采用现代流行语汇，点、面结合，塑造清新、活泼的立面形象。

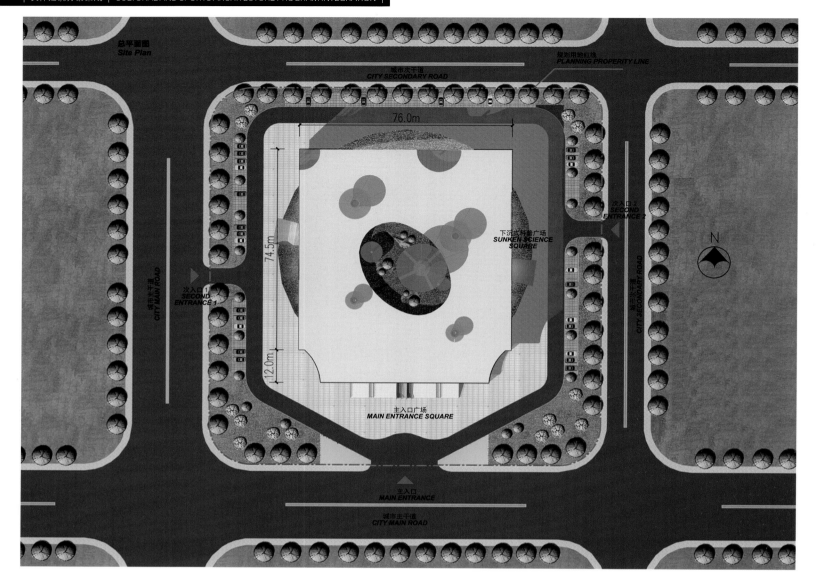

Function and Flow Lines
Space and Epidermis

Project Overview

The project is located in a modern city with a long history of thousands of years—Zhenjiang, which implies a profound historical accumulation, and is full of vitality of the city. The design walks in its past and present, trying to extract some typological elements, properly deformation, transformation, places, in order to complete its one of the best modern interpretations.

In addition, the designers strive to create a dialogue scene through the building: namely, the closely intertwined with the remembrance for the ancient, the modern perception and the imagination for the future; people in their ongoing dialogue feel the past, present and future.

Building Design

Plane functions and flow lines. The whole building plane was a square, four floors above ground and one underground floor, with a total gross floor area of about 14,500 square meters. The main floor of the plane is overhead space, and together constitutes the science sunken plaza with the underground floor. The main entrance of the venue is located in the south of the second floor, passing to the second main entrance from the floor by stairs and escalators. The east and west sides of the big step are the ticket office and souvenir shops; the first underground floor of the big step can be the equipment room. The main function of the second plane is short-term exhibition hall, science activity room and a number of permanent exhibition halls; the third plane functions are primarily for permanent exhibition and science report / Movie Theater; the forth plane is for the management office space, business research space, as well as the roof garden. Visit-able flow lines consists of two main components: the main entrance to the second floor of the south and sinking into science plaza stairs, and the elevator to the second floor and above; flow line of the logistics is mainly composed of body building with vertical transportation in the north and west sides to the forth floor.

Building space and epidermis. Architectural vertical space is designed to be: the basement and the first, second and forth floor of the building are 5m height, and the second floor is 6m, and building height is 21m. Six cylindrical layers of vertical space transportation body are together in series with each floor space. Another five vertical flared hollow vitreous interspersed in the each layer space, so interior space of each floor is rich and changeable; to bring the flow light and ventilation for the each layer space. In addition, the local space formed the inner courtyard over the hall, making the whole building to form highly mobile and diverse exhibition space. The center of the forth floor houses rooftop garden, providing superior space feeling for logistics office staff. Architectural epidermis design uses modern pop vocabulary and the combination of points and facet to create a fresh and lively façade image.

梦溪之水天上来
Mengxi Water from Sky

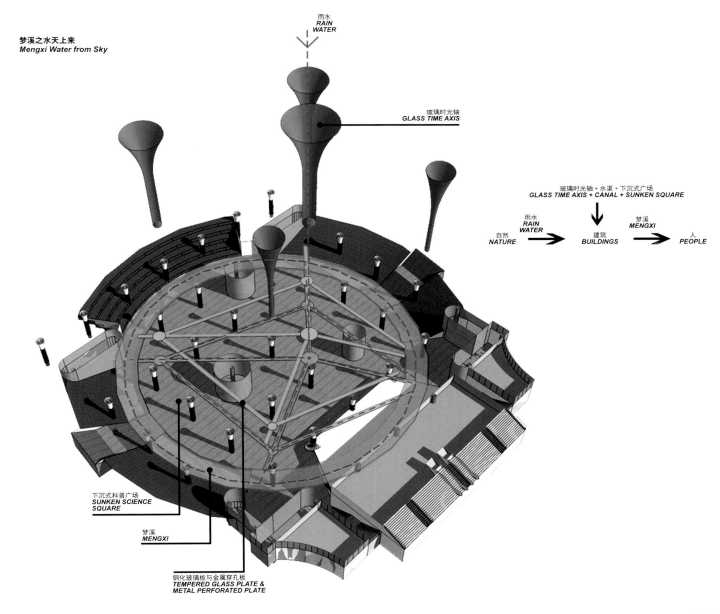

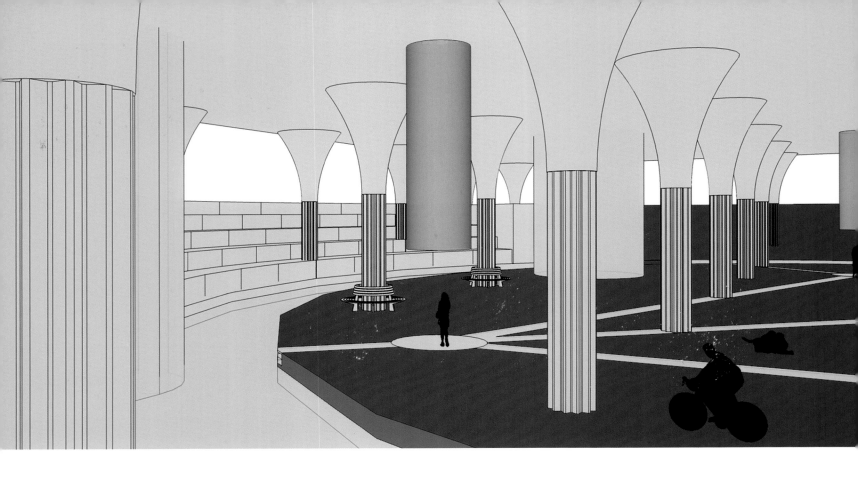

科技原理
Technology Principle

晴天 SUNNY　　　　　　　　　　雨天 RAINY

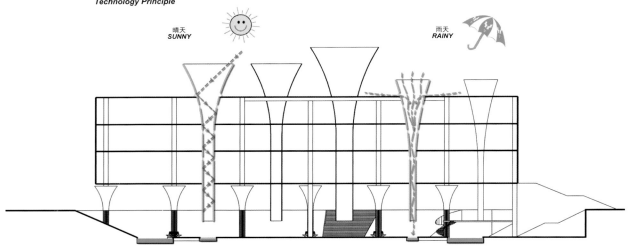

自然采光、排水原理示意：光的反射、折射原理；建筑内排水方式
NATURAL LIGHTING & DRAINAGE PRINCIPLE INDICATE: THE REFLECTION AND REFRACTION OF LIGHT; DRAINAGE PATTERNS WITHIN THE BUILDINGS

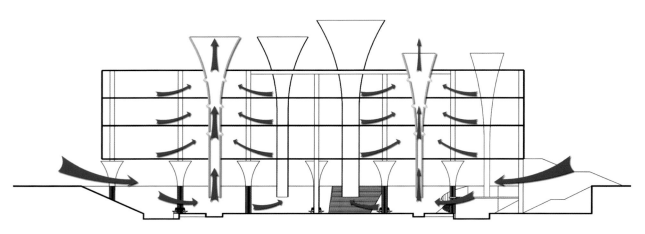

自然通风原理示意：烟囱效应
NATURAL VENTILATION PRINCIPLE INDICATE: CHIMNEY EFFECT

功能流线
Functional Flow Line

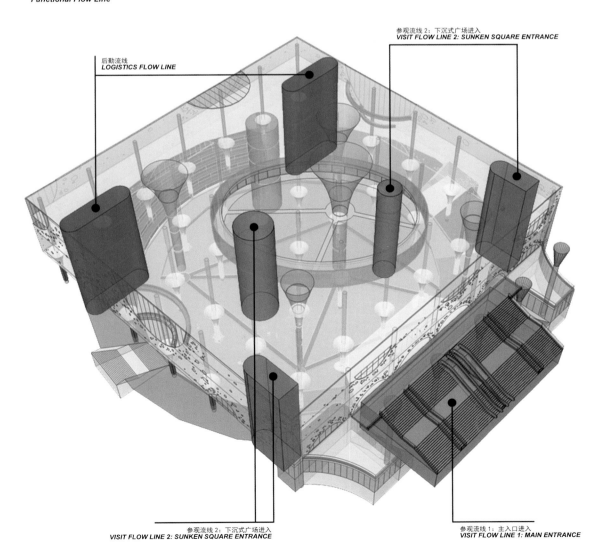

| 文体建筑方案集成 | CULTURAL AND SPORTS ARCHITECTURE PROGRAM INTEGRATION |

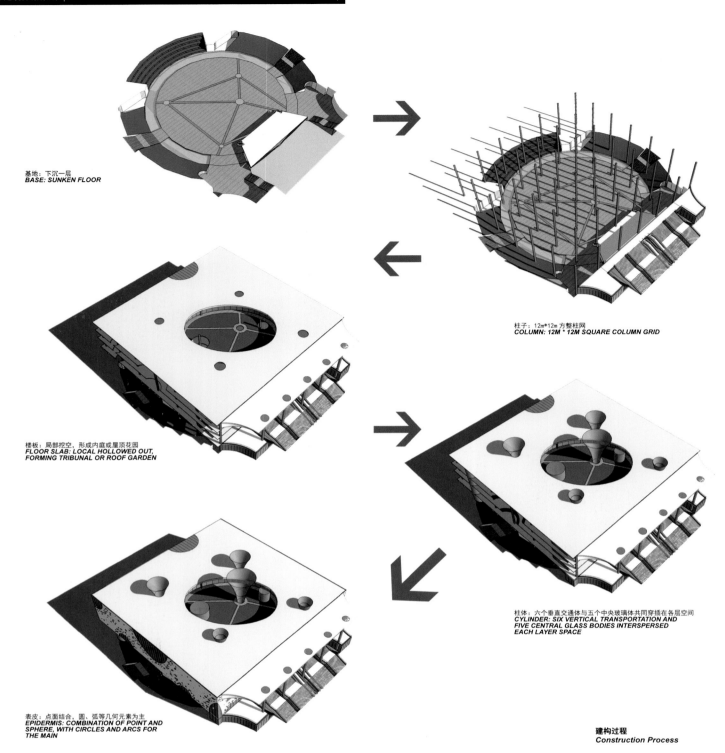

基地：下沉一层
BASE: SUNKEN FLOOR

柱子：12m*12m方整柱网
COLUMN: 12M * 12M SQUARE COLUMN GRID

楼板：局部挖空，形成内庭或屋顶花园
FLOOR SLAB: LOCAL HOLLOWED OUT, FORMING TRIBUNAL OR ROOF GARDEN

柱体：六个垂直交通体与五个中央玻璃体共同穿插在各层空间
CYLINDER: SIX VERTICAL TRANSPORTATION AND FIVE CENTRAL GLASS BODIES INTERSPERSED EACH LAYER SPACE

表皮：点面结合，圆、弧等几何元素为主
EPIDERMIS: COMBINATION OF POINT AND SPHERE, WITH CIRCLES AND ARCS FOR THE MAIN

建构过程
Construction Process

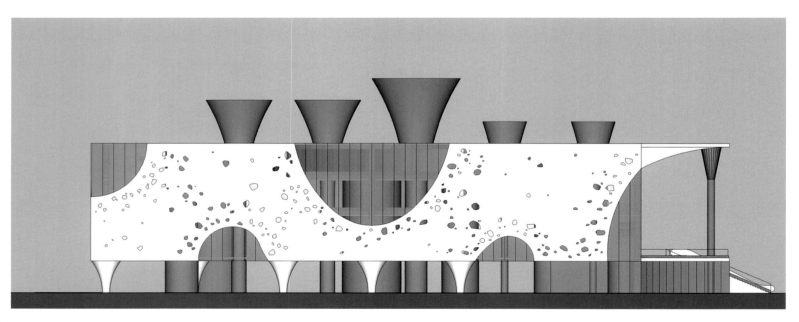

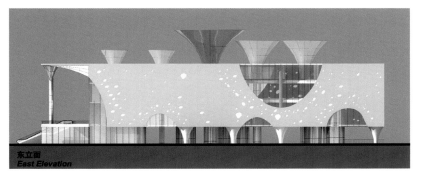
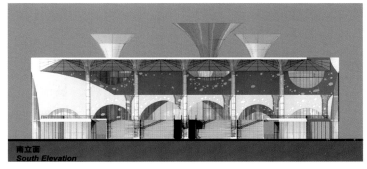
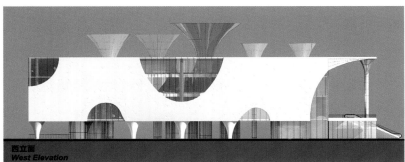
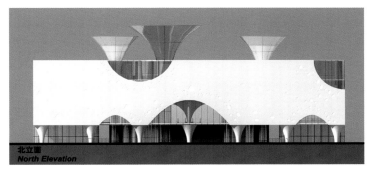

平面功能
Plane Function

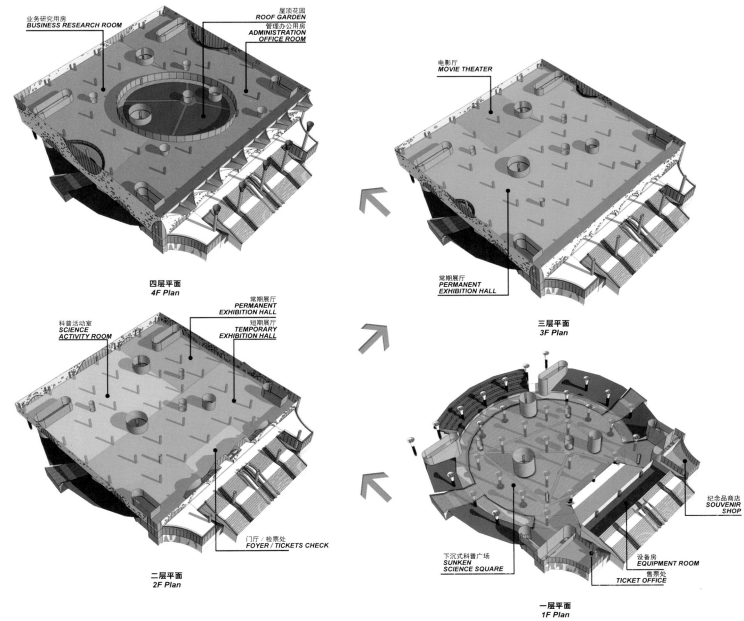

塞尔维亚贝尔格莱德科学促进中心
Center for Promotion of Science Belgrade, Republic of Serbia

设计单位：LAN Architecture
委托方：塞尔维亚共和国科技发展部
项目地址：塞尔维亚新贝尔格莱德
项目面积：26 200m²

Designed by: LAN Architecture
Client: Ministry of Science and Technological Development of the Republic of Serbia
Location: New Belgrade, Serbia
Area: 26,200m²

城市化
生态
程序

项目概况

"在已经被几个高楼大厦占领的附近，没有必要再建一座高楼，而是要建立一个新的城市空间。"贝尔格莱德科学促进中心便是在这一理念的激发下构建出来的。从历史角度来说，贝尔格莱德是一个独特的城市景象，它几乎被过往的历史所占据，然而，新的贝尔格莱德却也是这座城市抵抗和重生能力的证明，今天它代表的是一个城市新的面孔。

该中心的建设将成为一个契机，它将借助这个机会向世人揭示这一现实，并提出如何重塑一个不完整城市的解决方案。

建筑设计

该项目提出了三个主题：城市化主题、生态主题和程序主题，并且建筑物的形状和体量也是由这三个不同的参数产生的。

城市化主题：贝尔格莱德城市——镜子的另一面或视觉上重塑城市。烟囱是该项目的核心要素，并担任着城市视觉重造的重任，通过该要素将大众的目光重新连接到超出其接近物理或材料的城市元素上，使得科学促进中心看起来像一个动画的镜子，并通过这面镜子的反射看到贝尔格莱德曾经饱经历史的折磨和未来焕发的地理活力。"烟囱"的设计将成为附近建筑群的一个显目的、标志性的象征，它也将成为连接城市化与环境的枢纽，就像一面旗帜，展示出了城市和自然之间的和谐，以及新旧贝尔格莱德之间的对话。

生态主题：科学和能量。探索优化建筑体量，尤其是对空气交换的处理，是对建筑能源消耗优化处理的一个很有建设构思的问题；如何保证如太阳能、风能等自然资源在建筑形式之间进行能源传输，是该主题探索的主要方向。该建筑设计中所创建的"烟囱效应系统"——利用玻璃和混凝土进行组合，使得空气通过自然压力被吸入，所有的建筑空间可以通过放置在前院的专门的尺寸管连接到该系统中。清新的空气将通过位于公园的双重地面热交换器系统预先加热，并通过变成景观的结构元素的进气口，最后到达室内。据估计，一个大的太阳能烟囱可保证90%的建筑通风。

程序主题：感官经验。该建筑的立面本身作为城市机体的重要组成部分，而大楼的整个概念随立面的转换、原则的推移创造一种错觉，这种错觉是通过一侧的钢质三角轮廓和其他面上的木材创建的。

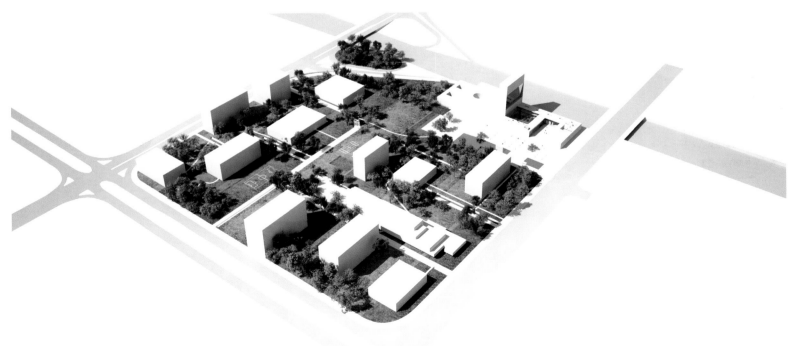

Urbanization
Ecology
Program

Project Overview

"Several high-rise buildings in the vicinity have been occupied, so there is no need to build a tall building, but to create a new urban space." Center for Promotion of Science Belgrade is stimulated and built under this concept. From a historical perspective, Belgrade is a unique urban phenomenon, literally inhabited by its history, however, new Belgrade is the proof of the city's capacity to resist and regenerate, and it represents today the modern aspect of the city.

The construction of the centre for promotion of science is a unique opportunity to reveal this reality, and to propose architecture as a solution of an incomplete urban process.

Building Design

The project proposes in one gesture three themes: urbanity, ecology and program. The buildings' shape and volume are generated by these three different parameters.

The urban theme: the city of Belgrade—the other side of the mirror / visually reinventing the city. The project's central element is the chimney, and it is a visual reinvention of the city. The chimney becomes a visual reinvention of the city, a reconnection through the gaze to urban elements beyond their physical or material proximity, so the Center for Promotion of Science looks like an animated mirror, and through its reflection you can see Belgrade's past torture of history and future vitality of geography. "Chimney" design has become a remarkable and iconic symbol in the surrounding group of buildings, and also becoming a hub for connecting urbanization and the environment, like a banner to demonstrate the harmony between the city and nature, as well as be a dialogue between the old and new Belgrade.

The ecological theme: Science and energy. Exploring the buildings volume, especially in the handling of air exchange, is an issue of well conceived construction for the optimization of energy consumption in a building; it is the main exploration of how to assure a transmission of energy sources such as sun and wind between natural resources and architecture. The "chimney effect system" created in the building design—through a combination of glass and concrete to suck up air by natural pressure. All of the buildings' spaces can be connected to this system through specially dimensioned ducts placed in the slab of the accessible forecourt. The fresh air will be pre-heated through a system of ground-coupled heat exchangers situated in the park and made visible through the air intake that becomes the structuring elements of the landscape mix with the interiors. It is considered that a big solar chimney assures 90% of the buildings ventilation.

The programmatic theme: the experience of the senses. Its façade inscribes itself as a vital part of the cities body, while the whole concept of the building goes by this principle and the façade creates an optical illusion. This illusion is created by a triangular profile with steel on one side and wood on the other sides.

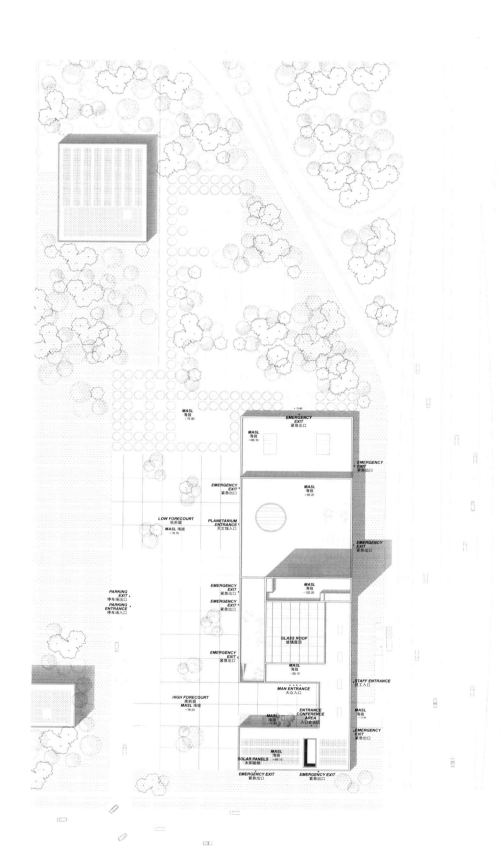

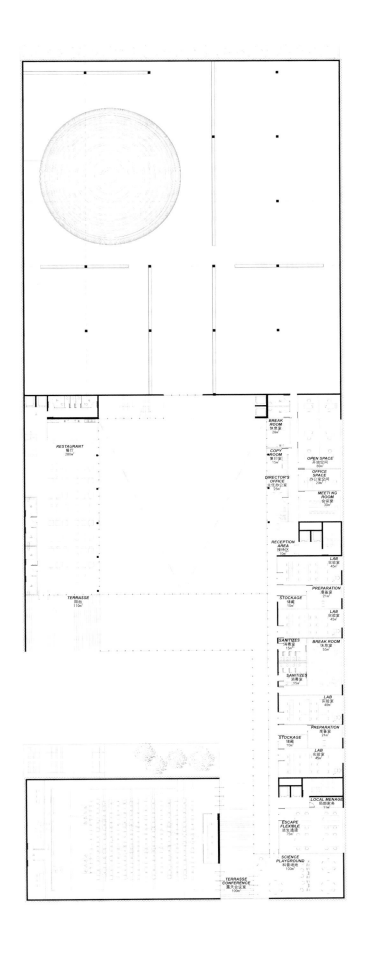
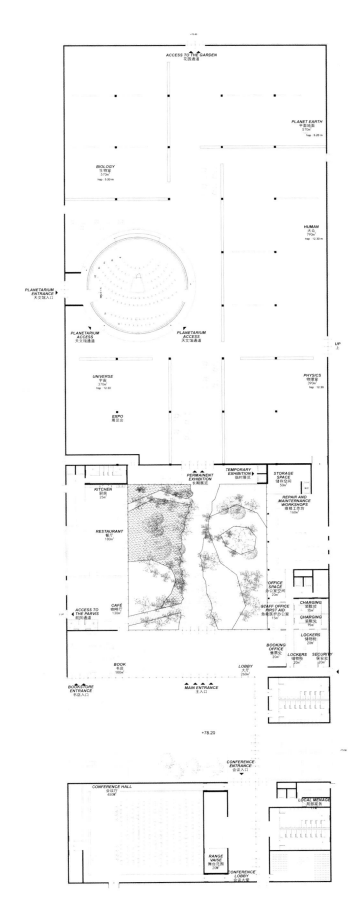

| 文体建筑方案集成 | CULTURAL AND SPORTS ARCHITECTURE PROGRAM INTEGRATION |

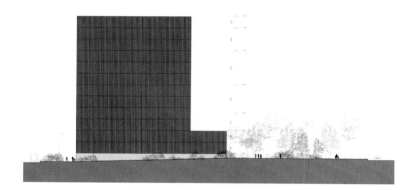
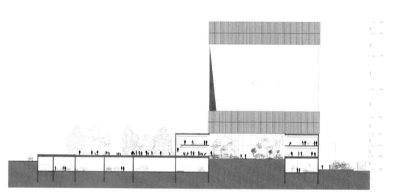
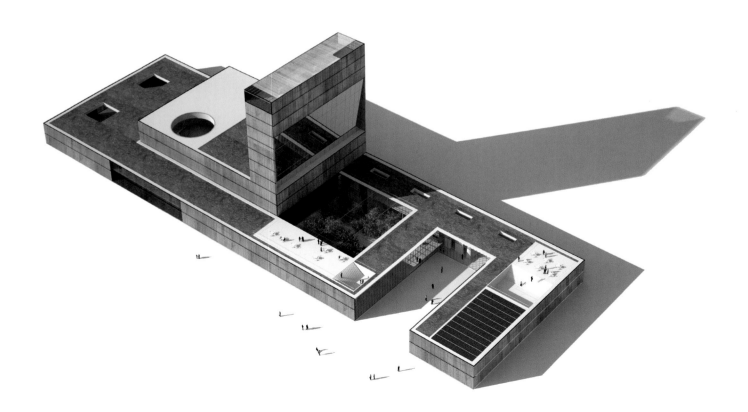

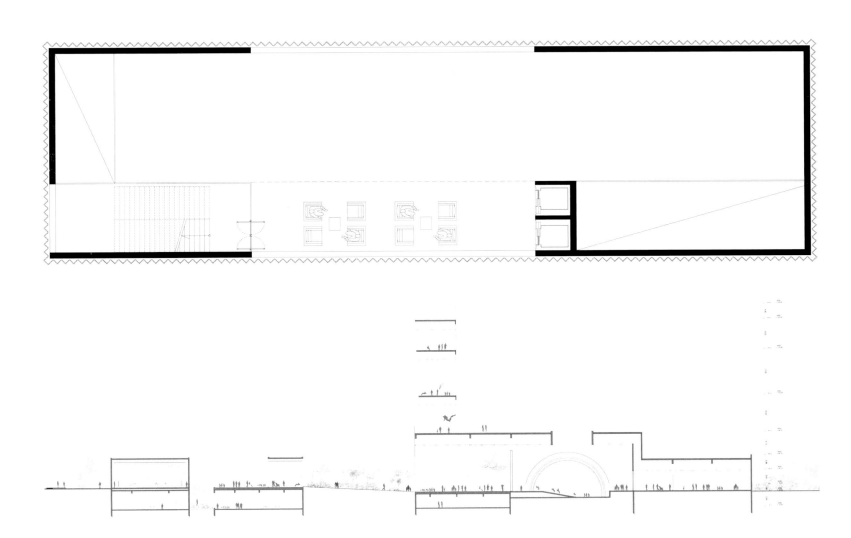

荷兰鹿特丹伊拉斯姆斯大学生活动中心
Erasmus Pavilion

设计单位：Powerhouse Company & DeZwarteHond	**Designed by**: Powerhouse Company & DeZwarteHond
委托方：鹿特丹伊拉斯姆斯大学	**Client**: Erasmus University Rotterdam
项目地址：荷兰鹿特丹	**Location**: Rotterdam, The Netherlands
建筑面积：1 800m²	**Building Area**: 1,800m²

项目概况

新鹿特丹伊拉斯姆斯大学生活动中心位于该大学区新总体规划的中心地带。这是 Powerhouse Company 与 De Zwarte Hond 事务所合作设计的项目，这座建筑被设置成为充满活力的校园中心交汇点，在这里，学术与业务交融，科学与文化交融。可持续性、可用性、透明度和亲密度是此学生活动中心的设计的主要目标。

建筑设计

此项目建筑设计的主要出发点在于它可以积极地改变外观，不仅适应天气变换和季节循环，也能根据内部活动的需要决定建筑自身的通透性和开放性。通过打开或关闭建筑外表面的动态百页，使用者可决定有多少自然光进入建筑，这减少了能量消耗，同时它也使人们能够自由调节建筑物的开敞通透度。横跨四个外立面的内表皮弧线是根据太阳路径来确定其曲度的，因此，内表皮可根据太阳方位来调节建筑内的自然光与热的量。建筑物的外观根据建筑内的事件和外部天气不断发生变化，它因为活动中心的各种人群活动而充满活力，呈现出有节奏的空间画面，使其在周边公共场所中表现出足够的存在感。

建筑表皮及其铝制百页与内部的木质天花吊顶结合，创造出生动的空间体验。建筑的透明性是由所有"暗空间"功能房间被放置于建筑物的核心决定的，这种"服务核心"有效提供了核心区周围分布的开放空间。在这个"服务核心"之上，我们放置了一个多用途礼堂，为讲座、表演和辩论提供空间。

屋顶的太阳能板，自然通风系统，以及灵活的分区使这座透明的建筑成为一个低能耗的标志性建筑。

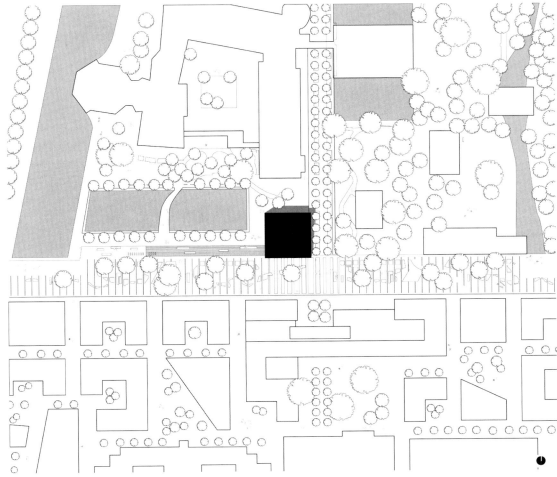

可变化外观
透明建筑
低能耗地标

| 文体建筑方案集成 | CULTURAL AND SPORTS ARCHITECTURE PROGRAM INTEGRATION |

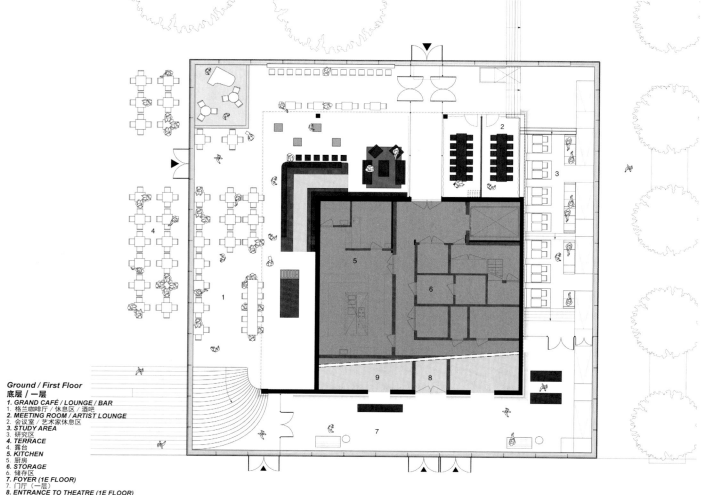

Ground / First Floor
底层 / 一层

1. **GRAND CAFÉ / LOUNGE / BAR**
   1. 格兰咖啡厅 / 休息区 / 酒吧
2. **MEETING ROOM / ARTIST LOUNGE**
   2. 会议室 / 艺术家休息区
3. **STUDY AREA**
   3. 研究区
4. **TERRACE**
   4. 露台
5. **KITCHEN**
   5. 厨房
6. **STORAGE**
   6. 储存区
7. **FOYER (1E FLOOR)**
   7. 门厅（一层）
8. **ENTRANCE TO THEATRE (1E FLOOR)**
   8. 剧院入口（一层）
9. **CLOCK ROOM (1E FLOOR)**
   9. 寄存处（一层）

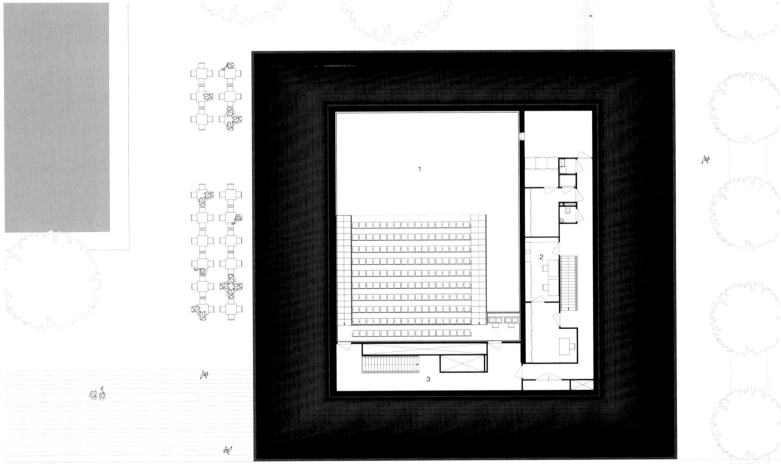

First / Second Floor
一层／二层
1. *THEATRE*
1. 剧院
2. *OFFICE (2ND FLOOR)*
2. 办公室（二层）
3. *TECHNICAL SPACE (2ND FLOOR)*
3. 技术空间（二层）

| 222–223 |

Changeable Appearance Transparent Building Low Consumption Landmarks

Project Overview

The new Erasmus University Rotterdam Student Center is located at the heart of a new master plan for the University area. Designed in combination with De Zwarte Hond, it is set to become a vibrant and central meeting point on the campus where research meets business and science meets culture. Sustainability, usability, transparency, and intimacy are the main objectives for the design of the Student Center.

Building Design

The design with the idea of creating a building that can actively change its façade—not only to adapt to the weather and the cycle of seasons, but also to allow an adjustable level of intimacy inside, depending on the events. Through opening or closing the dynamic lamellae, users can determine how much daylight comes in. This leads to a reduction of the energy consumption, but it also enables one to adjust the level of openness of the building.

The curved lines of the lamellae that sweep across the four façades are based on the path of the sun and moderate the amount of natural light and heat according to the sun orientations. The appearance of the building continuously changes depending on the events inside the building and the weather outside the building. It is animated by the vibrant rhythm of the Student Center's life, making it visible from the surrounding public spaces.

The façade and its mobile aluminum louvers combined with the wooden interior ceiling, creates a dynamic spatial experience. The transparency of the building is preserved by positioning all "dark space" programs in the core of the building. This "logistical core" efficiently serves the publicly accessible spaces distributed around it. On top of this logistical core we placed a multipurpose auditorium that offers room for lectures, performances, and debates.

The solar panels on the roof, the natural ventilation, and flexible zoning transform this transparent building into a low-consumption landmark.

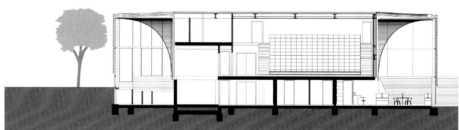

意大利佛罗伦萨音乐文化公园
Florence Music and Culture Park

设计单位：Andrea Maffei Architects	Designed by: Andrea Maffei Architects
委托方：Italy National Council for the Celebration of 150 years of Italian Republic	Client: Italy National Council for the Celebration of 150 years of Italian Republic
项目地址：意大利佛罗伦萨	Location: Florence, Italy
建筑面积：16 459m²	Building Area: 16,459m²
设计团队：Arata Isozaki & Associates M+T & Partners Ove Arup & Partenrs	Design Team: Arata Isozaki & Associates, M+T & Partners, Ove Arup & Partenrs

复合体建筑 有机屋顶 "鞋盒"礼堂

项目概况

该项目是由"对意大利统一150周年的部长委员会"所认定的,其目的是为整个国家的文化和科学价值主要作品的实现和完成基础设施项目的规划。

该剧院被设计成一个多功能的复合体,并与城市综合体进行交互,找到其适当的位置和配置,确认与周边区域的全面整合,以创造一个所有公民都能享受的地方开放城市和一个真正的文化中心。

建筑设计

该音乐公园包括一个拥有2 000个座位的剧院大厅,一个拥有1 000个座位的音乐会大厅及一个2 000个座位的露天剧院。该建筑前方有广场外部伴随,并且有一个小假山使之成为舞台之王。

此外,该建筑空间元素被设计成一组相互关联的体量,这些体量将用于表演和其他相关的活动,包括一个为服装间、指挥、合唱、乐队和歌手服务的体量,一个作为客房测试、办公和演员卫生间的体量,以及一个拥有所有商业空间的体量。而一楼的门厅有相当于真正广场的使命,通过吸引直接与它相连的大窗户来进行户外投射。系统中的所有体量被一个有机屋顶覆盖,通过该屋顶将整座综合大楼聚集起来并转化成一个体系进行运行。新礼堂有一个像鞋盒的方形形状,能产生最佳的声学性能效率。墙面及外部的路面铺满了当地的"塞茵那石",使之与意大利文艺复兴时期最伟大的纪念碑和谐平衡。

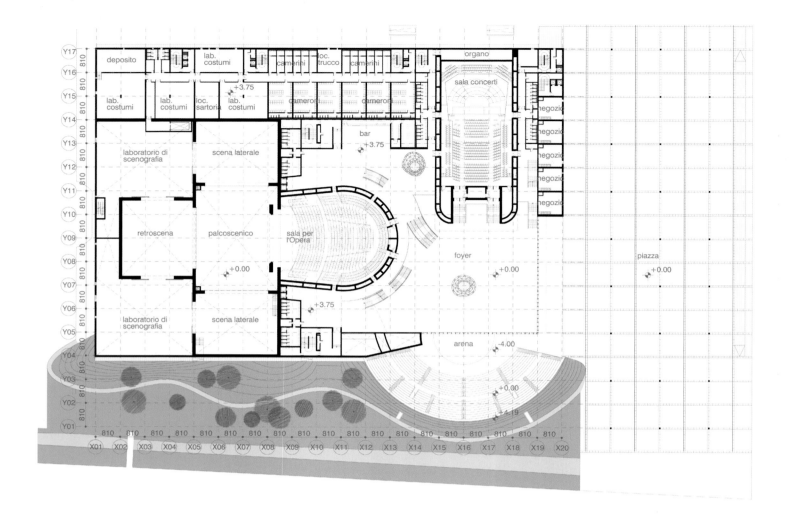

Complex Building Organic Roof "Shoebox" Hall

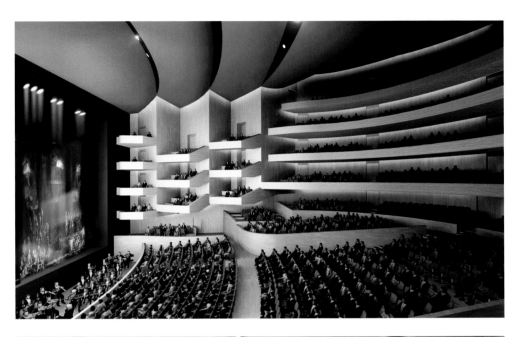

Project Overview

This work is a part of those identified by the "Committee of Ministers for the 150th anniversary of the Unification of Italy" for the planning of infrastructure projects aimed at the realization and completion of major works of cultural and scientific interest for the whole country. The theater was designed as a multi-functional complex to interact with an urban complex situation and find its proper place and configuration, confirming a full integration with the surrounding area to create a place opened to the city and a real cultural center that all citizens could enjoy.

Building Design

The project involves the construction of a 2000 seats hall for the opera, a 1,000 seats hall for concert and an outdoor arena with 2000 seats for opera and concert. The action is accompanied by external arrangements of the square in front and a small artificial hill that crowns the Arena.

We can depict the building as a set of interconnected volumes which are dedicated to those musical and theatrical performances. A volume that contains the services for dressing rooms, conductor and chorus, orchestra, singers; guest rooms test, offices, restrooms for actors and another volume with all the commercial space. The foyer has the vocation of real square that is projected outside through the large windows that bring in direct connection with it. The system volume described above is covered by an organic roof, and the roof has unified through an organic form that gathers and transforms them into an architectural plastic system. The new Auditorium has a square shape like shoe box, resulting from the best efficiency of acoustic performance. The walls as well as the pavement outside are covered by local "Pietra Serena" stone, to keep them in harmonious balance with the greatest monuments of the Italian Renaissance.

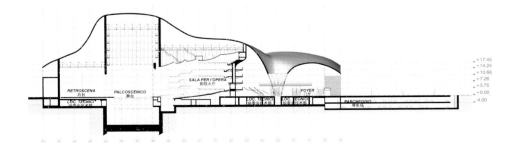

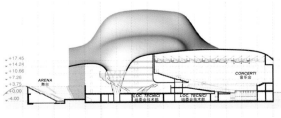

俄罗斯圣彼得堡阿里纳斯田径场综合楼
Complex of Track and Field Arenas

设计单位：STUDIO 44 ARCHITECTS
委托方：圣彼得堡城市规划建筑委员会
项目地址：俄罗斯圣彼得堡
建筑面积：96 720m²

Designed by: STUDIO 44 ARCHITECTS
Client: St. Petersburg Committee for City Planning and Architecture
Location: St. Petersburg, Russia
Building Area: 96,720m²

太阳系图像
单一柱座

项目概况

圣彼得堡，俄罗斯的"北方首都"，世界上最大的科学教育中心，是俄罗斯通往欧洲的窗口，阿里纳斯田径场综合楼就位于这座科学技术和工业高度发展的国际化城市中。项目的设计理念来源于太阳系的图像，它是由被遗弃的圣彼得堡体育和音乐会大楼改建而来的。新建综合楼建筑面积96 720m²，其中包括面积达24 300m²的可停放680辆车的停车场。

建筑设计

项目原有的综合楼作为该组合物的核心被保留，它被循环视距结构、多层次停车场所包围，而建筑新设计的透明圆顶在中庭空间和户外庭院上方；项目新建的9m高的单一柱座可容纳服务、零售等各种功能载体，也能够提供休闲游憩空间，诸如室内网球场、排球场、篮球场和五人制羽毛球球场等。

此外，该柱座的景观屋顶将作为散步和运动实践的公园区，也是田径场综合楼的室外网球场。这座围绕体育和音乐会大楼的废弃空间打造的新的阿里纳斯田径场综合楼为城市的居民增添新的功能和吸引力。

| 文体建筑方案集成 | CULTURAL AND SPORTS ARCHITECTURE PROGRAM INTEGRATION |

Site Plan
总平面图

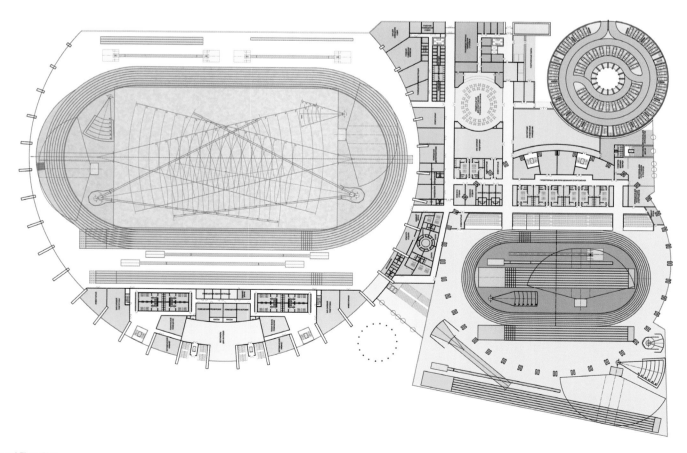

Ground Floor Plan
底层平面图

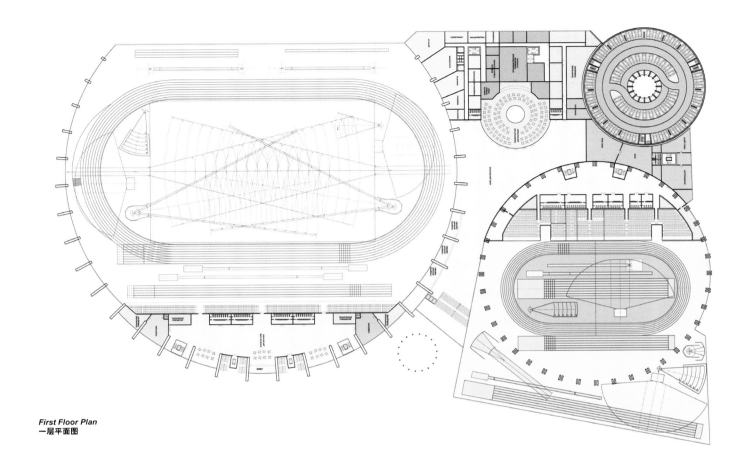

First Floor Plan
一层平面图

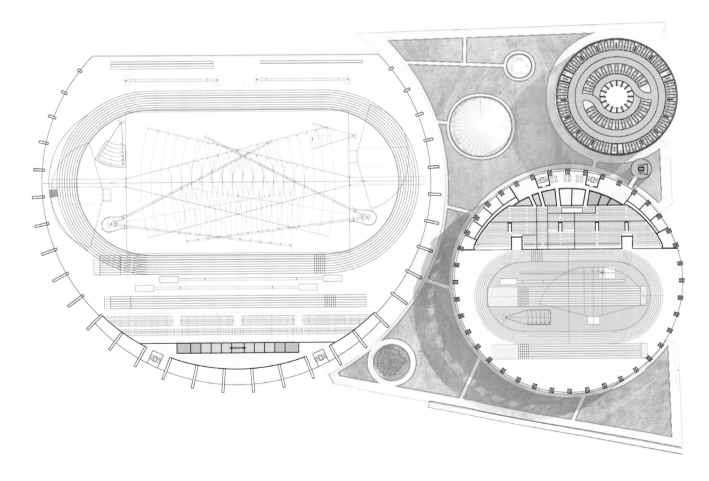

Second Floor Plan
二层平面图

First Level Plan
一层平面图

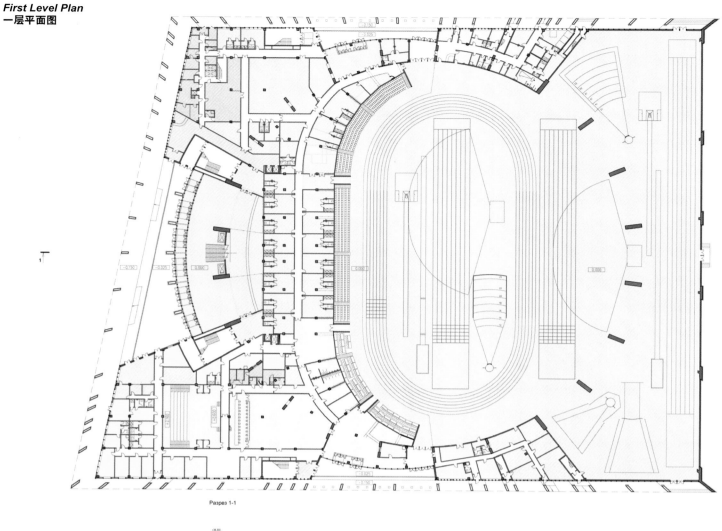

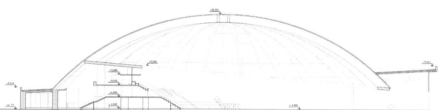

Solar System Image
Single Stylobate

Project Overview

St. Petersburg, Russia's "Northern Capital", the world's largest Science Education Center, is the window for Russia to Europe, besides, the Complex of Track and Field Arenas is located in the internationalization city with the highly developed science and technology and industry. The design concept of the project is based on the imagery of the Solar System, and it is rebuilt from the transformation of the abandoned territory near the Petersburg Sports and Concert Complex. The new building area is 96,720m^2, which includes the area of 24,300m^2 for outdoor car park for 680 cars.

Building Design

The original complex is remained as the core of the composition. It is surrounded by circular structures-stadia and multi-level car parks, while the transparent domes over the atrium spaces, outdoor courtyards; the new single stylobate with 9m height can accommodate various function carriers like service, retail, etc. and provide leisure and recreation spaces, such as indoor tennis courts, volleyball courts, basketball courts and futsal badminton courts.

Besides, the landscaped roofs of the stylobate act as park zones for promenades and sport practicing, i.e. outdoor tennis courts. Derelict spaces around the Sports and Concert Complex receive new functions and attractiveness for the inhabitants of the city.

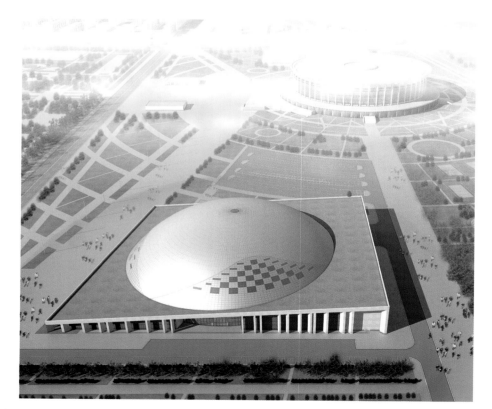

Second Level Plan
二层平面图

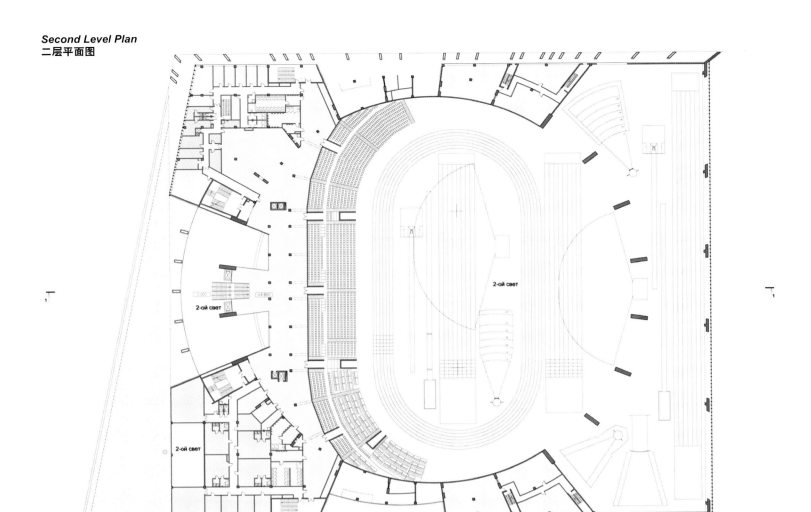

Third Level Plan
三层平面图

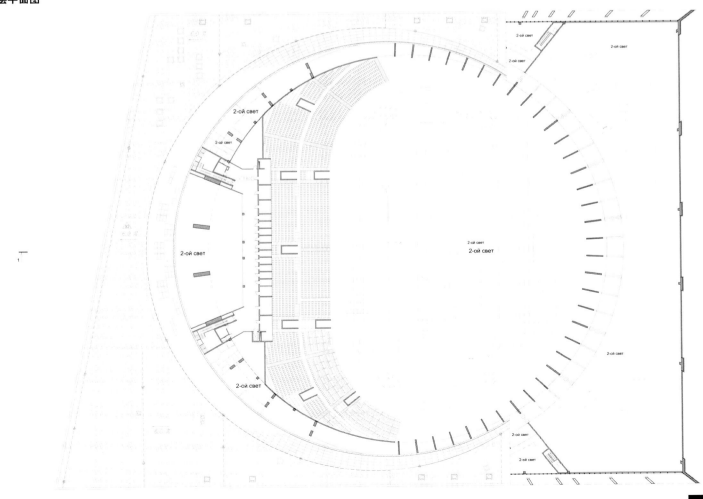

刚果共和国布拉柴维尔体育场
Brazzaville Stadium

设计单位：悉地国际有限公司　PTW Architects	Designed by: CCDI　PTW Architects
委托方：刚果重大项目委员会	Client: Congolese Major Projects Committee
项目地址：刚果共和国布拉柴维尔	Location: Brazzaville, Republic of the Congo
项目面积：903 767m²	Area: 903,767m²

世界一流体育中心
非运会主会场

项目概况

项目基地位于布拉柴维尔市区东北方向，用地南侧临近现有市政道路，与市区联系紧密；远处遥望刚果河，拥有着良好的景观资源优势。项目地块大致呈现东北高、西南低的缓坡地形态势。主体育场规划布置在整个用地势较为平缓的核心区域，建筑面积为 73 505m²，综合体育馆建筑面积为 30 895.2m²。

规划设计

布拉柴维尔体育场是 2015 年第 11 届非洲运动会的主会场，它包括一个 60 000 个座位的足球场、一个可以容纳超过 10 000 人的室内体育馆、一个拥有 2 000 个座位的游泳中心、其他配套设施及各种户外场地。

项目的设计目标是创建一个体育中心，可以在全球范围内被公认为是世界一流的设计、生态功能和高（精确）标准的体育设施；为非洲运动会提供设备齐全、全套设施场地和后勤；提高整体的印象和城市的标志性建筑的形象；建立有公共文化、体育、互动和休闲活动的适当的处所，使居民活动丰富化；创建延续和发展传统文化的处所。

建筑表面被分割成多个部分，各部分之间的狭缝可以渗入被过滤的光线，精心设计的百叶窗辅助空气流动。巨大的垂直叶片位于整个金色门面周边较低部分处，充当"鱼鳃"以便建筑进行呼吸。空气流入，使得看台及上层公共大堂区域变得凉爽。屋顶表层外观也折回建筑底层。这样提供了一个"阴影区"，保护建筑用户免受太阳及雨水的影响。屋顶照明的控制使用便于散开的自然光线透入建筑中，且不增加室内热量。

World-class Sports Center Main Venue for African Games

Project Overview

The base of the project is located in the northeast of Brazzaville. The land on the south side is near the existing municipal roads, closely connecting with the urban areas; seen the Congo River from the distant, the site has a good landscape resources advantage. The land of the project appears the gentle slope terrain trend with high in northeast and low in southwest. The plan and layout of the main stadium is arranged throughout the relatively flat core area in land terrain, with building area of 73,505m^2, comprehensive gymnasium floor area of 30,895.2m^2.

Planning Design

The Brazzaville Stadium is the main venue for 2015 African Games; it includes a 60,000-seat football stadium, a 10,000-seat indoor stadium, a 2000-seat aquatics center, ancillary facilities and various outdoor venues.

The objectives of the project are to create a sports center that can be recognized globally for its world-class design, ecological features and high (accurate) standard sport facilities; provide well-equipped, full-facility venues and logistic areas for the African Games; improve the overall impression and image of the city's iconic buildings; create appropriate premises where public culture, sport, interaction and leisure activities can be enriched for the inhabitants; create premise that perpetuates and develops traditional culture.

The overall building skin of the stadium is broken up into segments, slots in between each segment allow filtered light to penetrate and discreetly located louvres at high level assist air flow. Huge vertical blades on the lower section of the entire perimeter of the golden façade act as "gills" to allow the building to breathe. Air permeates to cool the stands and upper public concourse areas. The roof skin profile also folds back in towards the building at ground level. This provides a "shadow zone", which protects the building user from the sun and rain. Controlled use of roof-lights allows diffused natural light to enter the building without heat gains.

| 文体建筑方案集成 | CULTURAL AND SPORTS ARCHITECTURE PROGRAM INTEGRATION |

哥斯达黎加瓜纳卡斯特 Miravalles 教堂
Miravalles Church

设计单位：Fournier-Rojas Arquitectos（FoRo_ARQ）
项目地址：哥斯达黎加共和国瓜纳卡斯特
项目面积：750m²

Designed by: Fournier-Rojas Arquitectos (FoRo_ARQ)
Location: Guanacaste, Costa Rica
Area: 750m²

"鱼形泡沫"
宗教文化符号
永生

项目概况

Miravalles 教堂位于哥斯达黎加的瓜纳卡斯特省首府利韦里亚。在中世纪,教堂和寺庙都是基于"鱼形泡沫"理念而设计的。因此在本案中,也运用了符合"鱼形泡沫"中圆弧的部分,而这些在此项目中是正规的、象征性的元素。

建筑设计

项目正规的空间是基于"鱼形泡沫"设计的,椭圆形形状是由有着相同半径的两个圆相交而成的,以这样一种方式相交使每个圆的中心位于另一个圆的圆周上。它也呈现出了一种代表基督教的鱼。

屋顶、圣坛后面的墙壁、圣坛以及大型彩色玻璃窗都使用了这个形状。除了宗教象征,设计也使用了地方符号。为了做到这一点,设计师决定使用本地的黑石头将建筑固定于此。这座教堂的形状也回应了围绕城镇所在的大山的形状。整个建筑以倒影池为背景,象征着永生。来自屋顶的棚式建筑将雨水排入到这个倒影池中,产生水的运动和声音,从而引发信徒的冥想状态。

"Vesica Piscis" Religious Culture Symbols Eternal Life

Project Overview

Miravalles Church is located in the town of Fortuna in Guanacaste, Costa Rica. In Medieval times, churches and temples were designed based on the Vesica Piscis. Thus, in our case, sections of the circle such as those that conforms the Vesica Piscis are the formal and symbolic elements used in this project.

Building Design

The formal and spatial design of this project is based on the "Vesica Piscis", an oval shape which comes out of the intersection of two circles with the same radius, intersecting in such a way that the center of each circle lies on the circumference of the other. It represents a fish, which represents Christianity.

The roof, the wall behind the altar and the steps to the altar as well as the large stained glass windows use this shape. Besides religious symbolism, we use symbols of place. To accomplish this, the designers decided to use a local black stone to tie the building to place. The shape of this church also responds to the shapes of the mountains that surround the town where it's located. The whole building is set in a reflecting pool that symbolizes eternal life. Rainwater that sheds from the roof will drain into this pool, generating water movement and sound, thus provoking a state of meditation among churchgoers.

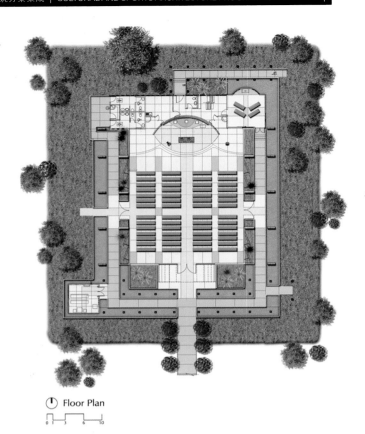

Floor Plan

韩国大邱 Gosan 公共图书馆
Daegu Gosan Public Library

设计单位：THEEAE LTD.	Designed by: THEEAE LTD. (The Evolved Architectural Eclectic Limited)
委托方：大邱广域市寿城区政府	Client: Daegu Metropolitan City Suseong-gu Government
项目地址：韩国大邱市	Location: Daegu Metropolitan City, Republic of Korea
项目面积：2 080m²	Area: 2,080m²

文化空间
可持续设计

项目概况

设计通过提供教室、展览区、讲坛、半私人阅读区及数字数据角等，让人们在从书本、杂志、文章等渠道中获取信息时有一个休息和放松的场所。当地社区的扩大方案也将成为新图书馆系统的另一个巨大改变。这不是一个戏剧化的形式，而是吸引公众到新图书馆空间来，并且徜徉于其中。

总平面图
Site Plan

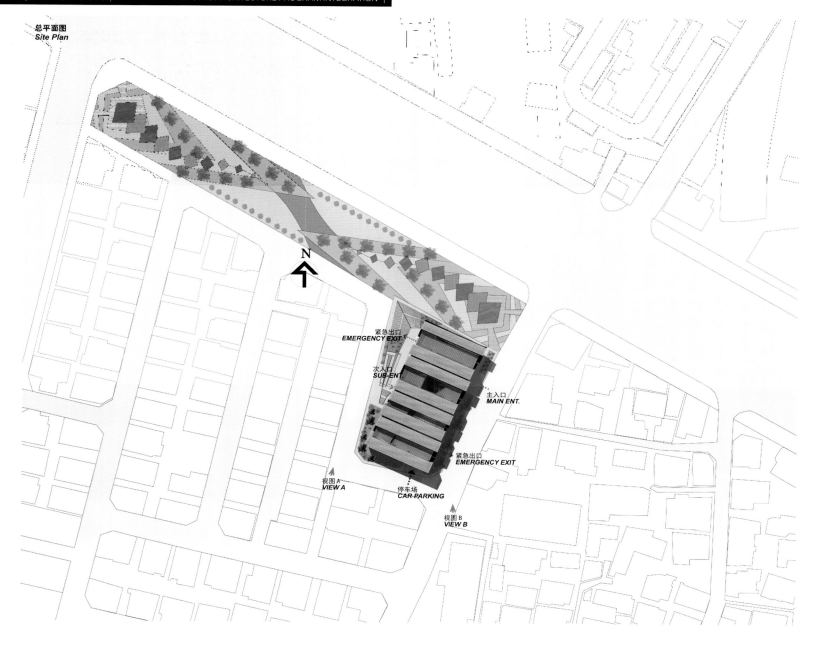

规划设计

新图书馆设计有一个带开放长廊的私人阅读区。基于这一想法,给为寻找特定书籍并想私下找地方静坐阅读的人们分配必需的空间,需要开发书架之间的距离。书架区按一定方向排列好的,形成一种建筑形式。然后,在水平方向上,它们之间也是彼此呈一定角度倾斜,提供足够多的入口进入书架及半私人开放阅读区域。如是,它的形状看起来就像书籍放在书架上这样一种形式自然形成。

项目所设计的空间都考虑到用户的体验,因此空间应该被阳光光线照亮。一大片长方形延伸空间经过每个书本区域,并且设计一个合适的开口,获得充足的阳光。同样,停车场入口按此方式设计。为了更好地利用阳光,项目将 Kalwall(透明墙系统)引入到主要设计元素中。这一系统不仅能获得来自太阳足够多的光线,同时,其抗热性也能很好地应对大邱市四季分明的气候。另外,它的建造方法简单,方便实现。

图书馆建筑结构系统为钢筋混凝土,其结构墙(书架墙)充当柱子,水泥楼板则充当构架系统。机械系统设计方面,考虑到通风性,将其设置在第三层。这样,室内空气流通将很容易被恢复,从高层将空气流向不易通风的低层。而且,从北至南的开口区域也将产生室内空间的自然通风系统。

局部剖面详图
Partial Section Detail

半透明墙
TRANSLUCENT WALL

当地社区节能问题，尤其是公共建筑，在设计方案中至关重要。我们设计建议使用KALWALL系统。通过减少人造光，实现其简单结构及高效节能材料。同时，它的低热透光度在冬季及夏季期间都能节约能源。

ENERGY SAVING IN THIS LOCAL COMMUNITY, ESPECIALLY AS A PUBLIC BUILDING, IS CRITICAL IN DESIGN PROPOSAL. OUR DESIGN PROPOSES THE USE OF KALWALL SYSTEM. ITS EASY CONSTRUCTION AND HIGHLY EFFICIENT ENERGY SAVING MATERIAL ARE USED BY REDUCING ARTIFICIAL LIGHT. ALSO, ITS LOW HEAT TRANSMITTANCE IS PROVEN TO SAVE ENERGY DURING WINTER AND SUMMER.

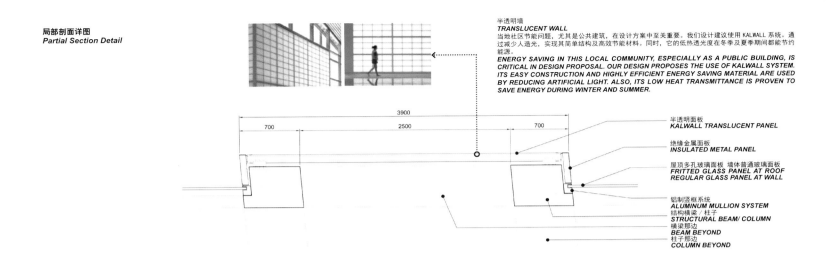

半透明面板 / KALWALL TRANSLUCENT PANEL
绝缘金属面板 / INSULATED METAL PANEL
屋顶多孔玻璃面板 墙体普通玻璃面板 / FRITTED GLASS PANEL AT ROOF REGULAR GLASS PANEL AT WALL
铝制竖框系统 / ALUMINUM MULLION SYSTEM
结构横梁/柱子 / STRUCTURAL BEAM/ COLUMN
横梁那边 / BEAM BEYOND
柱子那边 / COLUMN BEYOND

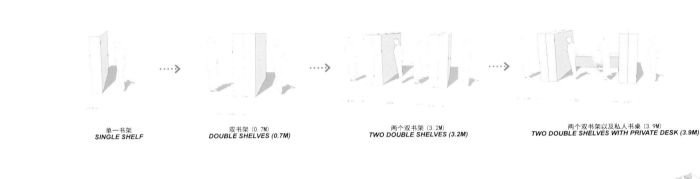

单一书架 / SINGLE SHELF
双书架 (0.7M) / DOUBLE SHELVES (0.7M)
两个双书架 (3.2M) / TWO DOUBLE SHELVES (3.2M)
两个双书架以及私人书桌 (3.9M) / TWO DOUBLE SHELVES WITH PRIVATE DESK (3.9M)

一个集中式站立型书架=两个双书架连续线性排列 / ONE MASS STANDING SHELF=TWO DOUBLE SHELVES CONTINUOUS LINEAR ARRAY
每个其他集中式书架线性排列 / EVERY OTHER MASS SHELF LINEAR ARRAY
每个集中式书架随意倾斜 / EACH MASS SHELF RANDOM INCLINATION
室内开放向导 / INTERIOR OPENING GUIDE MASS
室内开口最终设计 / FINAL DESIGN MASS OF INTERIOR OPENING

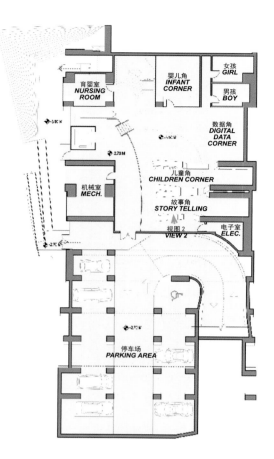

地下室平面图
Basement Floor Plan

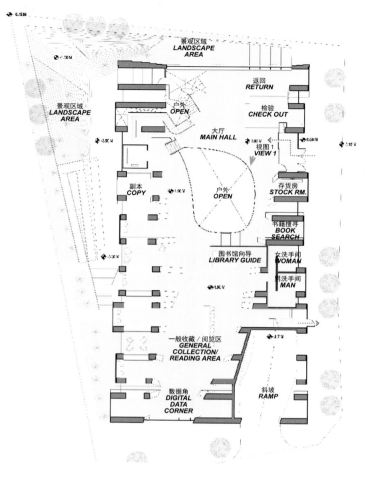

一层平面图
1st Floor Plan

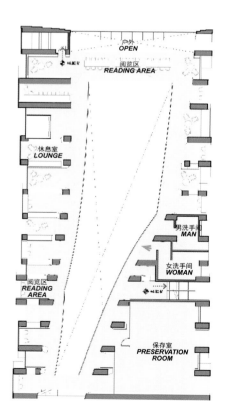

二层平面图
Level 2 Floor Plan

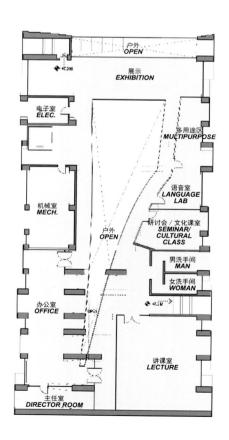

三层平面图
Level 3 Floor Plan

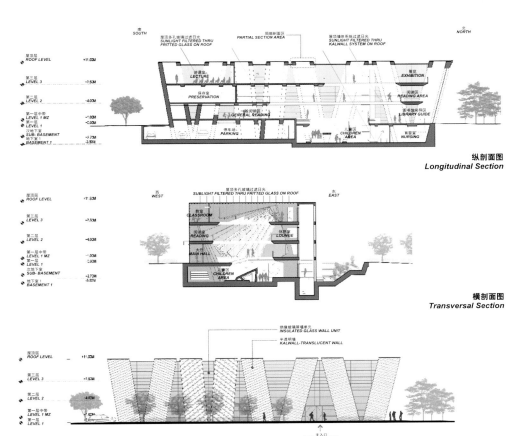

Longitudinal Section

Transversal Section

West Facade Elevation

Cultural Space Sustainable Design

Project Overview

The design is to agree that people are in need of finding places to sit and get relaxed for information from books, magazine, articles, etc. Thus, the expanded program for the local community by providing classrooms, exhibition area, lecture room, semi-private reading areas, digital data corner, etc. will be another major change in current library system. We believe not in a dramatic way but strongly the new design of space that should convey this change in a way to enhance and motivate public to come and enjoy the use of library.

Planning Design

The first ideas we aimed for this library was that it should have private reading area with opened arcade. In order to achieve it, the distance of book shelves each other was developed based on necessary space allocated for people looking for a specific book section and for people who want to take a time to sit and enjoy reading in private. By this arrangement, a unit of book shelves was created and it is arrayed in a direction to make a form of architecture. Then, it has been horizontally inclined each other to a certain degree by providing enough entrance, book shelves, semi-private open reading area one after another. By doing so, its shape seems more naturally formed as books on a shelf.

The space that we imagine for the library was all about user friendly, so the space should be well lighten by the sun. Hence, a mass of rectangular is stretched to pass through the each book sections to get a proper opening with lights from the sky. Also, parking lot entry was designed in the same manner. For the better use of this extensive sunlight, we brought Kalwall (transparent wall system) into our main design elements. Its transparency is to get enough brightness from the sun, and at the same time, its heat resistance also good for a climate like Daegu with four distinct seasons. In addition, its easy construction method would be even beneficial for the realization of the project.

The structure system is basically reinforced concrete structure. Its structural mega walls (bookshelves walls) act as columns. With floor slabs, they will act as truss system. Thus, for the mechanical design, we locate the space on level 3 floor for its easy ventilation. The interior air flow will easily resume from the top floor to circulate air into lower floor which return air will be located. Plus, the opening area from the north to the south will create natural ventilation of interior space.

挪威克里斯蒂安松歌剧院与文化中心
Opera and Culture House—Kristiansund, Norway

设计单位：Brisac Gonzalez Architects (uk) + Space Group (no)
委托方：Kristiansund Kommune
项目地址：挪威克里斯蒂安松
建筑表面积：15 000m²

Designed by: Brisac Gonzalez Architects (uk) + Space Group (no)
Client: Kristiansund Kommune
Location: Kristiansund, Norway
Surface Area: 15,000m²

**创新空间
解放文化**

项目概况

新的歌剧院及文化中心将歌剧院、图书馆、学校、文化设施及青年中心合为一体。这些综合设施以其前沿最深入的创新方式代表着文化。项目设计创造意外来处理空间、鼓舞空间、集体及个人共存空间，以激发实验、发明、异质性和自由，也就是解放文化。

规划设计

项目场地原有两个重要建筑，一个是19世纪的学校建筑，另一个是20世纪早期的"人之家"。本案设计综合三个不同的独立建筑，创造出一个多孔文化混合，既独立又相关，而且两个现存建筑将被彻底重新打造。一个带状的玻璃桥可兼做展览馆，连接"人之家"的图书馆和新建筑大楼的主门厅。

新建筑也将包含青年文化活动中心，一个地下通道将其连接至新礼堂的舞台层。建筑内部的公共循环路径作为集体空间，开放一系列冒险活动。从入口到餐厅的过程，通过红地毯环路迂回行进，立即展示出丰富的内部设置及其景观。

大厅礼堂布置很简单：让尽可能多的客人尽量靠近舞台，在舞台上，表演者能够看到一大群观众。多功能大厅设计灵活，兼具实用功能且美观大方。风景房间设计在新建筑最高层，它结合有餐厅、食堂和乐队排练室，共同创造出一个大型景观空间。而扩建的外部露台为其增添了灵活性与多变性。

新建筑的外观是帘状的，有选择性地显露出建筑内部的活动和场景。从外部看去，感觉就像是一件柔软的连衣裙，被小小的反光亮片覆盖着，而闪光处不断变化，不断流动。从内部看，透明的和半透明的光线均匀地照亮整个建筑。

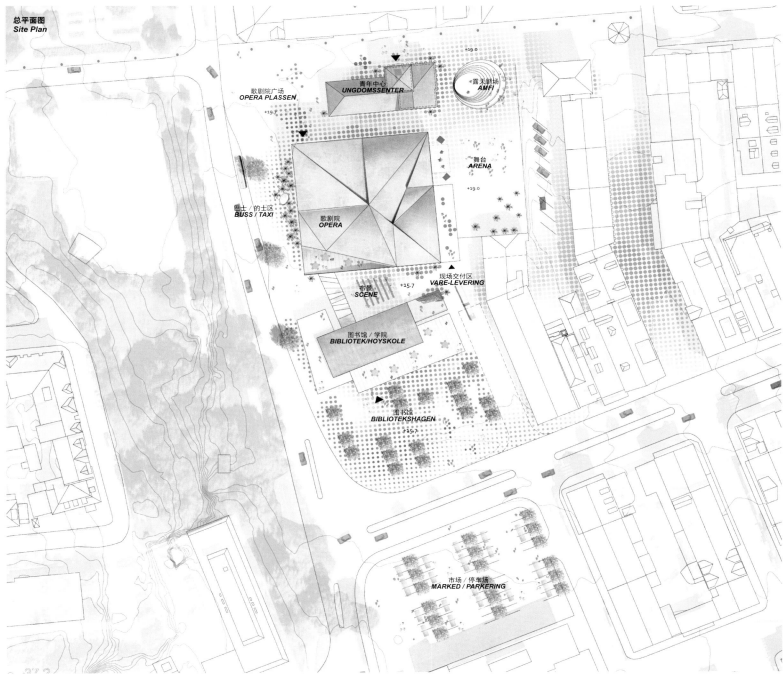

总平面图
Site Plan

| 文体建筑方案集成 | CULTURAL AND SPORTS ARCHITECTURE PROGRAM INTEGRATION |

平面图 1
Plan 1

平面图 2
Plan 2

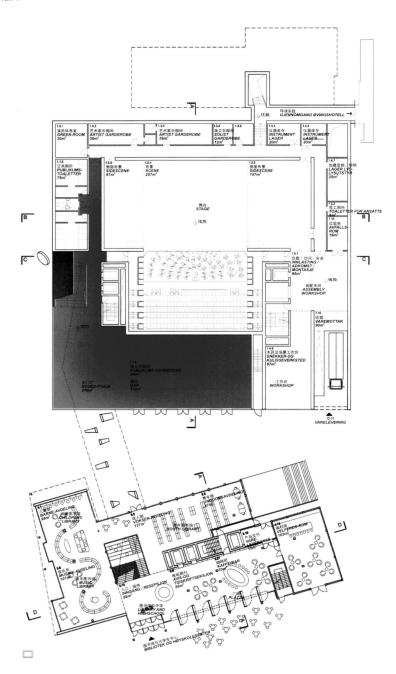

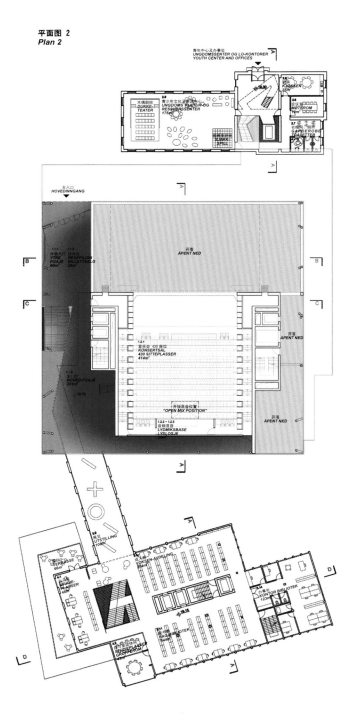

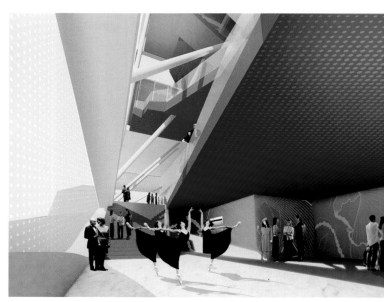

平面图 3
Plan 3

平面图 4
Plan 4

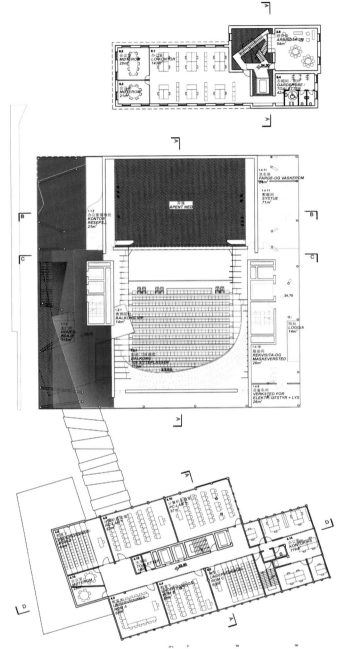
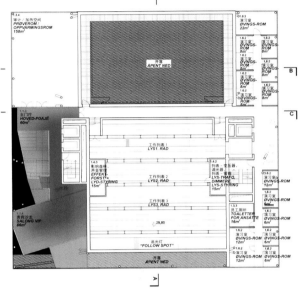

| 文体建筑方案集成 | CULTURAL AND SPORTS ARCHITECTURE PROGRAM INTEGRATION |

平面图 5
Plan 5

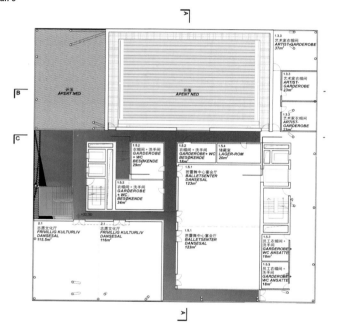

平面图 6
Plan 6

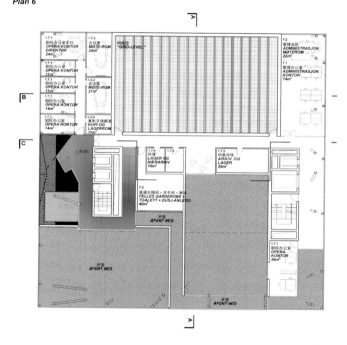

平面图 7
Plan 7

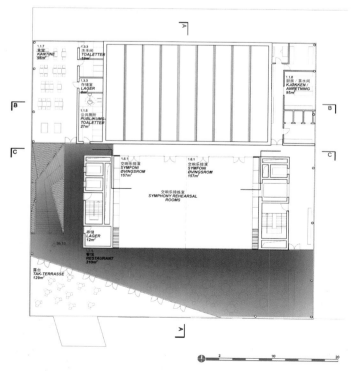

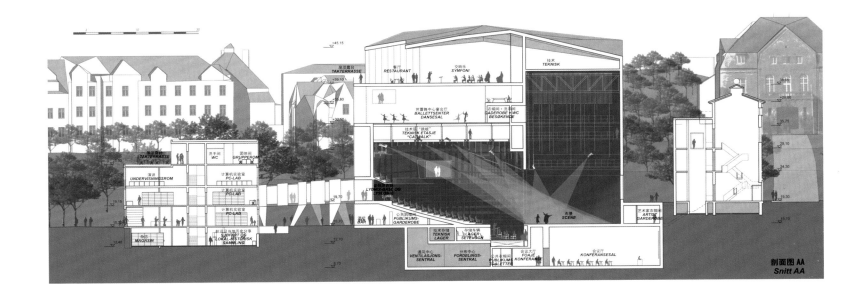

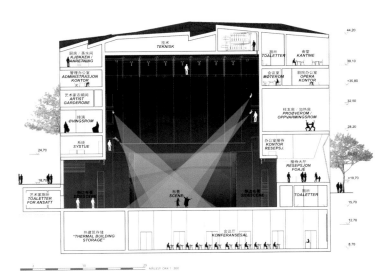

剖面图 BB
Snitt BB

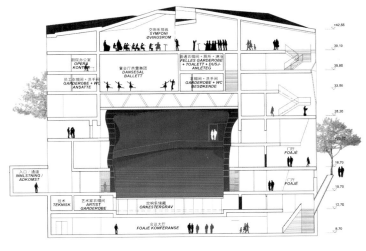

剖面图 CC
Snitt CC

Innovation Space Liberate Culture

Project Overview

The New Opera and culture house amalgamates opera, library, school, cultural facilities, and youth center. The complex represents culture in its most pervasive and innovative form. To create spaces for the unexpected, spaces to inspire us, spaces where the collective and individual can coexist, to inspire experimentation, invention, heterogeneity and freedom is to liberate culture.

Planning Design

There are already two significant buildings on the site—a 19th century school building and an early 20th century Folkets Hus (People's House). To engulf them by a third larger building would diminish their integrity. The design strategy creates a porous cultural compound of three very different free standing buildings that are autonomous yet connected. The two existing buildings will be fully refurbished and filled with a host of new activities. A ribbon-like glass bridge that doubles as exhibition gallery connects the library in the Folkets Hus to the main foyer of the new building. The school building will contain youth cultural activities. An underground passage will link it to the stage level of the new auditorium. Inside the new building the public circulation route serves as collective space open for an array of speculative activities. A procession of movement from entry to restaurant, via the red carpeted circuit that weaves it way through the building, reveals at once both the rich interior setting and views of Kristiansund.

The objective for the auditorium is simple: allow for the greatest number of guests to be as close as possible to the stage. From the stage, the performers will see an ocean of people, wall-to-wall. The multi-purpose hall, driven by the demand for flexibility, is often the victim to the paradoxes of big-small, invisible-exposed, functional-beautiful. On the top floor of the new building restaurant, canteen and orchestra rehearsal room can be combined to create a very large room with spectacular views. An expansive exterior terrace adds greater flexibility in a setting where different scenarios, be they planned, impromptu, or incidental can occur.

The façade of the new building is curtain-like, selectively revealing the activities within. From the exterior, the perception is of a soft dress, covered in small reflective sequins, shimmering, constantly changing, and flowing. From the inside, the lighting varies from translucent to transparent, providing an even light throughout the building.

East Elevation
东立面图

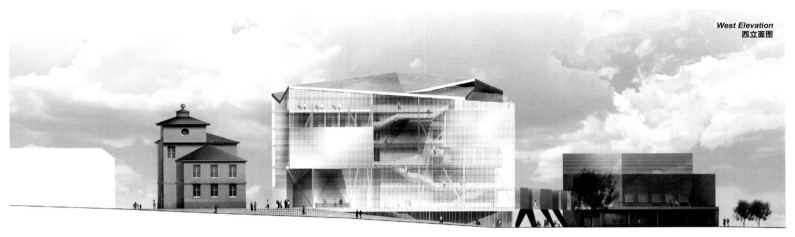

West Elevation
西立面图

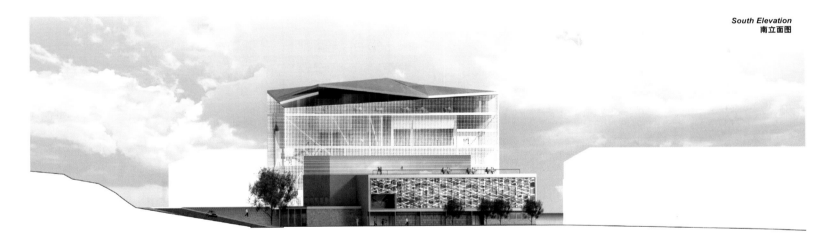

South Elevation
南立面图

North Elevation
北立面图

瑞典哥德堡世界文化博物馆
Museum of World Culture—Gothenburg, Sweden

设计单位：Brisac Gonzalez Architects
委托方：国有财产局
　　　　世界文化博物馆
项目地址：瑞典哥德堡市
总楼层面积：10 950m²

Designed by: Brisac Gonzalez Architects
Client: Statens Fastighetsverk
　　　　Museum of World Culture
Location: Gothenburg, Sweden
Total Floor Area: 10,950m²

公共文化平台
交流论坛

项目概况

该博物馆坐落于哥德堡市中心的一座山脚下，它是一个全新的重要的国家博物馆，项目包括展览馆、大学研究中心、图书馆、可以同时容纳200人的礼堂、餐厅及行政管理办公室。

规划设计

该博物馆为瑞典当地人学习收藏提供了一个全新的公共平台，同时这个平台也可作为国际级别的当地事件的交流论坛。

设计策略反复思考如何使东西翼两侧形成具有明显标志的区别。固定的西翼包含沿途的画廊空间以及办公室区域。开放的东翼朝向小山，以后的公共活动就在东翼进行和开展。而两者之间是一个峡谷地带，它包括有建筑服务，其间的公共环路在这三个区域内迂回行进。当你慢慢地往上走，从上面就可以逐渐清晰地看到下面的各个元素的姿态，它们共同创造出一系列的参考点，包括小山、中庭及博物馆的景色。

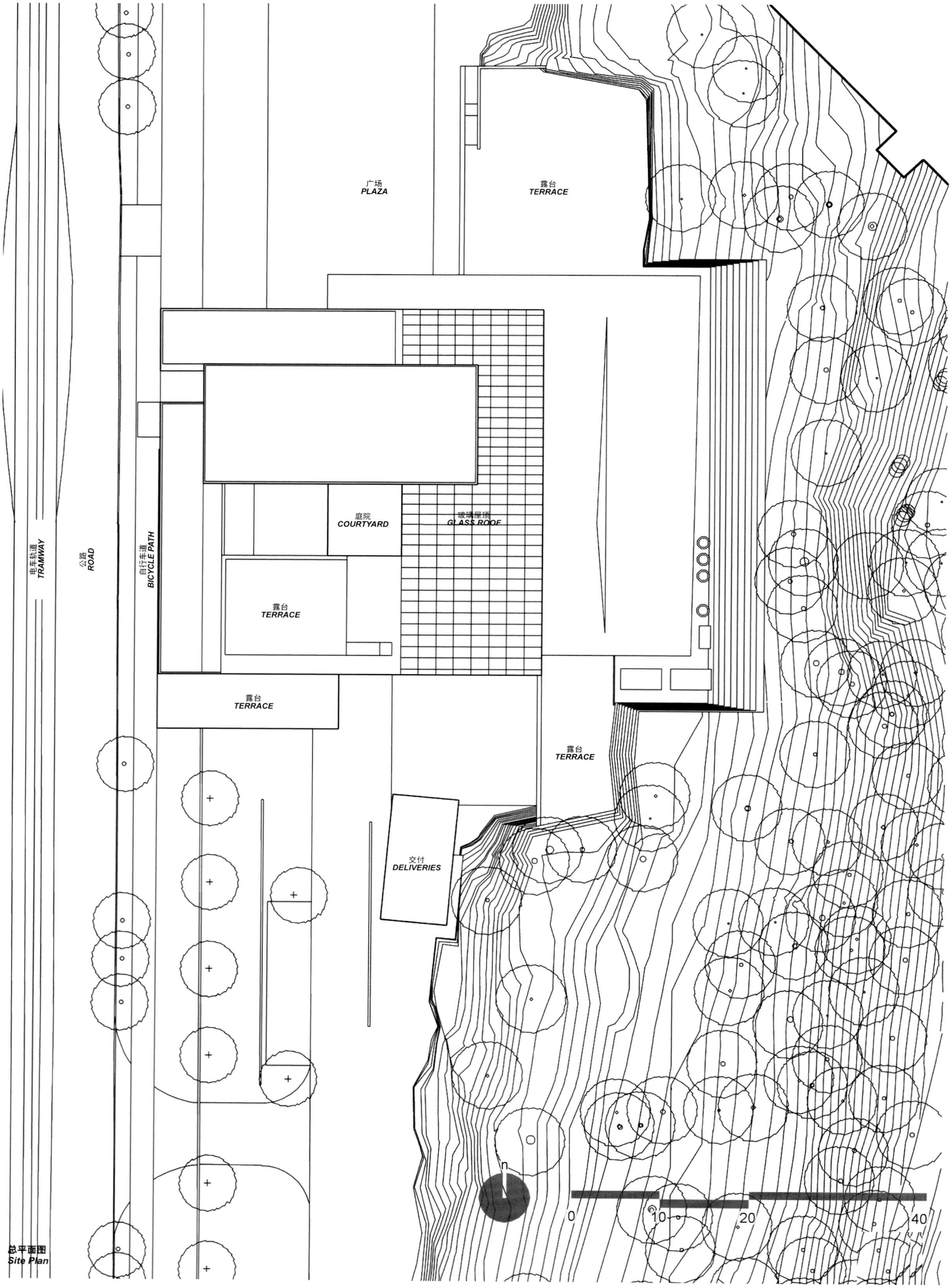

| 文体建筑方案集成 | CULTURAL AND SPORTS ARCHITECTURE PROGRAM INTEGRATION |

广场
PLAZA

露台
TERRACE

电车轨道
TRAMWAY

公路
ROAD

自行车道
BICYCLE PATH

庭院
COURTYARD

玻璃屋顶
GLASS ROOF

露台
TERRACE

露台
TERRACE

露台
TERRACE

交付
DELIVERIES

0   10   20   40

总平面图
Site Plan

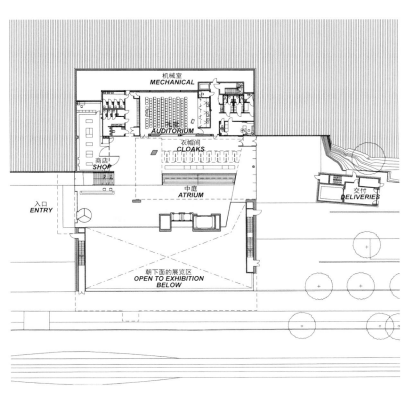

基层
Ground Floor

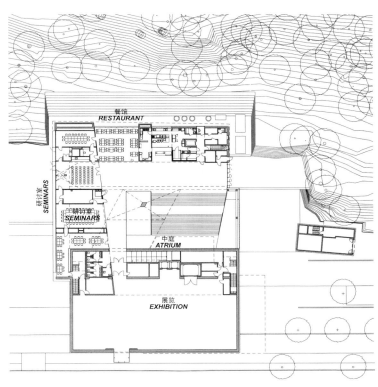

一层
First Floor

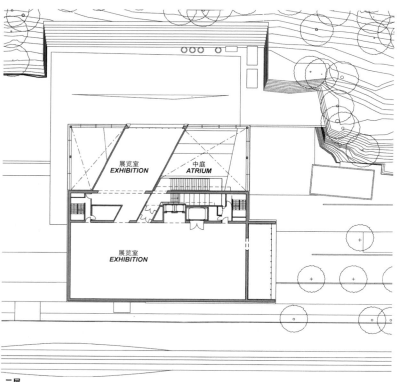

二层
Second Floor

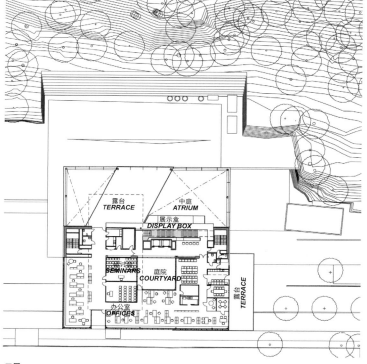

三层
Third Floor

Public Platform Exchange Forum

Project Overview

Situated at the foot of a hill in the city centre, the museum is a new major national museum including exhibition galleries, university research centre, library, 200-person auditorium, restaurant and administrative offices.

Planning Design

The museum provides a new public platform for the ethnographic collections of Sweden. It also serves as a new forum for international and local events.

The design strategy revolved around creating a clearly marked difference between a solid west wing, containing the gallery spaces and offices along the street, and an open east wing towards the hill, where public activities take place. Between the solid west and the open east is a canyon-like zone containing the building services, with public circulation weaving its way through the three areas. As one goes up the building, the elements which were seen from below are gradually perceived from above, creating a sequence of reference points throughout the building with alternating views of the hill, atrium and museum.

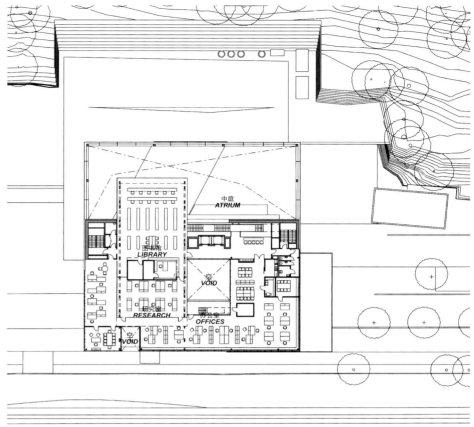

四层
Fourth Floor

横截面
Cross Section

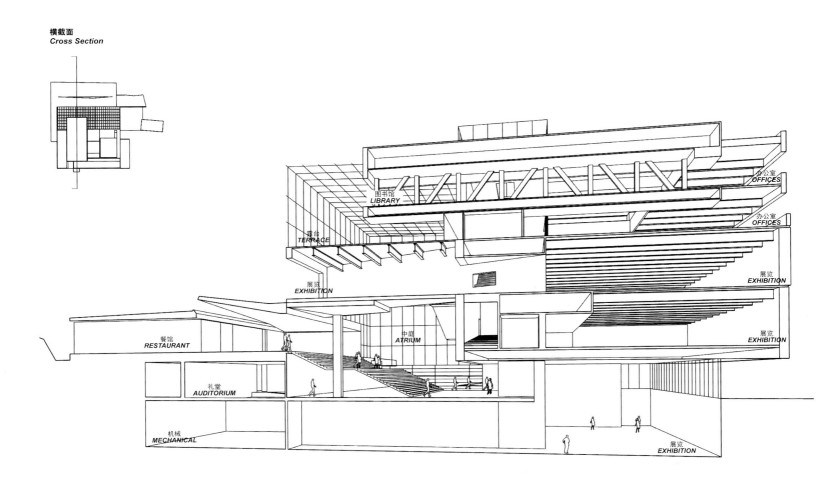

丹麦奥尔堡音乐厅
House of Music

设计单位：COOP HIMMELB(L)AU,
　　　　　　Wolf D. Prix & Partner ZT GmbH
委托方：丹麦奥尔堡北日德兰半岛音乐厅基金会
项目地址：丹麦奥尔堡市
建筑面积：20 257m²
图片版权：Duccio Malagamba

Designed by: COOP HIMMELB (L) AU,
　　　　　　　Wolf D. Prix & Partner ZT GmbH
Client: North Jutland House of Music Foundation, Aalborg, Denmark
Location: Aalborg, Denmark
Building Area: 20,257m²
Photos: Duccio Malagamba

交流平台
能源控制理念

项目概况

丹麦奥尔堡音乐厅既是一座文化建筑又是一座教育设施，其开放的结构促进了观众与艺术家、学生和老师之间的交流。从其外部形状就能解读出该建筑背后的理念——音乐学院拥抱着音乐厅。

建筑设计

U型的排演训练室被设置在建筑核心区域——能容纳1 300名游客的音乐厅附近。大气的门厅连接着这些空间，一个多层窗户区域开放至邻近的文化空间。门厅下面是三间不同大小的房子：私人大厅、韵律大厅、古典大厅。学生和游客透过窗户可以看到音乐大厅及排练房，还可以体验音乐活动，包括音乐会和预演。

音乐大厅——看台里面观众席流动及弯曲的形态与外部严密的立方形态对比鲜明。在乐队和露台的位置安排上，设计理念是以期提供最佳的音效及视觉效果。墙壁上非结晶石膏结构以及可调整高度的顶棚悬吊系统设计是基于声学专家的精确计算，这是确保达到最佳收听体验效果。

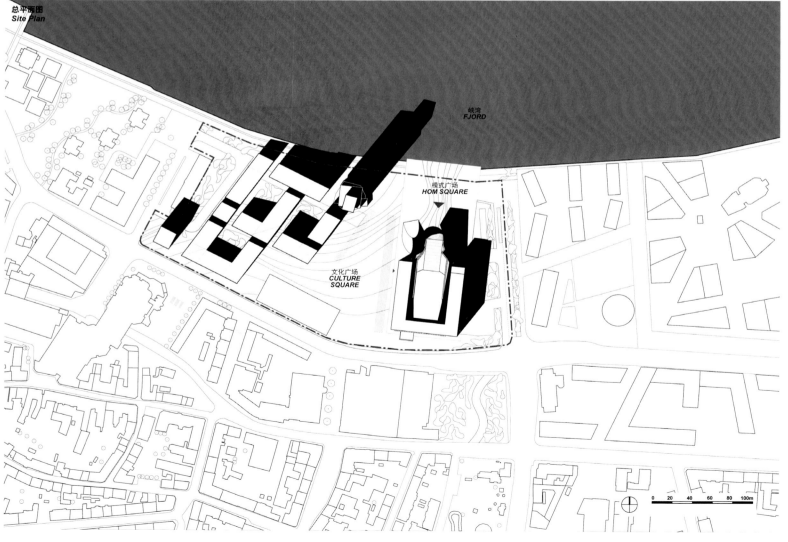

总平面图 / Site Plan

峡湾 / FJORD

模式广场 / HOM SQUARE

文化广场 / CULTURE SQUARE

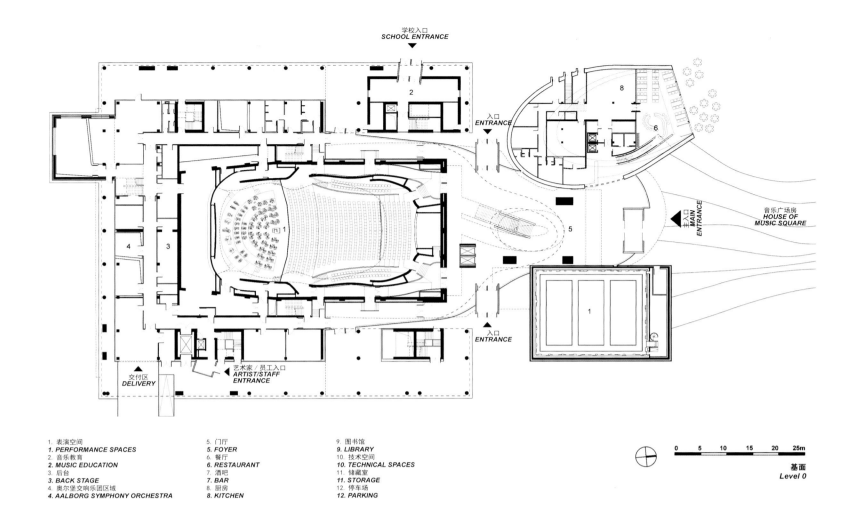

1. 表演空间	5. 门厅	9. 图书馆
1. PERFORMANCE SPACES	**5. FOYER**	**9. LIBRARY**
2. 音乐教育	6. 餐厅	10. 技术空间
2. MUSIC EDUCATION	**6. RESTAURANT**	**10. TECHNICAL SPACES**
3. 后台	7. 酒吧	11. 储藏室
3. BACK STAGE	**7. BAR**	**11. STORAGE**
4. 奥尔堡交响乐团区域	8. 厨房	12. 停车场
4. AALBORG SYMPHONY ORCHESTRA	**8. KITCHEN**	**12. PARKING**

基面
Level 0

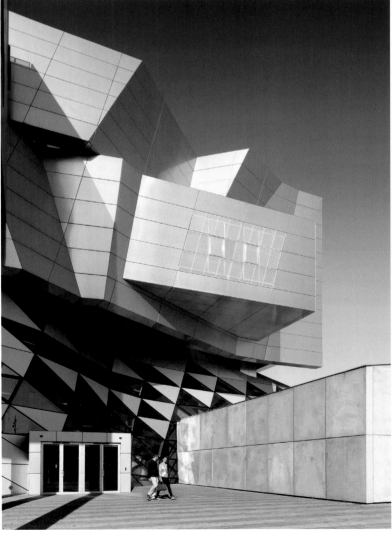

门厅——充当学生、艺术家、老师及游客的会面场所,五层楼高,带楼梯、阳台以及大型观景窗户,在这里可以举行各种活动。

能源控制理念——门厅利用大规模垂直空间的天然热浮力来通风,而不是使用风扇。混凝土地板下的供暖管道则用来冷却及加热。音乐大厅周围的混凝土墙则充当附加的热能存储容器。管道系统及通风口都安装有高效旋转热交换器。音乐厅的座位下面连接着具有低气流速度的高效通风系统。通过照明系统上面的天花板栅格可以引出空气,以防止室温升高。该建筑安装了一个房屋管理程序,可以调控整个建筑。当不需要使用时系统处于关闭状态,这样就能最大限度地减少能量的损耗。

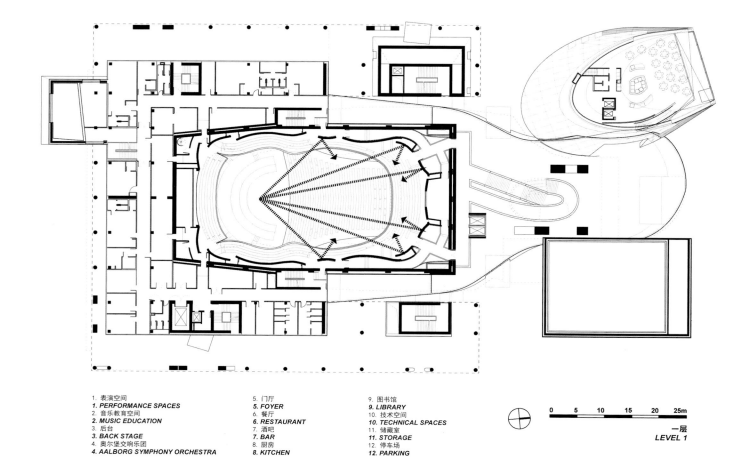

1. 表演空间
1. PERFORMANCE SPACES
2. 音乐教育空间
2. MUSIC EDUCATION
3. 后台
3. BACK STAGE
4. 奥尔堡交响乐团
4. AALBORG SYMPHONY ORCHESTRA
5. 门厅
5. FOYER
6. 餐厅
6. RESTAURANT
7. 酒吧
7. BAR
8. 厨房
8. KITCHEN
9. 图书馆
9. LIBRARY
10. 技术空间
10. TECHNICAL SPACES
11. 储藏室
11. STORAGE
12. 停车场
12. PARKING

一层
LEVEL 1

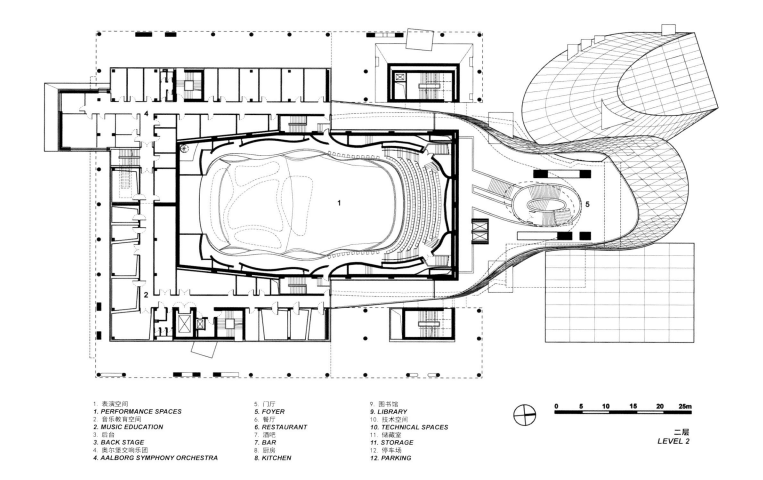

1. 表演空间	5. 门厅	9. 图书馆
1. PERFORMANCE SPACES	**5. FOYER**	**9. LIBRARY**
2. 音乐教育空间	6. 餐厅	10. 技术空间
2. MUSIC EDUCATION	**6. RESTAURANT**	**10. TECHNICAL SPACES**
3. 后台	7. 酒吧	11. 储藏室
3. BACK STAGE	**7. BAR**	**11. STORAGE**
4. 奥尔堡交响乐团	8. 厨房	12. 停车场
4. AALBORG SYMPHONY ORCHESTRA	**8. KITCHEN**	**12. PARKING**

二层
LEVEL 2

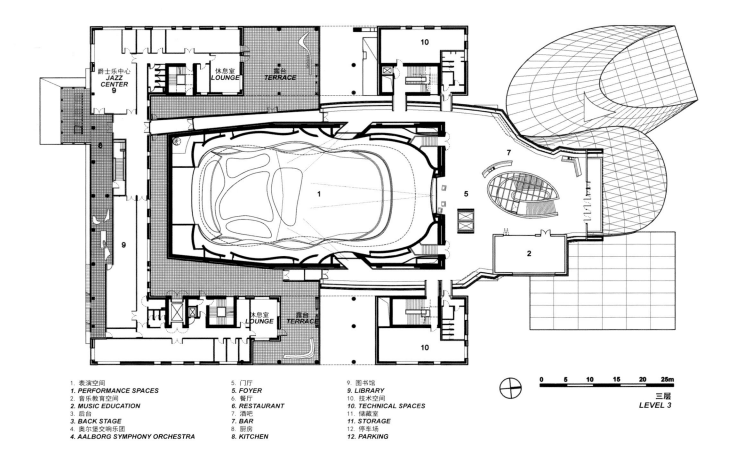

1. 表演空间 **1. PERFORMANCE SPACES** 2. 音乐教育空间 **2. MUSIC EDUCATION** 3. 后台 **3. BACK STAGE** 4. 奥尔堡交响乐团 **4. AALBORG SYMPHONY ORCHESTRA**	5. 门厅 **5. FOYER** 6. 餐厅 **6. RESTAURANT** 7. 酒吧 **7. BAR** 8. 厨房 **8. KITCHEN**	9. 图书馆 **9. LIBRARY** 10. 技术空间 **10. TECHNICAL SPACES** 11. 储藏室 **11. STORAGE** 12. 停车场 **12. PARKING**

三层 / LEVEL 3

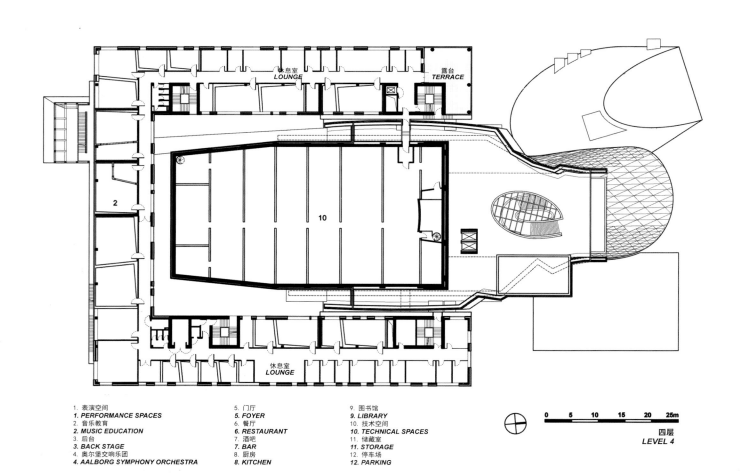

1. 表演空间 **1. PERFORMANCE SPACES** 2. 音乐教育 **2. MUSIC EDUCATION** 3. 后台 **3. BACK STAGE** 4. 奥尔堡交响乐团 **4. AALBORG SYMPHONY ORCHESTRA**	5. 门厅 **5. FOYER** 6. 餐厅 **6. RESTAURANT** 7. 酒吧 **7. BAR** 8. 厨房 **8. KITCHEN**	9. 图书馆 **9. LIBRARY** 10. 技术空间 **10. TECHNICAL SPACES** 11. 储藏室 **11. STORAGE** 12. 停车场 **12. PARKING**

四层 / LEVEL 4

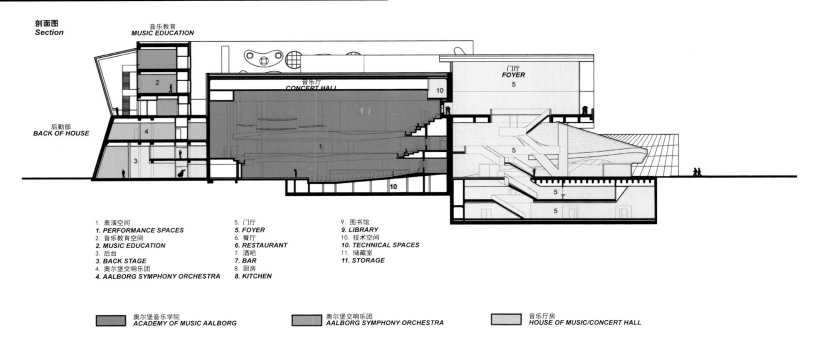

剖面图 / Section

1. 表演空间 / **1. PERFORMANCE SPACES**
2. 音乐教育空间 / **2. MUSIC EDUCATION**
3. 后台 / **3. BACK STAGE**
4. 奥尔堡交响乐团 / **4. AALBORG SYMPHONY ORCHESTRA**
5. 门厅 / **5. FOYER**
6. 餐厅 / **6. RESTAURANT**
7. 酒吧 / **7. BAR**
8. 厨房 / **8. KITCHEN**
9. 图书馆 / **9. LIBRARY**
10. 技术空间 / **10. TECHNICAL SPACES**
11. 储藏室 / **11. STORAGE**

奥尔堡音乐学院 / ACADEMY OF MUSIC AALBORG
奥尔堡交响乐团 / AALBORG SYMPHONY ORCHESTRA
音乐厅房 / HOUSE OF MUSIC/CONCERT HALL

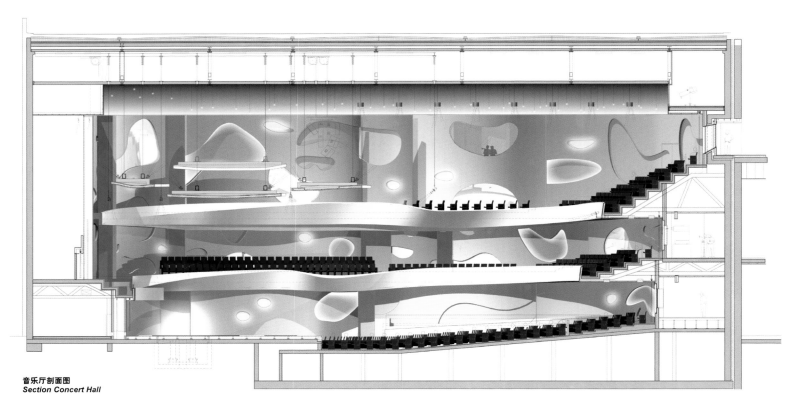

音乐厅剖面图 / Section Concert Hall

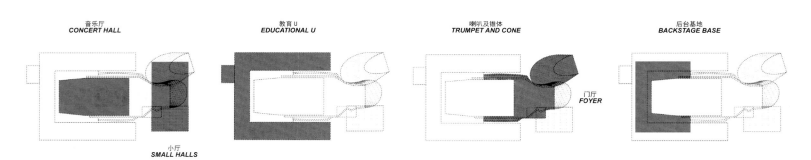

音乐厅 / CONCERT HALL　　教育 U / EDUCATIONAL U　　喇叭及锥体 / TRUMPET AND CONE　　后台基地 / BACKSTAGE BASE

小厅 / SMALL HALLS　　门厅 / FOYER

建筑部件 / Building Parts

Exchange Platform
Energy Control Concept

Project Overview

The "House of Music" in Aalborg, Denmark, as a combined school and concert hall: its open structure promotes the exchange between the audience and artists, and the students and teachers. The idea behind the building can already be read from the outer shape. The school embraces the concert hall.

Building Design

U-shaped rehearsal and training rooms are arranged around the core of the ensemble, a concert hall for about 1,300 visitors. A generous foyer connects these spaces and opens out with a multi-storey window area onto an adjacent cultural space and a fjord. Under the foyer, three more rooms of various sizes complement the space: the intimate hall, the rhythmic hall, and the classic hall. Through multiple observation windows, students and visitors can look into the concert hall from the foyer and the practice rooms and experience the musical events, including concerts and rehearsals.

The concert hall—The flowing shapes and curves of the auditorium inside stand in contrast to the strict, cubic outer shape. The seats in the orchestra and curved balconies are arranged in such a way that offers the best possible acoustics and views of the stage. The design of the amorphous plaster structures on the walls and the height-adjustable ceiling suspensions, based on the exact calculations of the specialist in acoustics, ensures for the optimal listening experience.

The foyer—The foyer serves as a meeting place for students, artists, teachers, and visitors. Five stories high with stairs, observation balconies, and large windows with views of the fjord, it can be used for a wide variety of activities.

Energy Control Concept—Instead of fans, the foyer uses the natural thermal buoyancy in the large vertical space for ventilation. Water-filled hypocaust pipes in the concrete floor slab are used for cooling in summer and heating in winter. The concrete walls around the concert hall act as an additional storage capacity for thermal energy. The piping and air vents are equipped with highly efficient rotating heat exchangers. Very efficient ventilation systems with low air velocities are attached under the seats in the concert hall. Air is extracted through a ceiling grid above the lighting system so that any heat produced does not cause a rise in the temperature in the room. The building is equipped with a building management program that controls the equipment in the building and ensures that no system is active when there is no need for it. In this way, energy consumption is minimized.

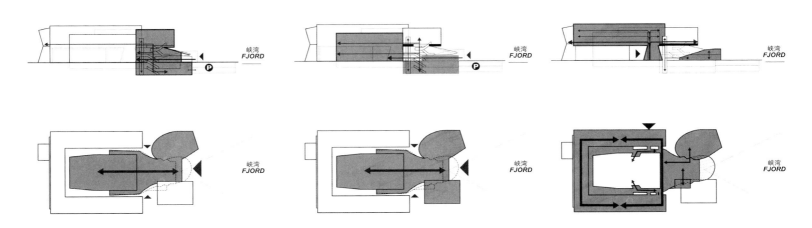

音乐房流通图
Circulation House of Music

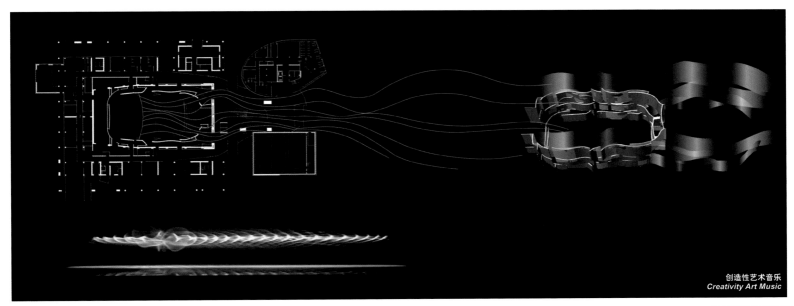

创造性艺术音乐
Creativity Art Music

公司简介 Company Profile

PTW Architects

澳大利亚PTW建筑设计公司创建于1889年，一直是建筑设计的革新先锋，建筑和规划设计一向追求最高标准。在建筑研究、发展以及确保所有设计项目的经济效益和成功实施方面，PTW始终处于领先地位。PTW建筑设计咨询（上海）有限公司为澳大利亚PTW建筑设计公司的分公司。PTW建筑师事务所于1889年在澳大利亚悉尼成立，其创始人是詹姆斯·佩多。在过去一个多世纪的不断实践中，PTW建筑师事务所不仅成为澳洲最古老的，而且是最大的也是最多元化的建筑公司之一。

PTW公司新近的建筑设计包括：水立方——北京2008年奥运会国家游泳中心、北京2008绿色奥运村、Pymont区的Darling岛公寓、St Margaret的改建、环形码头东部高级公寓、新建的皇家农业展览会场、悉尼国际游泳馆、Ryde水上运动中心、奥运村的总体规划和悉尼2000年奥运会运动村设计、上海黄浦区体育中心、上海黄金交易所办公楼、上海松江27号地块住宅区、昆明滇池盛高大城、杭州滨江风雅钱塘、苏州建屋美居酒店、南京天泓山庄等。

PTW Architects

PTW Architects has pioneered architectural innovation since the firm was established in Sydney in 1889 and has set the highest standards in design and planning. PTW currently employs approximately 135 people, has offices in Sydney - Australia, Beijing, Shanghai, Ho Chimin City and Hanoi. The firm is a leader in research and development and ensures maximum cost efficiency and performance success for all its projects.
Recent Projects by PTW include: Watercube for the Beijing 2008 Games National Swimming Centre, Darling Island Apartments - Pyrmont, St Margaret's Redevelopment – Surry Hills, Quay Grand and Bennelong Centre at East Circular Quay, New RAS Showground for the Olympic Coordination Authority, Sydney International Aquatic Centre, Ryde Aquatic Centre and Olympic Village master plan and Apartments for the 2000 Games, Colonnades – Milsons Point, The Forum Development – St Leonards, Angel Place Recital Hall and Offices, Royal Sun Alliance Centre – Auckland, and Sutherland Hospital - Caringbah.

Architekt Lukas Göbl Office for Explicit Architecture

Architekt Lukas Göbl Office for Explicit Architecture 致力于为各类型的建筑设计和城市规划项目提供从草图设计到设计构思原理、场址及材料的选择、现场施工监督等全方位的服务。其设计的独特性和个性源自对现实与理想的认识、大胆的构思、自由的思想以及精细的细部处理。在设计中，他们将传统绘画技术与现代传媒系统结合起来，综合考虑各方面的因素，使方案最优化。

Architekt Lukas Göbl Office for Explicit Architecture

The Office for Explicit Architecture works on all types of architecture and urban planning, providing comprehensive services in these areas. The spectrum ranges from the first decisive sketch and the formulation of the design philosophy to the careful choice of location and materials, meticulous planning, and project specification all the way to on-site construction supervision. The uniqueness and individuality of each of Explicit Architecture's works arises from the coexistence of vision and reality, boldness and precision, free spirit and exact detail. Paper and pen are just as good as digital design methods. Analog and digital systems both complement and enrich each other. Each of our designs goes through many different processes, with the goal of viewing it from all possible perspectives, thus allowing maximum optimization.

eLANDSCRIPT（译地）设计事务所

eLANDSCRIPT（译地）设计事务所成立于美国洛杉矶，是一家具有创新精神并致力于城市实践的年轻事务所。公司团队成员早年留学美国，并于美国及中国香港等地区的建筑设计、城市规划设计及景观设计事务所进行了多年的工作，积累了大量的实践经验和专业人脉资源。

事务所与国内外多家设计顾问团队保持长期设计及研究合作，其中包括香港大学可持续发展设计顾问团队、广州大学建筑设计研究院和重庆市风景园林规划研究院等。

eLANDSCRIPT

eLANDSCRIPT LLC was established in Los Angeles & Hong Kong, is an innovative spirit design firm which focuses on urban issues. Company team members studied in the U.S., and practiced in building design, urban design, and landscape planning firms in the U.S., Hong Kong and Mainland. eLANDSCRIPT has accumulated a lot of practical experiences and professional network of resources.
eLANDSCRIPT maintains a long-term design and research cooperation with a number of domestic and foreign research and design consultant teams, including the University of Hong Kong: Sustainable Design Consultant Team; Architectural Design & Research Institute of Guangzhou University; Chongqing Landscape Architecture Planning Institute. The eLANDSCRIPT team worked with multiple advanced Architecture, urban planning, and landscape and engineering firms.

LAN Architecture

LAN Architecture 由 Benoît Jallon 和 Umberto Napolitano 于2002年创立，LAN 从多个学科的角度来探索建筑设计，将社会、城市、功能和形式纳入考量范围，致力于为创新性和实际性的项目寻求现代且高雅的设计方案。其主要作品有：法国Chelles体育馆和市政大厅、法国Bure EDF档案中心、圣马洛法院、楠泰尔Minimum-security 监狱、Saint-Mesmes Marchesini总部等。

LAN获得了多个设计奖项，包括2004年法国文化和通信部颁发的NAJA、2009年芝加哥雅典娜博物馆国际建筑大奖、欧洲建筑艺术设计和城市研究城市中心Archi-Bau大奖、世界建筑米兰三年展特别奖、2010年AR国际房地产交易会未来项目奖、欧洲"40 Under 40"奖、2011年SAIE精选奖、AMO企业基金会特别奖。2011年，LAN事务所在LEAF奖中获得了保护环境的最佳可持续发展奖，还获得2011年 Prix de L'Equerre D'Argent提名。

LAN Architecture

LAN (Local Architecture Network) was created by Benoit Jallon and Umberto Napolitano in 2002, with the idea of exploring architecture as an area of activity at the intersection of several disciplines. This attitude has developed into a methodology enabling LAN to explore new territories and forge a vision encompassing social, urban, functional and formal questions. LAN's projects seek to find elegant, contemporary answers to creative and pragmatic concerns.
LAN has received several awards: the Nouveaux Albums de la Jeune Architecture (NAJA) prize awarded by the French Ministry of Culture and Communication (2004); the International Architecture Award from the Chicago Athenaeum and the European Urban Centre for Architecture, Art, Design and Urban Studies, the Archi-Bau Award, the Special Prize at the 12th World Triennale of Architecture, Sofa (2009); the AR Mipim Future Projects Award and the Europe 40 Under 40 Award (2010).

Maxwan architects + urbanists

Maxwan Architects + Urbanists 位于荷兰鹿特丹，是一家致力于现代建筑及城市设计的国际化公司。其注重客户、同行以及专业顾问间的协作，在每个项目中，都力求将文化、科学和美术影响融入到整体效果当中，创造出最具创意的城市规划或者最壮观迷人的建筑物。

公司坚持和可持续性、建筑工程、结构设计、成本管理、交通流通等诸多方面的顾问合作，以广泛的设计经验知识为客户提供值得信赖的、原创的、美观且可实行的设计方案。其许多项目获得社会的广泛认可，例如：鹿特丹的"Saw Tooth Tower"、"F&I办公大楼"等。

Maxwan Architects + Urbanists

Maxwan Architects + Urbanists is based in Rotterdam, the Netherlands. It is an international practice dedicated to contemporary architectural and urban design. Maxwan encourages collaboration between clients, colleagues and specialist consultants. In each of its projects, Maxwan seeks to merge cultural, technical and esthetical influences into an inextricable and beautiful whole, to make design innovative urban town plans and beautiful, intriguing buildings.
The firm collaborates with consultants in sustainability, building engineering, structural design, cost management, civil engineering, process management, traffic flows, etc. It is with this broad range of talent and perspective, experience with design, and love for design, that Maxwan would like to convince the clients that they are capable of designing original, beautiful and feasible plans for almost any request the clients could possibly have. Maxwan have designed widely recognized projects such as the "Saw Tooth Tower" in Rotterdam, the "Office Building F&I", etc.

M.S.B Arquitectura e Planeamento

M.S.B Arquitectura e Planeamento 是一家年轻的建筑设计事务所，于2004年由Miguel Mallaguerra、Susana Jesus和Bruno Martins建立于丰沙尔。MSB的设计师在项目设计上有丰富的经验，在建筑设计、城市规划和室内设计等领域颇有研究。近几年该事务所在住宅、酒店、设施、城市开发领域设计的作品展现了同时期建筑的魅力，在国内外比赛中受到认可。

事务所主张向世界开放的哲学，这一思想不仅表现在它们的沟通策略上，同时，也通过其与葡萄牙本国及国外伙伴的合作中彰显出来。在设计工作中，事务所十分注重多学科之间的团队合作，并将此理念视为成功的关键。

M.S.B Arquitectura e Planeamento

M.S.B Arquitectura e Planeamento was established in 2004 in Funchal by architects Miguel Mallaguerra, Susana Jesus and Bruno Martins. The MSBs are experienced in producing projects and studies directed to the areas of architecture, urban planning and interior design. The works produced inside this young work group reveal the fascination by contemporaneous architecture accompanying its time, having developed over recent years a number of projects in the area of Housing, Hospitality, equipment and Urban Development, noted for its quality recognized by the victories achieved in national and international competitions. The office maintains an open philosophy to the world, both in its communication policy, but also in how it works with its partners inside and outside of Portugal, privileging the work in multidisciplinary teams, and promoting ideas as the crucial factor in their projects. Dwelling, hotel industry, equipments and urban planning are distinct categories in which the MSBs already have a significant number of participations.

Archinauten

Archinauten 是 Wolfgang Mühlbacher 和 Andreas Dworschak 两人合作于 2000 年在奥地利林茨成立的联合建筑工作室。公司完成过很多项目的设计，包括从城市规划战略到室内设计，从单户住宅到文体、商业等公共建筑。

公司的许多建筑佣金是由获奖建筑比赛得来。公司将自己的工作如建筑作为整体一样看待，无论从初稿到成品。在创意、视觉和满足现实需求之间，Archinauten 能意识到自己的社会责任。

Archinauten 是一个年轻的、经验逐渐丰富的施工执行国际性激励的策划团队。随着建筑设计的发展、建筑的功能多样化要求，对话成为项目发展的一个至关重要的方面，因此策划团队的协调自然就成了工作的一个组成部分。

Archinauten

The collaboration between Wolfgang Mühlbacher and Andreas Dworschak led to the founding of a joint architecture studio based in Linz / Austria in 2000. Since then, they have worked on numerous projects under name of the architecture firm—Archinauten—ranging in scale from urban planning strategies to interior design. The palette of projects reaches from single-family homes to commercial and business objects all the way to public buildings such as cultural / sport centers and schools.

Many of the Archinauten's building commissions are the result of winning architectural competitions. They see their work as architects as being holistic—from the preliminary draft to the finished object. In between creativity, vision and realistically meeting needs, they are conscious of their social responsibility.

The Archinauten is a young, internationally inspired team of planners with experienced construction execution planners. Along with the architectural design, dialogue with the client and users about diverse requirements regarding the function of the building is an essential aspect of the project development. The coordination of the planning team that this requires is naturally a component of the work.

3GATTI Architecture Studio

3GATTI Architecture Studio 是一家将前沿建筑与视觉艺术结合起来的跨学科创新设计工作室，Francesco Gatti 是其创立者和主建筑师。工作室以罗马和上海为基地，由建筑师、设计师、艺术家、施工队等各类专业人才组成，在创新和工程领域通力合作，不断运营，使3GATTI 在过去的11 年里，已经成为国际领先的公司，专门从事公共、商业和文化建筑的综合项目。主要包括新南京汽车博物馆、上海"红墙"混合发展区、上海世博会场址 UBPA 区域的两个新展览馆、西安"书架酒店"以及沈阳红星购物商场等。

3GATTI Architecture Studio

3GATTI is interdisciplinary innovation design studio combined preface architecture and visual arts, and Francesco Gatti is the founder and the main architect. The Studio is based in Rome-Shanghai, and composed by architects, designers, artists, construction teams and other kinds of professional talents, making it become an international leading company over the past 11 years by innovation and engineering cooperation and continuous operation and specializing in the comprehensive program of public, business and culture construction. Mainly includes: The new Nanjing Automobile Museum, Shanghai "Redwall" Mixed Development Zone, two new exhibitions in Shanghai World Expo Site UBPA Area, Xi'an "Bookshelf Hotel" and Shenyang Red Star Shopping Mall.

Architects Collective / Architects Collective ZT-GmbH

Architects Collective ZT-GmbH（AC） 于 2006 年在维也纳由 Andreas Frauscher, Richard Klinger 及 Kurt Sattler 共同创立。自成立至今，事务所一直活跃于建筑、景观、城市规划等领域，是一家提供全方位设计服务的建筑事务所。

事务所汇聚了跨度超过 30 余年的设计团体的集体创作经验，其项目范围涵盖了广泛的建筑类型。AC 已设计并建造了若干超低能耗建筑并因此获得众多奖项。事务所的设计团队及各个专业的顾问共同合作并深化了一系列的新兴建筑技术以及节能设计理念。其三位合伙人分别侧重于对自小型高新技术结构的研发至大型建筑的设计、管理，以至城市规划设计等诸多学科。

Architects Collective ZT-GmbH

Architects Collective ZT-GmbH was founded in 2006 by Andreas Frauscher, Richard Klinger and Kurt Sattler in Vienna, and has worked internationally on projects in architecture, landscape and urban planning.

Designs are created and advanced by a collective idea finding process. Criteria for innovative technologies and energy-optimized concepts are developed within the team and by working with professional planners and consultants. The experience of the three agency partners ranges from fundamental research on small advanced structures to the planning and management of large complex buildings and master plans.

Architects Collective consists of 30 architects and engineers. The work of the office is continually represented in international publications and exhibitions.

CEBRA 建筑设计事务所

CEBRA，是一家丹麦的事务所，总部位于奥胡斯郡。事务所 2001 年由建筑师 Mikkel Frost, Carsten Primdahl 和 Kolja Nielsen 成立。客户遍及丹麦，同时事务所开始积极向海外项目拓展。事务所的项目涉及不同规模的项目，几乎是任何类型的项目，但不仅仅关注于它的面积大小，而是从工业设计到城市规划等等都包含其中。

事务所的设计思想是"项目相关"，即针对不同的项目设计师采用不同的、独特的方式。因为很多条件的变化，事务所的设计师们认为有必要保持灵活的设计思维，根据具体情况进行适应和发展，同时保持针对性及现代性。尽管如此，事务所的项目还是有共通的主题，或许可以说是 "CEBRA" 风格，这是一种属于自己的、贯穿于所有项目之中的特质。

CEBRA Architects

CEBRA is a multidisciplinary Danish architecture office founded in 2001 by architects Mikkel Frost, Carsten Primdahl and Kolja Nielsen. The headquarters is located in Aarhus. Its customers are over the whole Denmark, while the architects began to actively expand overseas projects. The projects involve different scales, almost any type of project. But it not only focuses on the size, but from industrial design to the urban planning, etc., which all is included.

Architects' design philosophy is "project-related", which is used a different unique way by the designers for different projects. Because of a lot changes in conditions, the Architects considers it is necessary to maintain a flexible design thinking for adapting and developing depend on the circumstances, while maintaining the relevance and modernity. Nevertheless, the projects have a common theme, and perhaps they can be said to "CEBRA" style, which is a part of their own qualities in being throughout all projects.

Belzberg Architects Group

Belzberg 建筑事务所自 1997 年成立以来，一直设在圣塔莫尼卡市。该公司已经获得了超过 35 个国家和地区的设计奖项，其中包括超过 15 个来自美国建筑师学会。他们的作品已经在超过 200 个出版物中起到了重要作用，遍及超过 25 个国家，包括频繁的定位在著名的期刊，如建筑实录、室内设计以及《纽约时报》。

Belzberg 建筑事务所已荣获洛杉矶市文化事务委员会设计荣誉奖和城市土地学会认可自己的设计作品和他们对城市规划作出的卓越贡献。芝加哥雅典娜博物馆建筑公认的几个项目，获得了多个著名的美国建筑奖项。在过去的 10 年中，Belzberg 建筑事务所已经成长为一个多元化的、专业的设计工作室，能生产大型、多样的可持续建筑解决方案和项目类型，同时保持个人和集体参与每个项目。2013 年，在洛杉矶的当代艺术博物馆选择了 Belzberg 建筑事务所的作品，被列入他们的节目 "新的 Sculpturalism：来自南加州的当代建筑"。在过去 25 年的南加州建筑中，这是已经成为多产的第一个激进形式的广泛的学术审查。

Belzberg Architects Group

Belzberg Architects has been located in the City of Santa Monica since its inception in 1997. The firm has earned over 35 national and local design awards including over 15 from the American Institute of Architects. Their work has been featured in over 200 publications throughout more than 25 countries including frequent features in notable periodicals such as Architectural Record, Interior Design, and the New York Times.

Belzberg Architects has been honored with the City of Los Angeles Cultural Affairs Committee Design Honor Award and the Urban Land Institute recognized their design work and their contribution to excellence in urban planning. The Chicago Athenaeum Museum of Architecture recognized several of their projects with multiple notable American Architecture Awards. In the past 10 years, Belzberg Architects has grown into a multi-faceted, professional design studio capable of producing large, and diverse sustainable building solutions and project types while maintaining a personal and intensive involvement with each project. In 2013, the Museum of Contemporary Art in Los Angeles has chosen the work of Belzberg Architects to be included in their show "A New Sculpturalism: Contemporary Architecture from Southern California". This is the first extensive, scholarly examination of the radical forms that have become prolific in Southern California architecture during the past twenty-five years.

美国 KDG 建筑设计咨询有限公司

美国 KDG 建筑设计咨询有限公司是集建筑设计、景观设计、城市规划、地产咨询等业务为一体的综合性设计机构。自 1993 年在美国洛杉矶地区成立以来，作品受到中美地产界和地方政府机构的普遍赞誉。KDG 信奉 "设计是创造价值的重要手段"，领先的设计理念和严谨的设计作风是其核心竞争力。

Kalarch Design Group

Kalarch Design Group is a multi-disciplinary design firm offering services in four core areas of practice: Architectural Design, Landscape Architecture, Urban Design & Planning, Development Consultation. Founded in 1993 in Los Angeles, California, KDG has developed into an experienced and respected practice among clients in China and US. KDG believes that "Design Creates Value". Leading design concept and rigorous design style are the core competitiveness of KDG.

阿特金斯

阿特金斯是世界领先的设计顾问公司之一，专业知识的广度与深度使之能够应对具有技术挑战性和时间紧迫性的项目。阿特金斯是国际化的多专业工程和建设顾问公司，能为各类开发建设项目提供一流专业服务，从摩天大楼设计到城市规划、铁路网络改造，以及防洪模型的编制，都能提供规划、设计、实施的全程解决方案。

作为阿特金斯集团远东区的全资子公司，阿特金斯中国于1994年正式进入中国市场。借助集团总部的强大支持，阿特金斯中国以提供多专业多学科的"一站式"全方位服务的核心优势区别于竞争对手，并通过国际经验和本地知识的有机结合，在中国近年来迅猛推进的城市化进程中取得了骄人的项目业绩。

其主要作品和奖项有：中央景城一期荣获2010年中国土木工程詹天佑奖住宅小区优秀规划奖；伊顿小镇荣获2010年中国土木工程詹天佑奖住宅小区金奖；上海办事处荣获LEED绿色建筑商业内部装修金奖等多个奖项。

Atkins

Atkins is one of the world's leading design consultancies. It has the breadth and depth of expertise to respond to the most technically challenging and time-critical projects and to facilitate the urgent transition to a low carbon economy. Atkins' vision is to be the world's best design consultant.

Whether it's the architectural concept for a new supertall tower, the upgrade of a rail network, master planning a new city or the improvement of a management process, designers plan, design and enable solutions.

With 75 years of history, 17,700 employees and over 200 offices worldwide, Atkins is the world's 13th largest global design firm (ENR 2011), the largest global architecture firm, the largest multidisciplinary consultancy in Europe and UK's largest engineering consultancy for the last 14 years. Atkins is listed on the London Stock Exchange and is a constituent of the FTSE 250 Index.

In 1994 Atkins established its first Asian office in Hong Kong followed by Singapore in 1996. Today Atkins also has offices in Hong Kong, Beijing, Shanghai, Chengdu, Chongqing and Ho Chi Minh and Sydney, all part of an integrated network that delivers innovative multidisciplinary projects and employs approximately 1,000 staff across the region from China to South East Asia and Australia.

Perkins Eastman

Perkins Eastman建筑设计事务所是一家知识型设计公司，提供独特的设计方案，创新的环境和优异的建筑设计、项目规划、室内设计、景观设计、城市设计和项目管理。

Perkins Eastman在六个城市设有办事机构，拥有500多名专业人士提供服务，包括：纽约州纽约市、北卡莱罗纳州夏洛特市、加利福尼亚州洛杉矶市、宾夕法尼亚州匹兹堡、旧金山以及安大略省多伦多市。根据主流刊物的评价，Perkins Eastman是纽约最大的建筑设计事务所；北美最大的十家建筑事务所之一；全球最大的20家建筑事务所之一。

Perkins Eastman已经在中国开展了近15年的业务，并于2007年在上海成立了一家外商独资企业（WFOE），以便更好地为地区的客户提供服务。该公司最近完成的项目包括中国科学院总部The Hall of Academics、山东一家大型新社区济南南城、北京一家大型新零售中心万柳购物中心以及众多整体计划和大型开发计划。Perkins Eastman将通过于2011年在越南胡志明市开设计划中的办事处拓展其在亚洲的业务。

纽约2010年12月9日电／美通社亚洲——国际顶级设计与建筑机构Perkins Eastman和Ehrenkrantz Eckstut & Kuhn Architects（简称"EE&K"）欣然宣布，双方已达成一致意见合并业务部门。此次合并是双方各自历史和长期相互尊重的自然产物，将显著增强双方业务部门的实力。合并后的国际机构将拥有近600名员工，从而实现更强大的能力和客户价值。这两家公司将合并它们位于纽约、华盛顿特区和中国的部门。

Perkins Eastman

Perkins Eastman is an international planning, design, and consulting firm that was founded in New York City in 1981. Today the firm has seven other offices across the US and five overseas. The firm has a professional staff of more than 750 made up of architects, interior designers, planners, urban designers, landscape architects, graphic designers, construction specification writers, construction administrators, economists, environmental analysts, traffic and transportation engineers, and several other professional disciplines.

In order to reflect the enhanced capabilities that have come from mergers with other organizations—most recently with the international architectural firm Ehrenkrantz Eckstut and Kuhn—these broader services are in many cases offered through our affiliate firms. EE&K brings creative thinking and big picture perspective to design problems of all scales. While Perkins Eastman is the parent firm with proven depth in 15 major project types, we believe a multi-brand firm is stronger and better illustrates the full range of our capabilities.

The firm has completed projects in 45 states and 30 countries. These have ranged from small buildings and interiors for non-profit organizations to large new healthcare and educational campuses, major mixed-use commercial developments, and entire new cities.

waltritsch a + u

waltritsch a + u于2001年建立于的里雅斯特，是一家从事于建筑设计、室内设计与城市规划的综合性设计公司。公司一直致力于提供专业技能、创意的解决方案和高效的专业服务，以达到与客户目标和志向相符的最终项目质量。公司将实用主义与乐观态度相结合，将研究和专业知识相结合，打造了最佳的应对既定任务特定、独特的解决方案。

waltritsch a + u一直保持与国际事务所的密切合作，其设计的项目和建筑在意大利、斯洛文尼亚、荷兰、德国、西班牙、葡萄牙、奥地利、瑞士、中国、韩国、新加坡等国家杂志上以及世界30多个国家的网站上出版，并且在纽约哥伦比亚大学、阿姆斯特丹贝尔拉格美术馆、威尼斯建筑双年展和2011东京世界建筑大会上展出，还曾获得多项重要奖项。

waltritsch a+u

waltritsch a+u was established in Trieste in 2001. The office works on architecture, interior design and urban planning. It offers professional skills, creative solutions and efficient professional services, in order to reach ultimate project quality in accordance to the client goal and ambition. Projects are approached mixing pragmatism with an optimistic stance, combining research and professional knowledge, to build specific and unique solutions as the best answer to the given task.

waltritsch a+u has also been involved in a series of partnerships with renewed international offices. Projects and buildings have been published in magazines and books in Italy, Slovenia, The Netherlands, Germany, Spain, Portugal, Croatia, Austria, Switzerland, China, Korea, Singapore, Malaysia etc., and in more than 30 countries worldwide on the web. Works have also received several prizes and been exhibited throughout the world in many occasions, such as Columbia University New York, Berlage Gallery Amsterdam, Biennale di Architettura Venezia and the World Congress of Architecture UIA 2011 held in Tokyo.

Powerhouse Company

Powerhouse Company是一家集建筑设计、城市规划及研究于一身的国际事务所，它由Charles Bessard和Nanne de Ru于2005年共同创立。Charles Bessard（1970年生于法国）负责位于哥本哈根、丹麦的事务所，Nanne de Ru（1976年生于荷兰）负责位于鹿特丹、荷兰的事务所。

建筑和设计的基础在于强大的创新过程和工艺的掌握。这也是为什么我们以我们的项目而自豪的原因。这些建筑和设计是与我们的客户、我们的合作伙伴和项目背景的集中的、服务至上的相互作用的结果。我们的方法是全面的：创意理念、细节、物质和空间的品质是我们项目的主要组成部分。这就是为什么我们的客户如此喜爱这些项目的原因。与此同时，我们认为，优化预算、日程编排和制作技术上完美的建筑也是美好的事情。

Powerhouse Company

Powerhouse Company is a international firms collected architectural design, urban planning and research in one, and was co-founded by Charles Bessard and Nanne de Ru in 2005. Charles Bessard (born in France in 1970) is responsible for the firms in Copenhagen and the Danish, while Nanne de Ru (born in the Netherlands in 1976) is responsible for the firms in Rotterdam and the Netherlands.

The foundations of architecture and design lie in the powerful process of creativity and in the mastering of craftsmanship. That is also why we are proud of our projects. They are the result of an intensive, service-oriented interaction with our clients, our partners and the project context. Our approach is comprehensive: the creative concept, the detailing, materiality and spatial qualities are intrinsic and integral parts of our projects. That's why our clients love them so much. At the same time, we think that optimizing budgets, orchestrating schedules and making technically perfect buildings are things of beauty, too.

WVA architects LTD

WVA architects LTD取名于术语"进行编织"，意为去交织，公司目标是编织同一个时代环境下的不同方面。这个过程是一个试图以非线性设计来创造建筑片段。通过跨越过去和未来的时间界限，在一个项目中编织多样文化，编织不同城市肌理来实现新的文化识别性，编织探索所有从新兴结构到可实施手段的可能。

每一个项目都是一个设计片段的机会，一个城市品质的序列来演化和形成城市肌理。交织、连接、伸缩尺度、空间节点都被重新考虑为达成新的解读，新的交织方式的资料。这也是韦瓦认为编织是最好的将所有这些要素连接的方式。

WVA architects LTD

Based in Beijing, with the beginning of the new post-modern age, WVA is a multi-disciplinary firm that offers service in the fields of architecture, urban planning, landscape and interior design.

Driven by the term "to weave", our aim is to intertwine different aspects of our contemporary environment, creating pieces of architecture through a nonlinear design process. Weaving various cultures and urban fabrics into a project sparks the potential for new cultural identities. WVA explores all possibilities of emergent structures and operational tools.

Each project is a new start, an opportunity to design a fragment, a sequence of urbanity that morphs and transforms the city fabric. Interactivity, connectivity, variations in scale, and articulation of spaces are rethought as date enabling new interpretations of interlacing them. In our experience weaving is the best link between these components.

OMA

大都会建筑事务所简称为 OMA（Office For Metropolitan Architecture），是荷兰建筑师雷姆·库哈斯在鹿特丹成立的。1975年创办，创办人除雷姆·库哈斯外，还有埃利亚·增西利斯（Elia Zenghelis）与 Madelon Vriesendorp 和 Zoe Zenghelis。创始人雷姆·库哈斯1944年生于鹿特丹，早年做过记者和电影剧本撰稿人，曾在伦敦建筑联合学院、美国康奈尔大学学习建筑。曾引起广泛争论的中国中央电视台新大楼设计方案就出自他之手。

目前 OMA 公司规模约为100名员工，总部位于鹿特丹，并在纽约市、北京都有办事处。专门承接建筑设计、市区规划和文化分析等工作，并以采纳 Arup 结构和雇用专门的机械工程师闻名，另有研究分公司称为 AMO。

OMA

OMA was founded in 1975 by Rem Koolhaas, Elia and Zoe Zenghelis and Madelon Vriesendorp as a collaborative office practicing architecture and urbanism. The office gained renown through a series of groundbreaking entries in major competitions. OMA is a leading international partnership practicing architecture, urbanism, and cultural analysis. OMA's buildings and masterplans around the world insist on intelligent forms while inventing new possibilities for content and everyday use. OMA is led by six partners—Rem Koolhaas, Ellen van Loon, Reinier de Graaf, Shohei Shigematsu, Iyad Alsaka and David Gianotten—and sustains an international practice with offices in Rotterdam, New York, Beijing, Hong Kong and Doha.

The counterpart to OMA's architectural practice is AMO, a research studio based in Rotterdam. While OMA remains dedicated to the realization of buildings and masterplans, AMO operates in areas beyond the traditional boundaries of architecture, including media, politics, sociology, renewable energy, technology, fashion, curating, publishing and graphic design.

AMO often works in parallel with OMA's clients to fertilize architecture with intelligence from this array of disciplines. In 2004, AMO was commissioned by the European Union to study its visual communication, and designed a colored "barcode" flag—combining the flags of all member states—that was used during the Austrian presidency of the EU.

Studio 44 Architects

Studio 44 Architects 是俄罗斯圣彼得堡最大的私人建筑公司，设计团队由包括建筑师、修复专家、结构设计师和工程师在内的120多位专家组成。

公司不仅在重修、整修和改造归为文化遗产的历史建筑有独特经验，还在充当总设计师和协调承办工作方面有广泛的经验，能为客户从概念规划和初期设计到相关预算方面提供一套完整的项目方案，并且能为客户提供一系列的服务，包括审批过程的技术支持、设计的状态检测、预算文件以及施工过程的建筑监督。

多年发展以来，公司已成功实施的项目包括：Ladozhsky 火车站、Atrium 商业中心、Linkor 商业中心、Grand Palace 购物长廊、哈萨克斯坦阿斯塔纳儿童休闲中心以及索契奥林匹克公园火车站等等。这些项目的成功实施，大大提高了 Studio 44 Architects 的职业声望，在社会各界赢得了一致好评，获得了超过60项的殊荣，主要包括：国家文学艺术奖、国家最高建筑奖——Chrystal Dedalus 奖和弗拉基米尔·塔特林奖，以及在圣彼得堡举办的建筑比赛和莫斯科 Zodchestvo 国际建筑节中均获得最高奖项。

Studio 44 Architects

Architectural studio "Studio 44" is one of the largest private architectural firms in St. Petersburg, Russia. The studio's team consists of over 120 highly qualified specialists in a variety of fields (architects, restoration experts, structural designers, engineers). "Studio 44" has unique experience in the reconstruction, restoration and adaptation of historic buildings classified as cultural heritage monuments. Also, it has an extensive experience in acting as general designer and coordinating contractor work. The company has both the human resources and the capacity to perform the whole cycle of project stages from concept planning and preliminary designs to drafting and to prepare the complete set of design and budget-related documentation. It provides a range of services including technical support during approvals and state inspection of design and budget documentation, as well as architectural supervision of the construction process.

The portfolio of "Studio 44" includes such successfully implemented projects as the Ladozhsky Railway Station, Atrium Business Center, Linkor business centers, the Grand Palace shopping gallery, Palace of Schoolchildren and Olympic Park Railway Station in Sochi. "Studio 44" enjoys a high professional reputation and holds over 60 prestigious awards including the State Prize for Literature and Arts, the top national awards in architecture: the Chrystal Dedalus and the Vladimir Tatlin Prize, as well as the top prizes of Architekton review competition held in St. Petersburg and Zodchestvo International Architectural Festival in Moscow.

Andrea Maffei Architects

Andrea Maffei Architects（AMA）拥有一些目前正在意大利开发的最重要的项目。包括 Citylife 摩天大楼和在2008年博洛尼亚的新高速列车站。这包括90 000m^2 的零售、写字楼及酒店空间，耗资近200万欧元。

AMA 参加了2007至2008年蒙特卡洛城市的海上扩张的比赛，沿着利伯斯金和亚历山大 Giraldi 的工作室，在海边与新住宅，将半岛形式的新住宅、酒店、办公室和博物馆的城市扩展到海边，投资超过50亿欧元。AMA 也获得了2009年贝加莫省新总部的竞赛。

现在 Andrea Maffei 签署了正在米兰建设中的 Citylife 摩天大楼的项目。它是由高达207m，共50层，总建筑面积约81 000m^2 的摩天楼建筑而成的。该建筑计划将在2014年年底竣工。

Andrea Maffei Architects

The company Andrea Maffei Architects (AMA) is leading some of the most important projects now under development in Italy. They include the Citylife skyscraper and the New high-speed trains station in Bologna (2008). This consists of 90,000 sq of retail, office and hotel spaces over the binaries, for a cost of almost 200 mil. Euros.

AMA participated in the big competition for the extension of the city of Monte Carlo on the sea (2007-2008), along with studio Libeskind and Alexandre Giraldi, to extend the city by the sea in the form of a peninsula with new residences, hotels, offices and museums with an investment of more than 5 billion Euros. AMA won also the competition for the new headquarters of the Province of Bergamo (2009).

Now Andrea Maffei signed the project of Citylife skyscraper in Milan under construction. It consists of a 207m high skyscraper, 50 stories, for about 81,000 sqm. of gross building area. The construction is planned to be completed at the end of 2014.

Saraiva + Associados

Saraiva + Associados 由公司主要负责人米格尔 Saraiva 建筑师于1996年在里斯本成立。S＋A 在建筑学、城市规划设计等领域发展它的活跃性，拥有超过80位有卓越的技能和专业知识的专业人士。技术精度、以客户为导向、完成期限和质量概念就是 S＋A 提供明显的竞争优势和显著的市场认可度的一些原则。

目前，S＋A 在阿尔及利亚（奥兰）、巴西（圣保罗）、中国（北京）、哈萨克斯坦（阿拉木图）、哥伦比亚（波哥大）、赤道几内亚（马拉博）、葡萄牙（里斯本和丰沙尔）、新加坡和阿联酋办事处（阿布扎比）设有办公室。而且，国际工作人员坚持与里斯本的创办人进行永久性的和密切的合作。

此外，通过动态出口活动，S＋A 已在安哥拉、佛得角、科特迪瓦、加蓬、加纳、几内亚、摩洛哥、莫桑比克、尼日利亚、俄罗斯、塞内加尔、阿塞拜疆、土耳其、和土库曼斯坦开展其活动。

Saraiva + Associados

Saraiva + Associados was established in 1996, in Lisbon, by Miguel Saraiva (Architect) who remains as the head of the company. S+A develops its activity in the areas of Architecture, Urban Planning and Design, with over 80 professionals with exceptional skills and expertise. Technical accuracy, customer orientation, accomplishment of deadlines and conceptual quality are some of the principles that provide a clear competitive advantage and noticeable market recognition of S+A.

Currently, S+A has offices in Algeria (Oran), Brazil (São Paulo), China (Beijing), Kazakhstan (Almaty), Colombia (Bogota), Equatorial Guinea (Malabo), Portugal (Lisbon and Funchal), Singapore and UAE (Abu Dhabi). The international staff, however, maintains a permanent and close collaboration with the founder office in Lisbon.

Furthermore, through a dynamic export activity, S+A has been developing its activity in Angola, Cape Verde, Ivory Coast, Gabon, Ghana, Guinea, Morocco, Mozambique, Nigeria, Russia, Senegal, Azerbaijan, Turkey and Turkmenistan.

OFIS Arhitekti

OFIS Arhitekti 是由 Rok Oman 和 Spela Videcnik 在1996年创建的一家位于斯洛文尼亚卢布尔雅那的事务所。两位合伙人都毕业于卢布尔雅那建筑学院，在随后的建筑实践中，他们多次参与文化及学术演讲。其代表项目包括位于斯洛文尼亚的蜂巢公寓、卢布尔雅那市立博物馆扩建、四季帐篷塔楼——梅赛德斯奔驰酒店设计方案等。

OFIS Arhitekti

OFIS Arhitekti is a firm of architects established in 1996 by Rok Oman and Špela Videčnik, both graduates of the Ljubljana School of Architecture and the London Architectural Association. Upon graduation they had already won several prominent competitions, such as Football Stadium Maribor and the Ljubljana City Museum extension and renovation. In 2001 they were awarded with the UK and Ireland's "Young architect of the year award".

The company is based in Ljubljana, Slovenia, but works internationally. They won a large business complex in Venice Marghera, Italy and a residential complex in Graz, Austria. However, it was by winning 180 apartments in Petit Ponts, Paris, their first large scale development abroad, which led them to open a branch office in France, 2007. This has been followed by a second large scale development with the construction of a football stadium for FC BATE in Borisov, Belarus, due for completion in 2012. They also have partner firm agreements in London, Paris and Moscow.

Neutelings Riedijk Architects

Neutelings Riedijk Architects 于 1987 年成立于荷兰鹿特丹，经过近 30 年的发展，该事务所已成为一个领先的国际事务所。事务所专注于公共、商业和文化建筑综合体等项目的设计，并在多年的实践中积累了丰富的经验。事务所将创新的设计理念融入到建筑设计中，并通过清晰的建筑形态来展现高品质的建筑形象。针对国际项目，事务所注重与当地建筑工程、造价估算和场地监督等方面的专业人员的交流与合作，从而形成一支能够承担多方面的挑战的团队。

Neutelings Riedijk Architects 的作品在国际范围内屡获大奖，其主要作品包括：安特卫普 Aan de Stroom 博物馆、海牙国际舞蹈与音乐厅、安特卫普 ACC 监控中心、Perm Tchaikovsky 歌剧院和芭蕾舞剧院、荷兰中央税务局等。

Neutelings Riedijk Architects

Neutelings Riedijk Architects was established in Rotterdam in 1987. It offers a strong commitment to design excellence: realizing high quality architecture through the development of powerful and innovative concepts into clear build form.

Over the last twenty five years Neutelings Riedijk Architects has established itself as a leading international practice, specializing in the design of complex projects for public, commercial and cultural buildings. The office has great experience in balancing the complex challenges of these projects to meet the ambitions of its clients.

For the international projects its design force is complemented by technical force through the association with local partners that specialize in architectural engineering, cost calculation and site supervision. This arrangement allows the Company to raise a team of specialists in order to meet the specific requirements of each project. The work of Neutelings Riedijk Architects has gained world wide appreciation through several awards and numerous publications in the international press.

ATELIER BRÜCKNER

ATELIER BRÜCKNER（布鲁克纳事务所）从内容出发进行空间的构思和设计。精彩内容的实现、气氛的营造、空间意象的创造都遵循着一个剧情，观众通过一条叙事主线被引入到一个故事中，这就是场景设计的范围。场景设计的使命在于将空间造型及其内部所包含的信息联合起来作为一件完整的艺术品：造型与内容成为整体。这一场景化的完整艺术作品旨在通过和谐地处理形式与内容的关系，通过客观地使用空间、图形、灯光及多媒体设计元素，来创造一种新的构成方式。

事务所的这一思维和工作方式，借鉴那条建筑界著名的规则，可以归纳为"内容决定形式"：带有场景设计特征的空间造型。

ATELIER BRÜCKNER

ATELIER BRÜCKNER embarks the spatial conception and design starting from content. To achieve wonderful content and create an atmosphere and space imagery all follow a story, bringing the audience into the story through a main narrative line, which is the range of scene design. The mission of scene design is to make the information containing in the space shape and inside unite as a complete artwork, so the shape and content become a whole. The scenarized complete artwork aims to harmoniously deal with the relationship between form and content, through objective using space, graphics, lighting and multimedia design elements to create a new way of composition.

The thinking and working styles of the Firm are learned the rules of the famous piece of the construction industry, which can be summarized as "content determines form": spatial modeling with scene design features.

Trahan Architects

Trahan 建筑事务所，由 Victor F 领导的一个一流的公司。Trey Trahan，美国建筑师协会会员，因作品曾获得国际赞誉，该作品立刻具有强烈的个性、历史基础和崇高审美。每个项目共享尖端技术及与安静和体验美感结合的创新的较高水平。Trahan 的极简主义美学作品植根于路易斯安那州的深刻理解及周边地区，而且通过国家的最先进的执行变得完全现代化。

自 1992 年成立以来，Trahan 建筑事务所已深入到每个项目的细节——从客户和用户到场地和程序——具有强烈的点和个性。该公司首先理解客户端、场地、程序和预算的具体要求。每个建筑项目驳回先入为主的原则或设计，完全符合客户的需求和期望，并且每个项目团队与客户保持一致联系。

与 Trahan 建筑事务所合作的广泛的客户包括大学和金融机构；教会、零售和工业设施；体育场馆；国家博物馆体系；医疗设施；联邦、州和当地政府。该公司在中国北京已经赢得了三个重要的国际设计竞赛，分别为制药和生物工程产业基地；医疗研究机构；新城设计。

Trey Trahan 是唯一的美国建筑师于 2005 年获得了著名的新兴建筑评论奖，并应邀参加了建筑联盟的第 25 周年纪念"新兴之声"系列讲座。他是三个国家 AIA 荣誉奖的获得者，并且该公司已经获得了众多国家和地区的设计奖项，有 9 人是 St. Jean Vianney 天主教会的。Trahan 建筑事务所的作品已经被超过 14 个国家的刊物上刊登了。

Trahan Architects

Trahan Architects, an award-winning firm led by Victor F. Trey Trahan, FAIA, has received international acclaim for work that is at once intensely personal, historically grounded, and aesthetically sublime. Each project shares a high level of technological sophistication and innovation coupled with a sense of quiet and experiential beauty. Trahan's aesthetically minimalist work is rooted in a profound understanding of Louisiana and the surrounding area, yet becomes completely contemporary through its state-of-the-art execution.

Much of the firm's material palette of light and concrete is informed by the region's architectural history, such as the eighteenth-century use of bousillage (a primitive form of concrete that is native to the area); these age-old materials, combined with Trahan Architects intense research into the latest systems and construction techniques, leads to work that feels simultaneously rooted and contemporary. Trahan's architecture avoids pastiche, cliché and easy symbolism; each building has a depth and an element of surprise that inspire constant discovery.

Since its founding in 1992, Trahan Architects has approached the details of each project—from client and user to site and program—with an intense level of focus and individuality. The firm begins by developing an understanding of the specific requirements of client, site, program, and budget. Dismissing preconceived formulas or designs, each building project is tailored exactly to the clients' needs and expectations, and each project team stays in constant contact with the client.

The broad range of clients working with Trahan Architects includes universities and financial institutions; ecclesiastical, retail, and industrial facilities; sports arenas; a state museum system; medical facilities; and buildings for federal, state, and local government offices. Internationally, the firm has won three premier international design competitions in Beijing, China, for a Pharmaceutical and Bioengineering Industry Base; for a medical research facility; and for the design of a new city.

Trey Trahan was the only American architect to receive the prestigious Architectural Review Award for Emerging Architecture in 2005, and was invited to participate in the Architectural League's 25th Anniversary "Emerging Voices" lecture series. He is the recipient of three National AIA Honor Awards and the firm has received numerous state and regional design awards, nine of them for the St. Jean Vianney Catholic Church. Trahan Architects' work has been published in more than 14 countries.

amphibianArc

amphibianArc（双栖弧建筑设计公司）由王弄极先生于 1992 年创建，总部在洛杉矶，并在上海设有办事处。在实践中完美融合艺术表现和解决问题的技巧一向是公司设计贯彻的信条。公司一直专注于"液态建筑"概念的实践，在作品中不仅包括流行于近代建筑舞台的曲线形体，更包含了华夏文明中表意文字的逻辑。

近年来，双栖弧在美国和中国设计了众多建筑项目，公建项目包括屡获殊荣的北京天文馆（美国建筑学会 AIA/LA 奖、中国詹天佑土木工程奖、中国建国 60 周年建筑创意大奖等）、宜昌新区重点片区规划（2013 年宜昌新区重点片区规划国际竞赛二等奖）、宁波鄞州公交城运枢纽中心等；商业开发项目包括有蚌埠百乐门国际文化经贸广场、杭州万象天成运河汇城市综合体等；住宅项目有空间书法（2013 芝加哥雅典娜美国建筑奖）等。

在过去的 10 年中，中国市场方兴未艾，蓬勃发展，amphibianArc 以其一贯的建筑热情，参与了多次中国公共建筑的竞标。其投标方案，实验性和实用性兼备，思路独特，形式新颖。公司作品被各地权威刊物竞相刊登，包括美国《建筑实录》、《洛杉矶建筑师》、英国《世界建筑》、《Dezeen》及网络杂志《Designboom》等。公司也凭借不菲的成绩、优秀的实力和良好的知名度受邀参加了 2014 洛杉矶 SKYLINE 展览、2013 北京 DADA 建筑设计及装置作品展、2012 戛纳国际房地产建筑大会建筑评论未来奖提名典礼等一系列展览和活动。

amphibianArc

In 1992, Nonchi Wang, the founder of amphibianArc, received an Honorable Mention in the NARA/TOTO World Architecture Triennale. The theme of this conceptual competition was the "Symbiosis of History and the Future". The jury consisted of Toyo Ito, Kisho Kurokawa and Kenneth Frampton. Except for the competition theme and encouragement from the jury to break boundaries, no program or requirements were given. amphibianArc represents a philosophy of design and architecture that is informed by the idea of evolution in the formation of human spirit and habitation.

amphibianArc gains understanding of human habitation in the context and temporal scale of human evolution. It is our goal to create buildings that not only reshape the lived reality but also inspire minds that will invent the future. In last two decades, amphibianArc has been focusing on the idea of Liquid Architecture, an architectural discourse adapting to the ever-shifting human consciousness towards the digital realm in the computer age.

Due to the founder Nonchi Wang's Chinese cultural heritage, Liquid Architecture encompasses not only the curvilinear forms prevalent in contemporary architectural scene, but also ideogramic methodology that is the foundation of Chinese word making. Adopting ideogramic forms into its formal repertoire, amphibianArc is able to institute narratives into design, creating narrative architecture.

LAB Architecture Studio

LAB Architecture Studio（澳大利亚 LAB 建筑师事务）于 1994 年由唐纳德·贝茨先生和彼得·戴维森先生创建，是一个国际性的建筑设计机构，主要从事建筑设计与城市设计。LAB 总部位于澳大利亚墨尔本，之后相继在上海、伦敦、迪拜、新德里设立办事处。完成的项目包括一系列可持续总体规划、城市设计、市政建筑与文化建筑设计、商业综合体、写字楼、酒店、度假村与住宅开发设计项目。

LAB 建筑、事务所着重于新概念城市设计与当代建筑设计，并在这些设计中展现可持续城市规划及革新建筑设计的新理念。LAB 是一家探索型的事务所，在设计中，通过对建筑形式、公共领域、公共空间社交活动的研究分析，得出全新的功能结构和秩序策略。事务所的设计中一贯鼓励体现当代和当地的时代烙印、并将区域差异、传统习惯、市场情况，、建设条例和技术等因素充分考虑到设计应用中。

LAB Architecture Studio

LAB architecture studio is a specialist architecture and urban design practice internationally recognized for the originality of its distinctive building designs and creating sustainable urban planning proposals. The studio has developed to become an international practice undertaking design and construction of projects in China, S.E. Asia, India, the Middle East, Europe, the UK and the Australia-Pacific region. This global engagement offers clients a unique perspective on current development trends and emerging contemporary design ideas.

The studio was founded by Donald Bates and Peter Davidson in 1994, and the Melbourne studio was established in 1997 following the LAB's success in the Federation Square competition. Melbourne is the focal office for the studio's international operations, with supplementary offices in Shanghai, London and Dubai.

As a specialist design practice working in a number of global locations LAB has the specific capability of producing concept options and developing design proposals within short time frames, communicating ideas to clients through effective graphic expression or concept models, and providing design leadership within collaborative project teams. The studio importantly understands the local differences and necessary considerations of tradition, market conditions, regulations and construction techniques in the developing of any design project.

NL Architects

NL Architects 总部位于阿姆斯特丹，由 Pieter Bannenberg、Walter van Dijk、Kamiel Klaasse 和 Mark Linnemann 于 1997 年建立。在 NL 看来，当今的建筑与郊区存在的问题及发展策略是密不可分的，这也是他们构建创新、前瞻性项目的关键。NL 认为建筑是以一种实验性形式展现的活动，融合了经济、文化、技术和环境等因素，城市则是一个平衡了经济发展与环境保护的生态系统。

事务所的精选项目包括格罗宁根广场、阿姆斯特丹菲英 K 街区、韩国环形住宅、赞斯塔德 A8ernA、乌特勒支篮球吧、乌特勒支 WOS8。其获得的奖项包括：2006 年城市公共空间欧洲地区大奖、2005 年密斯·凡·德罗大奖、2004 年 N.A.I 大奖、2003 年里特维德奖以及 2001 年鹿特丹设计奖。

NL Architects

Amsterdam-based NL Architects was founded in 1997 by Pieter Bannenberg, Walter van Dijk, Kamiel Klaasse and Mark Linnemann. In the view of NL, architecture today cannot be separated from suburban issues and strategies, which is their key to develop innovative and forward-looking devices and arrangements. Architecture is presented as a field of experimental activity, at the crossroads of economic, programmatic, technical and environmental thinking. The city is perceived as an ecosystem, an environment where logical systems of urban growth and natural factors, consumption and production, flux and stasis are balanced; and where recycling is set up as a method of stable and sustainable functioning, whether it has to do with waste, energy, materials or even architecture.

Its selected works include Groninger Forum (Groningen, 2006–2011), Funen Blok K (Amsterdam, 2010), Loop House (Korea, 2006), A8ernA (Zaanstad, 2006), Basket Bar (Utrecht, 2003), WOS 8 (Utrecht, 1997).NL Architects participated in many exhibitions around the world: Out There, Architecture Biennale Venice (Venice, 2008), New Trends of Architecture in Europe and Asia-Pacific (Shanghai, 2007), Sign as Surface (traveling Exhibition USA, 2003), Fresh Facts, Biennale Venice (Venice, 2002), NL Lounge, Dutch Pavilion Biennale Venice (Venice, 2000).

Among the numerous awards there are the European Price for Urban Public Space (2006), Mies van der Rohe award (2005), N.A.I award (2004), Rietveld Price (2003), and the Rotterdam design price (2001).

深圳市瀚旅建筑设计顾问有限公司

深圳市瀚旅建筑设计顾问有限公司成立于 1995 年，国家建筑行业建筑工程综合甲级设计资质，并通过 ISO9001 国际质量体系认证。业务范围包括规划设计、建筑设计、景观设计工程顾问，建筑涵盖办公、商业、教育、大型住宅、综合体等各种类型，创作了众多有代表性的优秀设计作品。瀚旅以蓬勃的创造力、执着的建筑理想、扎实技术上的创新以及超前的服务意识，不断推出创造性的设计理念和设计手法，获得了业内的广泛认可。曾获得"提升城市品质金房奖"、"全球生态人居最佳范例奖"、"深圳市优秀勘察设计奖"、"深圳市房屋建筑工程优秀施工图评选金奖"、"最佳雇主"等多项荣誉。

瀚旅拥有雄厚的技术实力，致力于各类课题的研究分析和软件开发，通过大量研发成果和设计创作的融合，筑成了瀚旅稳健发展的基础。瀚旅将 BIM 技术应用于项目设计的各个阶段，并在绿色建筑领域不断实践与创新。与国内外多家优秀设计公司进行广泛的交流和合作，不断与时俱进。

瀚旅力求给客户提供"信得过的全方位顾问式"专业设计服务。瀚旅对每一个项目除常规方案、初步设计和施工图外，还在全程的各个设计阶段增加优化设计环节。以客户的视角针对项目的定位、品质、空间及造价成本等因素，进行建筑、结构、设备等各专业的方案对比及推荐，在与客户的互动中共同寻找品质和造价的平衡，实现建筑理想和商业目标的双赢。

瀚旅是国内较早建立现代管理模式的民营建筑设计企业，采用与国际接轨的建筑设计行业管理模式。企业管理和项目管理实现职业化、科学化、精细化和网络化，项目设计的策划、组织、监管、反馈等机制以每一个设计项目为中心，围绕致力于优秀作品和提供客户信得过的全方位顾问式服务展开，确保每一个项目的设计和服务都能体现瀚旅的最高水准。

Shenzhen H&L Architects and Engineers Co. Ltd.

Shenzhen H&L architects and Engineers Limited Company was founded in 1995, with an integrated Class-A Engineering Design Qualification certificate, as well as ISO9001 Quality System Certification and International Accreditation. The business scope in H&L covers the planning design, architectural design, landscape design and engineering consultant. H&L has rich experience in many architecture types which include large scale residential developments, offices, commercial, education and mixed-use developments etc., creating many exemplary and outstanding design works. The company has been widely recognized by the industry for its vigorous creativity, persistent architecture ideas, innovation in strong skills and advanced sense of service, continuously producing creative design ideas and design techniques. H&L have also been awarded "The Golden House for the Promotion of Urban Quality", "Best Practice in Global Ecology for Human Settlement" and "Shenzhen Design Excellence award" among many others.

H&L with its strong technical ability has always been dedicated to subject research and software development, forming a stable and developed foundation through the integration of abundant research achievements and design creation. H&L exchanges extensively and cooperates with many excellent design companies, and will continue to move forward with the times.

H&L make every effort to provide clients with a professional "Credible Full Scope Consultant Service". H&L provides an optimized process for each design stage for the entire project in addition to primary design and construction drawings. To achieve an ideal win-win for architecture and commercial objectives, H&L will made recommendation after analyzing each projects of architecture, construction and engineering to meet the customer's visual angle.

H&L applies an international architecture design and industrial system, and is one of the first private enterprises to set up a modern management model. H&L's enterprise management and project management are realized through professional, scientific, precise and networking. The planning, organizing, supervision and feedback system gives priority to each project. H&L provides the reliable "Full Scope Consultant Service" based on an excellent design dedicated to our client, ensuring the design and service for each project incorporates the highest standards of H&L. The management model responds to the enterprise culture after years of exploration and practice in H&L; it has gained exclusive experience from the two completely unique fields of "Innovation" and "Standardization".

绿舍都会

绿舍都会（SURE Architecture）2006 年成立于英国伦敦，是一家专注可持续城市更新与生态建筑的研究与实践的国际化设计公司。现有伦敦公司作为公司总部辐射全球，并分别在中国北京和香港成立区域分公司主持亚洲及太平洋地区业务。我们的服务范围是项目策划定位、城市规划、建筑设计、景观设计、室内设计、幕墙设计、生态顾问和项目管理。

绿舍都会拥有一支来自不同背景且很有实力的设计和研究国际化团队。公司把多元化、个性化的技巧和经验集中起来。共享富有创造力天赋和专业技术人才是为我们取得国际和地性成就的钥匙。绿舍都会鼓励工作和内部文化中的个性化，通过对不同文化的角度和思维模式，因地制宜并求同存异，创造既有地方特色而又能与国际同步的产品及服务，打造拥有国际化工作模式、设计经验及理论、服务理念和管理模式的设计公司。

Sure Architecture

Sure Architecture (Sustainable Urban Regeneration and Eco-Architecture) was established in London, UK in 2006. It is a design and academic research architecture practice. It establishes close cooperation relation with the experts from Tsinghua University, the University of Sheffield and other professional institutions. It tries to stimulate architectural practice and creation by academic study. Sure Architecture has a professional design from different background and the research team, which makes the international education background and the working experience become the biggest advantage and foothold for the company.

MA2

MA2 是一个年轻的设计公司，致力于追求创新结构和应急策略。公司曾参加过许多国际设计竞赛，其卓越设计受到广大消费者的认可。MA2 因釜山塔楼综合楼与韩国建立了关系，而且该项目在国际建筑师协会主办的比赛中获得了三等奖，还得到韩国许多其他的建议。MA2 由 Micheal Arellanes 二世创立，收集并调查了许多基因组形成和架构变异的作品。

该办公室的理念植根于创新设计和技术的进步。公司在不断地发展，在数字化时代尝试架构进步的新程序和技术。先进的建筑设计是 MA2 获得收益的基础。思想和新技术进步在其建筑话语中尤感兴趣，并熟练地贯穿其作品。随着复杂性和设计能力的行业增长，MA2 将不断追求其最先进的和最新的操作。理论和研究是 MA2 工作的一部分，并且不断发展。

公司的使命是将建筑理论转变为建筑形式。由于出现的方法论和新发现的进步正在形成，MA2 将为建筑环境积极引进创新提案，努力成为建筑设计中的领导者。

MA2

MA2 is a young design office dedicated to the pursuit of innovative structures and emergent strategies. It has competed in many international design competitions and received recognition for design excellence. MA2 has built a relationship with the Republic of Korea for its work in the Busan Tower Complex which was awarded 3rd in a UIA sponsored competition, along with many other proposals for Korea. MA2 is founded by Michael Arellanes II, a Columbia University Alumni and author of Mutagenesis, a collection of work investigating genomic formations and mutation in architecture.

The office philosophy is rooted in the progress of innovative design and technology. MA2 is constantly evolving, trying new procedures and techniques for the advancement of architecture in the digital age. Advanced Architectural Design is the foundation of which MA2 proceeds. Evolutionary thought and new technological advancements in the architectural discourse is of most interest and practiced throughout its work. As the profession grows in its complexity and design capability, MA2 will be seeking after its most progressive and current execution. Theory and research are part of the work at MA2 for its importance in the ongoing development of exuberant environments.

The mission of MA2 is to transition architectural theory to built form. As advancements occur in methodologies and new discoveries are being made, MA2 will be active in bringing innovation to proposals for the built environment. Research is an important part of the design office, by doing so MA2 can be a leader in what is possible for architectural design.

TheeAe LTD.

TheeAe 是由建筑折中学派演变而来的缩写词，这一名字本身就代表了创造独特建筑的理念和精神。TheeAe LTD. 自创立以来，就秉持着这一理念，通过新的设计元素，带给人们愉悦的空间体验。除了追求创造性，该团队还致力于满足客户在环境和功能方面的需求，其项目既兼顾了实用性和美观性，也还综合考虑了功能、成本以及可持续性。

公司领导人吴勋重具有十多年在世界各地参与各类项目的经验。他认为，创造性就是一种在大自然的随意性中引入秩序的能力，也正是在这一理念的引导下，他积极寻求有机的设计方法，使建筑随着时间的流逝、功能和用途的变化而产生相应的演变，从而形成一种新的建筑，这样的建筑语言也就是"TheeAe"。

TheeAe LTD.

TheeAe is abbreviation of the evolved of architectural eclectic. Its name is ideas and dedication to create unique architecture. Since its establishment, TheeAe has adhered to this concept to bring people joyful spatial experience with new design elements. Besides the pursuit of creativity, the team devotes to satisfy the environmental and functional requirements of clients. All of the projects have well mingled practicality with aesthetics, functional programs, cost and sustainability.

The leader of TheeAe—Woo-Hyuncho has been working with a variety of projects around the world for more than 10 years. As he believes "the creativity is the ability to introduce order into the randomness of nature", his design seeks organic approach which the buildings evolve as they are developed by the change of function and use as time goes by. This eventually will produce the architecture as a new creation. That is the language called "TheeAe".

UNStudio

UNStudio 由本·范·伯克尔和卡洛琳·博斯于 1988 年组建，是一家专门从事建筑设计、城市开发和基础工程建设的建筑设计事务所。公司名 UNStudio 代表的是 United Network Studio，强调团体的协作与配合。事务所致力于在设计、技术、专业知识和管理等方面不断提升自身质量，以在建筑领域做出应有的贡献。

2009 年亚洲 UNStudio 建立，其第一个办事处设在中国上海，该办事处由最开始致力于杭州来福士广场项目的设计，逐渐扩展成为一个全方位服务的设计公司，有着全面、专业的跨国建筑师团队。

UNStudio 拥有一系列长期目标，务求定义并指引事务所在建筑界的表现。我们致力于为建筑领域做出应有贡献，能够在设计、技术、专业知识和管理等方面不断提升自身质量，跻身公共网络项目的专家行列。我们的作品无不体现出环境可持续发展、市场需求与客户意愿的完美结合。我们希望能够打造出令事务所与客户共同满意的项目。

拥有 20 多年国际设计项目经验的 UNStudio，不断扩展与全球各地的国际型咨询人员、伙伴及顾问之间的合作，并提升自身能力。该国际性的人际网络，以及我们在荷兰阿姆斯特丹及中国上海市中心设立的办事处，让我们在世界各地均得以有效率地运作。我们在亚洲、欧洲及北美洲已有 70 多个已建成的设计项目，并在现有的项目基础上继续拓展包括中国、韩国、台湾、意大利、德国及美国在内其他国家的国际业务。

UNStudio

UNStudio, founded in 1988 by Ben van Berkel and Caroline Bos, is a Dutch architectural design studio specializing in architecture, urban development and infrastructural projects. The name, UNStudio, stands for United Network Studio, referring to the collaborative nature of the practice. In 2009 UNStudio Asia was established, with its first office located in Shanghai, China. UNStudio Asia is a full daughter of UNStudio and is intricately connected to UNStudio Amsterdam. Initially serving to facilitate the design process for the Raffles City project in Hangzhou, UNStudio Asia has expanded into a full-service design office with a multinational team of all-round and specialist architects.

Throughout more than 20 years of international project experience, UNStudio has continually expanded its capabilities through prolonged collaboration with an extended network of international consultants, partners, and advisors across the globe. This network, combined with our centrally located offices in Amsterdam and Shanghai, enables us to work efficiently anywhere in the world. With already over seventy projects in Asia, Europe, and North America, the studio continues to expand its global presence with recent commissions in among others China, South-Korea, Taiwan, Italy, Germany and the USA.

同济大学建筑设计研究院

同济大学建筑设计研究院成立于 1958 年，是全国知名的集团化管理的特大型甲级设计单位。持有国家建设部颁发的建筑、市政、桥梁、公路、岩土、地质、风景园林、环境污染防治、人防、文物保护等多项设计资质及国家计委颁发的工程咨询证书，是目前国内设计资质涵盖面最广的设计单位之一。

经过近 50 多年的积累和进取，该院拥有了雄厚的设计实力、丰富的人力资源、先进的设计手段。全院现有职工 933 人，一级注册建筑师 112 名，一级注册结构工程师 146 名。自 1986 年以来共有近 200 项设计作品获奖。

作为一所国际著名高校设计单位，设计院非常重视建筑教育。培养了一大批硕士生、博士生。此外，研究院具备丰富的对外交流合作经验，曾成功的与来自美国、加拿大、德国、法国、西班牙等国的著名事务所合作，并互派员工交流学习。

按 ISO9001 标准建立的质量保证体系通过中国 SAC 和美国 RAB 双重认证。自 2001 年起，与中国人民保险公司签订了每年累计赔偿一亿元人民币的工程设计险合同。有能力提供顶尖的设计产品和一流的咨询服务。

Architectural Design and Research Institute of Tongji University

Architectural Design and Research Institute of Tongji University was founded in 1958. It is a large Class A unit which is a well-known collectivize management throughout the country. It has the construction, municipal, bridges, highways, geotechnical, geology, landscape architecture, environmental pollution control, air defense, heritage and many other design qualifications issued by the Ministry of Construction and the engineering consulting qualification certificate issued by the State Planning Commission, and it is currently one of the design qualification units with the most extensive covers.

After nearly 50 years of accumulation and enterprising, the Institute has a strong design strength, rich human resources and advanced design methods, in which has 933 staff, including 112 first-class registered architects and 146 first-class registered structural engineers. Since 1986, a total of nearly 200 design works have won awards.

As an internationally renowned university design unit, the Institute attaches great importance to building education to train a large number of master students and doctoral students. In addition, our Institute has extensive experience in foreign exchanges and cooperation, and had successfully cooperated with the renowned firms from the United States, Canada, Germany, France, Spain and other countries, and sent staff for the exchange and study.

The quality assurance system built according to the ISO9001 standard has won the dual certifications of Chinese SAC and the United States RAB. Since 2001, it has signed a contract with the Chinese People's Insurance Company on the engineering design of a total of 100 million RMB of annual compensation, so it has the ability to provide top-class design products and world-class consulting services.

Maryann Thompson Architects

Maryann Thompson Architects（玛丽安·汤普森建筑事务所）由玛丽安·汤普森于 2000 年 10 月创办。在此之前，玛丽安曾是汤普森·玫瑰建筑事务所的高级/始创公司的合作伙伴。

玛丽安·汤普森建筑师事务所是一家剑桥的建筑设计公司，为公共和私人客户提供范围广泛的服务。他们专注于可持续的、区域驱动的架构，试图提高他们工作场地的现象特质。他们的建筑调查以一个丰富而周到的内外部之间的边缘的关注为中心，利用光作为媒介，并采用温暖的自然材料，以突出归属感。

事务所有15位员工有着不同的背景，包括建筑、园林建筑、绿色建筑、规划、室内设计和视觉艺术。玛丽安·汤普森事务所有 LEED 认可的专业人士，而玛丽安在建筑和园林建筑中拥有学历，为大家带来跨学科的实践方法，其中的景观问题是他们设计思想的核心。他们认为整合规划、计划和设计以一个整体性的、建立共识的方式来完成项目。他们的项目从大型机构建筑的规模到小规模的住宅范围内变动。

Maryann Thompson Architects

Maryann Thompson Architects was founded by Maryann Thompson in October of 2000. Prior to that Maryann had been a senior/founding partner in the firm Thompson and Rose Architects.

Maryann Thompson Architects is a Cambridge-based architecture firm that offers a wide range of services to public and private clients. We specialize in architecture that is sustainable, regionally driven and that attempts to heighten the phenomenological qualities of the site in which we work. Our architectural investigations revolve around such concerns as the creation of a rich and thoughtful edge between inside and outside, utilizing light as a medium, and employing warm, natural materials in order to accentuate a sense of place.

The firm's staff of 15 comes from diverse backgrounds, including architecture, landscape architecture, green architecture, planning, interior design and the visual arts. Maryann Thompson Architects has LEED Accredited professionals on staff, and Maryann carries degrees in both architecture and landscape architecture, bringing to the practice an interdisciplinary approach where issues of the landscape are central to our design thinking. We believe in integrating planning, programming and design in a holistic, consensus-building approach to projects. Our projects range in scale from large institutional buildings to small-scale residential work.

澳大利亚柏涛（墨尔本）建筑设计公司

澳大利亚柏涛（墨尔本）建筑设计有限公司是澳大利亚最大的建筑设计公司之一。柏涛咨询（深圳）有限公司为澳大利亚柏涛（墨尔本）建筑设计亚洲公司在华设计机构。

一个多世纪以来，柏涛一直走在世界建筑设计行业的前列。杰出、实用和经济是柏涛公司设计的原则；设计上的创新和技术上的更新是公司的宗旨；技术上的可靠和设计的独特更是公司长期的声誉之所系。庞大的技术资源，广泛、丰富的经验，使柏涛公司能够承担各种规模、各种类型的区域规划设计和各类建筑的设计。

柏涛建筑设计集团除在澳大利亚几个大城市外，还在东南亚、欧洲和美国设有分支机构，设计业务遍布世界各地。柏涛墨尔本公司集中了一大批优秀的建筑师和相关专业人员，除开展通常的建筑设计业务之外，还有体育、医疗、居住建筑设有专门的研究机构。现在，柏涛亚洲分公司已拥有一百余名高素质的外籍及中国建筑师，业务范围开始向中国其他地域延伸，已承接了沈阳、武汉、长沙等地的多项设计任务。公司主要作品有澳大利亚国家网球中心、澳大利亚墨尔本奥林匹克公园及自行车赛馆、ESSO 澳洲总部、墨尔本水底世界水族馆、马来西亚运动中心，以及众多的建于澳洲本地与国外的酒店、商业大厦、政府大厦、写字楼工程、居住区建筑、医疗设施等。

Peddle Thorp Melbourne Pty Ltd.

Peddle Thorp Melbourne Pty Ltd. has been one of the largest architectural design firms in Australia. For over one hundred years, Peddle Thorp has always been at the forefront of the architectural world.

Pre-eminent, practical and efficient design works are the principles; design innovation and technical renovation and treated tenets; established reputation has been based on technical reliability and unique design. With superior technical resources, advanced computer skills and comprehensive design expertise, Peddle Thorp has been specializing in regional planning and architectural design at various scales and in various sectors. Peddle Thorp Group has practiced worldwide with offices in Australia, Asia, Europe and American. Supported by outstanding architects and experts, Peddle Thorp provides professional design service for mixed-use developments, hospitality, workplace, sports, health, education, retail and various residential projects. Major works are Australian National Tennis Center, Melbourne Olympic Park and Velodrome, ESSO Headquarters, Melbourne Underwater World Aquarium, Malaysia Sports Center, Portofino Residential and a series of other high class developments internationally.

扎哈·哈迪德建筑师事务所

扎哈·哈迪德建筑师事务所的创始人扎哈·哈迪德于 2004 年获得普利兹克建筑奖，她的作品、理论研究和学术成就在国际上广受关注。她的每一个动态创新项目都是基于 30 多年的革新实验和研究，都涉足都市生活、建筑和设计三个内部关联的领域。

哈迪德现在与资深设计师 Patrik Schumacher 共事。她的兴趣在于建筑、风景和地质三者交界的区域，而该事务所则关注将自然地质和人造系统整合在一起，因此需要运用前沿科技进行试验，这一过程的结果就是经常得到意想不到的动态建筑形式。

在前沿研究和设计调查领域，扎哈·哈迪德建筑师事务所一直保持着全球领先的地位。通过与该领域处于领先位置的艺术家、设计师、工程师和客户之间的合作丰富了事务所的知识储备。同时，高科技的运用帮助事务所实现了流线型的、综合的动态建筑结构。

Zaha Hadid Architects

Zaha Hadid, founder of Zaha Hadid Architects, was awarded the Pritzker Architecture Prize (considered to be the Nobel Prize of architecture) in 2004. Her practice integrates natural topography and human-made systems, leading to experimentation with cutting-edge technologies. Such a process often results in unexpected and dynamic architectural forms.

Zaha Hadid Architects continues to be a global leader in pioneering research and design investigation. Collaborations with corporations that lead their industries have advanced the practice's diversity and knowledge, while the implementation of state-of-the-art technologies have aided the realization of fluid, dynamic and therefore complex architectural structures.

In the fields of cutting-edge research and design survey, Zaha Hadid Architects has kept a worldwide leading position. By cooperating with the artists, designers, engineers and customers who are in leading position in the field to enrich the Architects' knowledge reserves, meanwhile, using the high-tech help the Architects to achieve the streamlined, integrated and dynamic architectural structures.

FOUNDRY OF SPACE

FOUNDRY OF SPACE（FOS）是基于曼谷设计的办公室实务建筑和城市规划。他们定位自己是建筑师和文化分析之间的衔接，其中的社会经济、政治、环境及其他相关因素是在他们的关键设计参数之间。

FOS 的魅力主要集中在每个项目的内容和环境中。通过他们对建设方案和创新设计符合转型过程的全面研究，他们认为，他们的架构不仅能够提高人居建筑的环境质量，而且还能激发任何工作、玩乐、休息或者对他们建筑的体验。

同时，FOS 一直致力于为建筑和城市结构之间的相互渗透，以刺激新的内容和现有环境之间的建设性的共存。FOS 的服务领域包括建筑、室内、总体规划和研发。

FOUNDRY OF SPACE

FOUNDRY OF SPACE is a Bangkok-based design office practicing architecture and urbanism. We position ourselves at the convergence between architects and cultural analysts where socio-economic, political, and environmental and other relevant factors are among our key design parameters.

FOS's fascination is primarily focused on contents and context of each project. Through our comprehensive research in building programmes and transformation process towards logically innovative design, we believe that our architecture is not only able to enhance quality of built environment for people to live in, but also to inspire whoever works, plays, rests or has any kind of experiences with our architecture.

Simultaneously, FOS consistently strives for mutual permeation between architecture and its urban fabric in order to stimulate the constructive coexistence between the new contents and the existing context. FOS's areas of services include Architecture, Interior, Master planning, and Research & Development.

北京殊舍建筑设计有限公司

北京殊舍建筑设计有限公司于 2009 年在北京成立，是一个由众多对建筑充满激情的年轻人组成的建筑研究及实践的设计团队。殊舍建筑以理性的思维考量项目设计所面临的个方面因素，通过分析与沟通，提出最可行的设计方案。殊舍建筑的作品涵盖城市设计、住宅规划、公共建筑、景观设计和室内设计等。

殊舍建筑强调建筑的唯一性和逻辑性，他们认为每个建筑都有自己特有的内涵、外观及空间。项目的限制条件、地理位置、地形特征、文脉历史、气候状况、周边环境以及项目的使用者和建造者都决定了这个建筑的特性。在设计过程中，殊舍建筑都会综合考虑以上因素，通过整体规划、空间组织以及细节推敲解决相关问题。

Beijing Shushe Architectural Design Co., Ltd.

Shushe Architecture was founded in Beijing in 2009. Shushe Architecture is a young and passionate team devotes in architecture research and design. Taking the rational considerations of every aspects of the problem, find the best way out through analysis and communication. Shushe's works cover the area of urban design, residential planning, public architecture design, landscape design and interior design.

"Shu" means special, but not in the way of look or space or concept, is about the only solution under the particular conditions, which makes it special. The firm speaks for the architecture of its unique and logic, and their own character, form and space. All the limitations, location, topography, history, weather, surrounding environment, users and designers together decide its character. They take consideration of all these mentioned factors through every project, name the unsolved problems, and then take them down from the whole picture to space, finally the details.

Steven Holl Architects

Steven Holl Architects（史蒂芬·霍尔建筑事务所）是一所由史蒂芬·霍尔（Steven Holl）于1976年创立的建筑设计事务所，业务范围包括建筑设计与城市规划。事务所的设计师们擅长有关艺术和高等教育类型的建筑设计，包括赫尔辛基当代美术馆、纽约普拉特学院设计学院楼、爱荷华大学艺术与艺术史学院楼、西雅图圣伊格内修斯小教堂。

史蒂芬·霍尔建筑师事务所事务所在国际享有盛誉，其作品以高质量的设计多次获奖、出版及展览，北京当代MOMA荣获由国际高层建筑与城市住宅协会（CTBUH）所颁发的"2009年世界最佳高层建筑"大奖和2008年美国建筑师协会纽约分会可持续设计大奖以及2009年西班牙对外银行（BBVA）基金会知识前沿奖，尼尔森·阿特金斯美术馆获得2008年美国建筑师协会纽约分会建筑荣誉大奖。史蒂芬·霍尔被授予"美国建筑师协会金奖"，这个奖项认可对建筑界做出持久贡献的个人。

Steven Holl Architects

Steven Holl Architects is a 40-person innovative architecture and urban design office working globally as one office from two locations; New York City and Beijing. Steven Holl leads the office with senior partner Chris McVoy and junior partner Noah Yaffe. Steven Holl Architects is internationally-honored with architecture's most prestigious awards, publications and exhibitions for excellence in design. Steven Holl Architects has realized architectural works nationally and overseas, with extensive experience in the arts (including museum, gallery, and exhibition design), campus and educational facilities, and residential work. Other projects include retail design, office design, public utilities, and master planning.

Steven Holl Architects has been recognized with architecture's most prestigious awards and prizes. Most recently, Steven Holl Architects' Cite de l'Ocean et du Surf received a 2011 Emirates Glass LEAF Award, and the Horizontal Skyscraper won a 2011 AIA National Honor Award. The Knut Hamsun Center received a 2010 AIA NY Honor Award, and the Herning Museum of Contemporary Art received a 2010 RIBA International award. Linked Hybrid was named Best Tall Building Overall 2009 by the CTBUH, and received the AIA NY 2008 Honor Award. Steven Holl Architects was also awarded the AIA 2008 Institute Honor Award and a Leaf New Built Award 2007 for The Nelson-Atkins Museum of Art (Kansas City).

盖里建筑师事务所

弗兰克·欧文·盖里（Frank Owen Gehry），1989年普利兹克建筑奖获得者，以设计具有奇特不规则曲线造型雕塑般外观的建筑而著称。由于深受洛杉矶城市文化特质及当地激进艺术家的影响，他于1962年建立了自己得公司——Gehry Partners。

盖里早期的建筑锐意探讨铁丝网、波形板、加工粗糙的金属板等廉价材料在建筑上的运用，并采取拼贴、混杂、模糊边界、去中心化、非等级化等各种手段，挑战人们既定的建筑价值观和被捆缚的想象力。明确地讲，盖里的设计风格源自于晚期现代主义，其中最著名的建筑是位于西班牙毕尔巴鄂的有着钛金属屋顶的古根海姆美术馆。近期于香港完成的The Opus，是他首次于亚洲设计的住宅建筑项目。

Gehry Partners, LLP

Frank Gehry established his practice in Los Angeles, California in 1962. The Gehry partnership, Gehry Partners, LLP, was formed in 2001 and currently supports a staff of over 120 people. Gehry Partners, LLP is a full service architectural firm with extensive international experience in the design and construction of academic, museum, theater, performance, commercial and residential projects. Gehry Partners employs a large number of senior architects who have extensive experience in the technical development of building systems and construction documents, and who are highly qualified in the management of complex projects.

Every project undertaken by Gehry Partners is designed personally and directly by Frank Gehry. All of the resources of the firm and the extensive experience of the firm's partners are available to assist in the design effort and to carry this effort forward through technical development and construction administration. The firm relies on the use of Digital Project, a sophisticated 3D computer modeling program originally created for use by the aerospace industry, to thoroughly document designs and to rationalize the bidding, fabrication, and construction processes.

The partners in Gehry Partners, LLP are: Frank Gehry, Brian Aamoth, John Bowers, Anand Devarajan, Jennifer Ehrman, Berta Gehry, Meaghan Lloyd, David Nam, Tensho Takemori, and Laurence Tighe & Craig Webb.

Arch Group

建筑办事处Arch集团于2007年由建筑师Michael Krymov和Alexey Goryainov在莫斯科建立的。办事处活动的主要范围是设计公共的室内设计，以及建筑和不同用途的结构。

Arch集团进行设计作品的整个范围从第一协商和概念开发到所有部分的详细设计文档的发展，以及施工管理。集团许可各类设计作品，建筑师Michael Krymov和Alexey Goryainov是建筑师联盟成员。该团队结合了各地20名充满热情的专业人士，其中包括16位建筑师。

集团的目标是揭示该项目的最大审美和功能性潜力，并以客户的需求为准确保其正确实施。我们认为优质的建筑不应该意味着对客户极端的财政负担。高品质的建筑是一位建筑师为达到适当目的的专业方法和兴趣的结果，它是基于对细节的关注，从而精心设计文档。

Arch集团的项目实施获得了专业奖项，并得到出版社的认可。截至目前已经实现超过了25个显著项目。

Arch Group

The architectural bureau Arch Group was set up by architects Michael Krymov and Alexey Goryainov in Moscow in 2007. The main area of activity of the bureau is designing public interiors, as well as buildings and structures for different purposes.

Arch Group performs the whole scope of design works from the first consultations and concept development to the development of detailed design documentation in all sections, as well as the construction supervision. Arch Group has the license for all types of design works; architects Michael Krymov and Alexey Goryainov are members of the Union of Architects. The team is combined with around 20 enthusiastic professionals, including 16 architects. Our aim is to reveal the maximum aesthetic and functional potential of the project and to ensure its proper implementation with a view to the customer's needs. We think that quality architecture should not imply an extreme financial burden for the customer. High quality architecture is a result of an architect's professional approach and interest in achieving proper result; it is based on the attention to details resulting in elaborate design documentation.

Their projects and implementation captured professional awards and got recognition of the press. As of now more than 25 significant projects have been realized.

法国AS建筑工作室

法国AS建筑工作室于1973年在巴黎创立，工作室从创立时就把团队合作精神作为工作哲学，并在发展过程中始终保持开放的姿态，如今已成为汇集了25位不同国籍的优秀建筑师、城市规划师、室内设计师及成本评估师等的专业团队。

法国AS建筑工作室将建筑定义为一种以社会为依托的艺术，一种以人生为框架的创作。工作室摒弃单纯的个人主义，立足于团队合作与知识分享，注重团队成员之间的交流与合作，甚至以争论来实现个人技能向团体创作潜力的转换。

工作室赢得了多项荣誉，包括：阿拉伯世界研究中心获得银尺奖及阿卡·汗建筑奖、巴黎Dom Remy路住宅楼获得Losange奖、上海惠生生化园区获得金钢奖、马Sainte-Marguerite医院获得金钥匙奖、法国Saint-Etienne市Casino集团总部设计获2007年SIMI大奖（颁发给办公及企业总部杰出设计项目的专门奖项）等众多奖项。

Architecture-Studio

Founded in Paris in 1973, Architecture-Studio has regarded team spirit as its work philosophy since its foundation. Team of AS. has progressively assembled itself. Now it brings together professional architects, urban planners, interior designers and cost valuators of 25 natinalities.

Architecture-Studio defines architecture as "an art committed with society, the construction of the surroundings of mankind". Its foundations lie on work group and shared knowledge, with the will to go beyond individuality for the benefit of dialogue and confrontation. Thus, the addition of individual knowledge turns into wide creative potential.

The awards by Architecture-Studio include the Aga Khan Award for the Arab World Institute in Paris, the Losange d' Argent for a housing complex on rue Domrémy in Paris and the Acier d', Award for Wison Chemical Head-Office in Shanghai, Sainte-Marguerite Hotel, etc.

Synthesis Design + Architecture

Synthesis Design + Architecture（SDA）是一家新兴的现代设计公司，在建筑设计、基础设施建设、室内设计、装置设计、展览设计、家具设计等领域积累了多年的专业实践经验，其卓越的设计工作已获得国际认可。

该团队由多学科的专业设计人员组成，包括注册设计师和建筑设计者，以及在美国、英国、丹麦、葡萄牙、台湾接受了专业教育与培训的计算机专家。基于全方位的设计理念，该公司的设计作品融合了世界性的设计元素，并以简单明了的色调和线性的整体结构为特色，引领了建筑设计的基本走向。

Synthesis Design + Architecture

Synthesis Design + Architecture (SDA) is an emerging modern design company, with years of professional experience specializing in architectural design, infrastructure construction, interior design, equipment design, exhibition design, furniture design and other fields, and its superior design works have been won international recognition.

The team is comprised of multi-disciplinary design professionals, including registered architects and building designers, as well as computer experts received specialized education and training in the United States, Britain, Denmark, Portugal, and Taiwan. Based on a full range of design concepts, the company's design works combine worldwide design elements, with the simple colors and linear overall structure for the features to lead the basic trend of architectural design.

非常建筑

非常建筑于1993年由张永和、鲁力佳创立，现已发展成国际知名、中国领先的建筑设计事务所。在近20年的创作实践中，非常建筑在国内外建筑、文化艺术和设计领域产生了很大影响，荣获了众多国内外重要奖项，其作品涵盖了城市规划、城市设计、公共建筑、文化设施、景观设计、室内设计等各方面，并在国内外被广泛出版和展览。

公司在2006年荣获美国《商业周刊》中国设计奖，河北教育出版社办公楼荣获2004年度中国建筑艺术年鉴学术奖，柿子林会所荣获2004年度"WA中国建筑奖"优胜奖，广东清溪"坡地住宅群"住宅设计获1996年度美国PA进步建筑奖。

其完成的主要项目包括四川安仁十年大事记馆、京兆尹素食餐厅、2010年上海世博会上海企业联合馆、吉首大学综合教学楼及黄永玉美术馆、柿子林会所等。建筑、艺术和设计领域前瞻性的视野、丰富的经验、良好的专业素养和独特的创造力，使非常建筑成为建筑行业的一支中坚力量。

Atelier FCJZ

Atelier FCJZ was established by Yung Ho Chang and Lulijia in 1993. The leading company in China has developed into an internationally renowned architectural design practise. With 20 years experience, Atelier FCJZ has a significant influence in the fields of architecture home and abroad, cultural arts and design and won many important awards. The works of Atelier FCJZ include urban planning, urban design, public architecture, cultural facilities, landscape design, etc.

Atelier FCJZ was awarded Chinese Design Awards of Business Week in 2006. Heibei Education Press Office Building by Atelier FCJZ was awarded Almanac of Chinese Architectural Arts 2004; Persimmon Woods Club House got honorable prize of "WA Chinese Architecture Award" 2004; "Housing Development on Slopes" (Qingxi, Guangdong) got Progressive Architecture Awards.

The main projects by Atelier FCJZ include Anren Hall of Decennium Events, Jingzhaoyi Vegetarian Diet Restaurant, 2010 Shanghai Corporation Pavilion of World Expo, and Persimmon Woods Club House, etc. With, prospective view, rich experience, excellent professions and distinctive innovation in architecture, arts and design, Atelier FCJZ has become the dominant force in building industry.

Urban Office Architecture

Urban Office Architecture（UOA）专注于卓越的和可持续性的设计，努力提议更多超越场地、预算和风格限制的潜能。UOA的解决方案是基于尖端技术和富有远见的信仰，相信建筑是类比和鼓舞人心的本质过程，能够让艺术成为一个司空见惯的必然。

每个项目由该公司创建人卡罗·恩佐和中高层管理人员从头到尾进行监督的。UOA以每个被受用的项目为骄傲和荣誉。我们认为，只要直接参与公司的主要创始人16年获得国家和国际经验，公司可以提供卓越的设计并满足客户的需求。UOA通过所有项目的阶段范围，从可行性研究到施工管理和项目清算进行参与。

UOA在世界各地的比赛和项目中已经荣获了多项奖项，其中包括著名的Europan 06、马里博尔、斯洛文尼亚和新泽西州特伦顿罗宾斯小学。UOA在本领域演讲和访谈中起到了重要作用，已在全球各地出版了许多书籍和杂志，包括在迪拜、台湾、德国、美国、意大利、法国、中国和印度。

Urban Office Architecture

Urban Office Architecture(UOA)focuses on design excellence and sustainability in an effort to suggest more potential beyond the limitations of site, budget and style. UOA's solutions are based on cutting edge technologies and the visionary belief that architecture is a process of analogical and inspirational nature, which is able to make art a quotidian necessity.

Each project is supervised by the Firm Founding Principal Carlo Enzo and senior staff at all stages from beginning to end. Urban Office Architecture (UOA) takes pride and honor in each project is hired for. We feel that it is only through the direct involvement of the Firm Founding Principal that the 16 years of gained national and international experience the firm can be used to deliver design excellence and meet the need of our clients. UOA involvement extends through all project phases from Feasibility study to Construction Administration and Project Closeouts.

UOA has been awarded numerous prizes for competitions and projects around the world, including the prestigious Europan 06, Maribor, Slovenia and the Robbins Elementary School in Trenton, NJ. UOA has also been featured in lectures and interviews and has been published in many books and magazines around the globe, including Dubai, Taiwan, Germany, USA, Italy, France, China and India.

Tuncer Cakmakli Architects

Tuncer Cakmakli Architects 是由 Tuncer Cakmakli 组织和领导的一家专业的设计公司。大到城市规划，小到门把手的设计，该公司都能为客户开发项目或建筑项目提供最富价值和收益的设计及规划服务。

公司有能力针对各客户的实际需求，为其量身定制一个团队，将团队知识网络中分散的知识和技术充分利用起来，广泛采用数字和物理模型来检测和检查设计决策，对使用者未来空间体验可能造成的影响，力求为其打造一个最符合要求和期望的解决方案。因此，公司一直广受客户信任和好评，其客户不仅包括德国和土耳其市政，还涉及有瑞士、奥地利、丹麦、匈牙利、西班牙和以色列等国外政府部分。另外，公司的主要作品有：Kosbaş Kocaeli 免税区项目、布尔萨蔬果鱼类批发市场项目等。

Tuncer Cakmakli Architects

Tuncer Cakmakli Architects is organized and led by Tuncer Cakmakli. From urban planning all the way down to the design of a door handle, it offers complete design and planning services that add value and profitability to the clients' development or construction projects.

The firm is able to assemble a team that is right for each project, as necessary, to draw on the vast knowledge and expertise that is distributed throughout its network. Digital as well as physical models are extensively used to test and scrutinize the implications of design decisions for the future experience of the users of a space, finding the best solution for particular needs and expectations. Therefore, the firm is trusted and well received by its clients. Among its repeat clients are municipalities in both Germany and Turkey, as well as the foreign offices of the governments of Switzerland, Austria, Denmark, Hungary, Spain, and Israel. The works of the firm include the Kosbaş Kocaeli Free Zone and the Wholesale Greengrocers' and Fishmongers' Markets for the City of Bursa.

上海同为建筑设计有限公司

上海同为建筑设计有限公司成立于2003年，由旅居加拿大的华人城市设计师李长君先生在上海创立。该公司始终坚持设计品质，不断研究先进规划设计思想，结合当代中国的建设需要、时代特征、人文特质，以埋头创作、不事张扬的精神，积极参与城市设计、城市规划、建筑设计、景观设计的创作探索。

在业务发展过程中，秉承"兼容并蓄、大道同为"的企业文化理念，注重整合各类社会资源形成了全方位的综合性的设计服务能力。其作品多次获得各种奖项，先后获得了美国、俄罗斯、日本的设计奖项三项，国内设计奖项七项，各类项目投标中标。

Shanghai Tongwei Architectural Design Co., Ltd.

Shanghai Tongwei Architectural Design Company is established by Li Changjun—an ethnic Chinese urban designer—in Shanghai, 2003. The firm pursues quality design and constantly study advanced planning & design concepts. In accordance to the construction needs, characteristics of the times and humanistic characteristics, the firm takes an active role in urban design, urban planning, architectural design and landscape design, with its spirits of unpretentiousness and perseverance.

In business development, the company adheres to its enterprise culture concept of "Versatile & Tolerant", creating all-round comprehensive design service by integrating various social resources. Its works have successively received three awards from America, Russia and Japan, as well as seven national awards and several biddings.

New Wave Architecture

New Wave Architecture 建立于2006年，位于伊朗德黑兰，公司共有70名员工，是一家创新型建筑设计公司。该公司致力于探索全球性的建筑语言，在新兴理念、高标准的创新空间美学、人性化设计等方面探索新的途径，以实现创新又富有挑战的现代设计。公司凭借在校园和教育类建筑的卓越表现，以及在医疗和保健设施类建筑的专业设计，获得了国内外多项建筑大奖和荣誉。

事务所的主要作品有：Broojen健康管理中心、Chelgerd健康管理中心、扎黑丹大学展览厅、中心礼堂和图书馆、艺术和建筑系大楼、人类科学系大楼、Hirmand文化综合体、Dalgan文化综合体、德黑兰Shareiati医院、德黑兰内分泌和新陈代谢中心、伊朗Abali Onlooker住宅、伊朗纳米科技研究中心等。

New Wave Architecture

New Wave Architecture (Lida Almassian/Shahin Heidari) founded in 2006 is a 70-person innovative architecture design firm in Tehran, Iran. It has been nationally and internationally honored with architecture's prestigious awards, publications, competitions and citation for design excellence with extensive experience in the campus and educational faculties, specialized hospitals and health care facilities. Other projects include retail design, residential works and recreational facilities. Over 150 projects have been designed, accomplished or due to be completed.

New Wave Architecture seeks for global language of architecture to approach an innovative and challenging contemporary movement. It explores the new ways of emerging ideas, demanding and distinctive spaces regarding the aesthetic aspects, humanity and global communication.

Fournier-Rojas ARQUITECTOS

Fournier-Rojas 建筑师事务所（Foro 建筑师事务所）成立于 1975 年，是一家建筑、规划、室内、平面艺术公司，室内设计，对现代的、热带的、生物气候的、环保的设计有兴趣，其大作设计是由美国受过教育的建筑师 Alvaro Rojas 和哥斯达黎加的艺术家与建筑师 Sylvia Fournier 领导的。

Foro 建筑师事务所是一家小型建筑公司，由于人工和自然或现有的新的人工景观，在伦理上和创造上致力于整体景观的设计，并且在环保设计和场地、时间以及具体架构的范围内，该公司已深深植根于此地。

Fournier_Rojas ARQUITECTOS

Fournier-Rojas ARQUITECTOS (FoRo ARCHITECS) was found 1975, an architecture-planning-interiors-graphic arts firm interested in modern-tropical-bioclimatic-environmentally responsible design led by USA educated architect Alvaro Rojas and Costa Rican artist and architect Sylvia Fournier.
FoRo ARQuitectos is a small architecture firm ethically and creatively dedicated to the design of Integral Landscape resulting from the encounter of the artificial and the natural and/or the existing and the new artificial landscapes, within the boundaries of environmentally responsible design and site and time specific architectures that are deeply rooted in place.

OPORTO OFFICE FOR DESIGN AND ARCHITECTURE

OPORTO OFFICE FOR DESIGN AND ARCHITECTURE（OODA）是一家基于 Portugueses 的建筑设计事务所，致力于创造并提出创新的项目和想法。它试图将先进的设计理念结合先进的技术——思想，探索当代的建筑，以积极主动的方式，结合具有竞争优势理论框架的方法来进行实践。

从外观设计上和景观的复杂点而言，OODA 在一些世界上最重要的建筑中作为项目经理、技术协调员和资深设计师是有经验的。其全球的人才和国际经验都显示在 OODA 团队过去工作过的项目中。

OPORTO OFFICE FOR DESIGN AND ARCHITECTURE

OPORTO OFFICE FOR DESIGN AND ARCHITECTURE(OODA) is a Portugueses-based architectural design office dedicated to create and suggest innovative projects and ideas. It tries to combine sophisticated design philosophy with advanced technology-thoughts in exploring the contemporary architecture, in a pro-active way, combined with a competitive advantage theorical approach to the practice.
OODA have experience as project managers, technical coordinators and senior designers in some of the most important buildings in the world from the design and complexity point of view. Its global talent and international experience are shown some of those projects OODA team have worked in the past.

Henning Larsen 建筑师事务所

Henning Larsen 建筑师事务所于 1959 年由建筑师 Henning Larsen 创立，是一家根植于斯堪的纳维亚文化的国际建筑公司。Henning Larsen 建筑师事务所由首席执行官 Mette Kynne Frandsen 以及设计总监 Louis Becker 和 Peer Teglgaard Jeppesen 负责管理，已在哥本哈根、慕尼黑、贝鲁特以及利雅得建立了办事处。

Henning Larsen 建筑师事务所在环境友好型和综合节能型设计等方面卓有建树，以创造怡人、可持续的项目为目标，给当地的人们、社会及文化创造持久的价值。设计师对项目负有高度的社会责任感，注重与客户、业主以及合作者的交流，积累了从建议书草案到细节设计、监督和施工管理等各方面的建筑知识和理论。

Henning Larsen Architects

Henning Larsen Architects is an international architecture firm with strong Scandinavian roots based in Copenhagen, Denmark, founded in 1959 by noted Danish architect and namesake Henning Larsen. The management of Henning Larsen Architects consists of Mette Kynne Frandsen (CEO), Louis Becker (International Design Director) and Peer Teglgaard Jeppesen (Scandinavian Design Director). And we have opened offices in Copenhagen, Munich, Beirut, and Riyadh.
Henning Larsen Architects is renowned for their contributions on environment friendly and comprehensive energy conservation designs. Our goal is to create vibrant, sustainable buildings that reach beyond themselves and become of durable value to the user and to the society and culture that they are built into. Our designers have a high sense of social responsibility for the projects, and focus on the exchange with the clients, owners and collaborators. We have acquired a comprehensive knowledge of the many aspects of building—from sketch proposals to detailed design, building owner consulting and construction management.

蓝天组

1968 年由沃尔夫·德·普瑞克斯和海默特·斯维茨斯基在奥地利维也纳设立了建筑设计事务所，取名蓝天组，其业务范围涉及建筑设计、城市规划、平面和艺术设计多个服务领域。经过 20 年的奋斗，公司在 1988 年于美国洛杉矶成立了境外第一家分支事务所。

蓝天组的主要项目有：法国默伦塞纳城市总体规划、荷兰 Groninger 博物馆东馆、瑞士 2002 年世博会比尔市艺术长廊会议中心、1998 年竣工的德国德莱斯顿 UFA 影院中心、德国慕尼黑宝马中心、洛杉矶视觉与表演艺术高中、法国里昂汇聚博物馆、丹麦奥尔堡音乐之屋、德国法兰克福欧洲央行新总部和韩国釜山电影中心等。

在过去的 30 年间，蓝天组获得了许多国际竞赛大奖：1982 年柏林建筑艺术促进大奖；1988 年维也纳建筑奖；1992 年 Erich Schelling 建筑大奖；1989-1991 年连续三年得 P.A. 大奖；1999 年奥地利城市建筑大奖；2001 年欧洲钢结构设计大奖；2005 年，凭借阿克隆艺术馆设计方案赢得美国建筑奖；2008 年凭借 BMW 展厅与阿克隆艺术馆赢得 RIBA 大奖等。

COOP HIMMELB(L)AU

COOP HIMMELB(L)AU was founded by Wolf D. Prix, and Helmut Swiczinsky in Vienna, Austria, in 1968, and is active in architecture design, urban planning, plane design and art design. In 1988, a second studio was opened in Los Angeles, USA over the struggle of two decades.
COOP HIMMELB(L)AU's most well-known projects include: the master plan for the City of Melun-Sénart in France; the Groninger Museum, East Pavilion in Groningen (1994) in the Netherlands; the design for the EXPO.02—Forum Arteplage in Biel, Switzerland (2002); the multifunctional UFA Cinema Center in Dresden, Germany (1998); the BMW Welt (2007) in Munich, Germany; the Central Los Angeles Area High School #9 of Visual and Performing Arts in Los Angeles, USA (2008); the Musée des Confluences in Lyon, France (2014); the House of Music in Aalborg, Denmark (2014); the European Central Bank's new headquarters in Frankfurt am Main, Germany (2014); the Busan Cinema Center in Busan, South Korea (2011); etc.
Over the course of the past three decades, COOP HIMMELB(L)AU has received numerous international awards. These include: the Förderungspreis für Baukunst, Berlin (1982), the Award of the City of Vienna for Architecture (1988), the Erich Schelling Architektur Preis (1992), the Progressive Architecture/P. A. Award (1989, 1990, and 1991), the Großer Österreichischer Staatspreis (1999) as well as the European Steel Design Award (2001). In 2005, for the design of the Akron Art Museum, our studio received the American Architecture Award. In 2008 COOP HIMMELB(L)AU received the RIBA International Award for the Akron Art Museum. In the same year the RIBA European Award and the World Architecture Festival Award: Production was given for the project BMW Welt, etc.

Mecanoo Architecten

Mecanoo Architecten（荷兰麦肯诺建筑事务所）于 1984 年正式在代尔夫特成立，是由来自 25 个民族的 120 位创意专业人士的高度跨学科的工作人员组成，包括建筑师、室内设计师、城市规划师、景观设计师和建筑技术师。

该公司是由它的原始创办架构师——教授 Houben、技术总监 Aart Fransen 及执行董事 Peter Haasbroek 执导，他们加入了合作伙伴 Francesco Veenstra、Ellen van der Wal、Paul Ketelaars 和 Dick Van Gameren。他们获得了 30 多年的丰富经验，加之结构化的规划过程结果在设计方面实现了专业技术和对细节的注重。

Mecanoo Architecten

Mecanoo, officially founded in Delft in 1984, is made up of a highly multidisciplinary staff of over 120 creative professionals from 25 nationalities and includes architects, interior designers, urban planners, landscape architects and architectural technicians.
The company is directed by its original founding architect, Prof. ir. Francine M.J. Houben, technical director Aart Fransen and executive director Peter Haasbroek who are joined by partners Francesco Veenstra, Ellen van der Wal, Paul Ketelaars and Dick van Gameren. The extensive experience gained over 30 years, together with structured planning processes results in designs that are realized with technical expertise and great attention to detail.

Doparum 建筑师事务所有限公司

Tamás Dobrosi 及他的同事在完成匈牙利布达佩斯 Károly Kós 协会专业研究生课程"Journeyman school"项目之后，于 2009 年创立了 Doparum 建筑师事务所。并得到了著名匈牙利有机建筑大师 Imre Makovecz 的全力支持。

事务所不仅渴望参与"未来建筑"设计，而且也渴望参与"回顾过去"建筑设计。通过传递所拥有的文化及精神知识，事务所及其年轻成员的基本目标是促进匈牙利建筑的独特本质及其发展。

Doparum Architects Ltd

After finishing "Journeyman school", a professional postgraduate course of Károly Kós Association in Budapest, Hungary, Doparum Architects was founded by Tamás Dobrosi and his colleagues in 2009, with the wholehearted support of Imre Makovecz, the internationally acclaimed master of hungarian organic architecture. Eager to take part not just in "building the future", but in "remembering the past" as well, the basic goal of the office and its young members, by passing on the cultural and spiritual knowledge they possess, is to contribute to the unique nature and development of Hungarian architecture.

彼特鲁格建筑设计事务所

彼特·鲁格建筑设计事务所（Peter Ruge Architekten）立足于德国柏林与中国杭州，是一家从事建筑设计、城市规划的国际性建筑设计事务所。从欧洲到亚洲，从提出方案到具体实施，主持建筑师彼特·鲁格具有超过20年的从业经验。彼特在各地均有讲座，目前在德绍建筑研究所从事可持续建筑设计教学工作。

事务所主要设计特长在于从低能耗可持续发展的角度进行现有旧房、新式生态建筑，以及整体区域进行改造与设计规划。事务所依旧欢迎客户给予机会与挑战：中国长兴市被动式住宅、中国杭州旧工厂改造项目（在建）、以及德国柏林住宅升级改造项目。

在所获得的众多奖项中，彼特·鲁格建筑设计事务所在绿色环保低能耗与可持续发展生态建筑设计领域尤为突出。彼特·鲁格设计的三个项目均已获得国家认证可持续性建筑奖。

Peter Ruge Architekten

Peter Ruge Architekten is an international architectural and urban design firm with offices in Berlin, Germany and Hangzhou, China. Peter Ruge runs the company and has over 20 years experience realizing buildings and urban developments across Europe and Asia. He has lectured at various locations. Currently he teaches Sustainable Design at Hochschule Anhalt / Dessau Institute of Architecture.

For Peter Ruge Architekten a wide range of prospects has opened up concerning sustainable optimization of existing portfolios, residential developments and city planning. With expertise in planning and realisation, Peter Ruge Architekten continues to readily welcome the opportunity to take on client's challenges: a Passvie House Project in Changxing, China, the industrial conversion of a redundant factory into modern workspaces in Hangzhou (under construction), and upgrading residential portfolios in Berlin.

In addition to various awards, three projects by Peter Ruge have been awarded with a national certification for sustainable architecture.

Brisac Gonzalez 建筑师事务所

Brisac Gonzalez 建筑师事务所是一家多语种的跨国建筑公司，于1999年由 Cecile Brisac 及 Edgar Gonzalez 在伦敦创立。

公司实践已赢得了世界各地的众多奖项和竞赛，目前正在进行的有英国、法国、斯堪的纳维亚、俄罗斯和中东等地的项目。通过在许多地方从事各种建筑类型项目，他们不断充实实践建筑知识。他们关心的是最广泛意义上一切建筑学内容，无论是城市或室内设计、文化、商业建筑，或社会学问题。他们通过与客户合作，挑战传统从而创造出优秀的建筑物和更好的生活及工作场所，为建成环境营造更大的价值。

公司的工作并不是生产重复性的设计，而是从严谨的工作方法中进化而来，包括与专业顾问合作、综合研究项目场地，以及广泛使用三维研究，最重要的是反馈客户的需求。哥德堡及 Le Prisme 世界文化博物馆、法国多功能音乐厅等项目使得 Brisac Gonzalez 建筑师事务所获得客户的广泛关注，也赢得了国际赞誉。

Brisac Gonzalez Architects

Brisac Gonzalez is a multi-lingual and multi-national architecture firm established in London in 1999 by Cecile Brisac and Edgar Gonzalez.

The practice has won numerous awards and competitions across the world, and is currently working on projects in UK, France, Scandinavia, Russia, and the Middle East. By working on a variety of building types in numerous locations we continuously enrich the knowledge in our practice and the buildings we produce. We are concerned with all things architectural in the broadest sense, be it urban or interior design, cultural, commercial buildings, or sociological matters. We work by collaborating with our clients and challenging the conventional to create outstanding architecture, better places to live and work, and greater value in the built environment.

Instead of producing repetitive designs, our work evolves from rigorous working methods that include collaboration with specialist consultants, comprehensive research of our projects sites extensive use of three-dimensional studies, and most importantly responding to our client needs. Projects such as the Museum of World Culture in Gothenburg and Le Prisme, a multi-purpose concert hall in France brought Brisac Gonzalez to the attention of a wide field of clients and gave the office international acclaim.